Mast Cell Proteases in Immunology and Biology

CLINICAL ALLERGY AND IMMUNOLOGY

Series Editor

MICHAEL A. KALINER, M.D.

Medical Director
Institute for Asthma and Allergy
Washington, D.C.

1. Sinusitis: Pathophysiology and Treatment, *edited by Howard M. Druce*
2. Eosinophils in Allergy and Inflammation, *edited by Gerald J. Gleich and A. Barry Kay*
3. Molecular and Cellular Biology of the Allergic Response, *edited by Arnold I. Levinson and Yvonne Paterson*
4. Neuropeptides in Respiratory Medicine, *edited by Michael A. Kaliner, Peter J. Barnes, Gert H. H. Kunkel, and James N. Baraniuk*
5. Provocation Testing in Clinical Practice, *edited by Sheldon L. Spector*
6. Mast Cell Proteases in Immunology and Biology, *edited by George H. Caughey*

ADDITIONAL VOLUMES IN PREPARATION

Mast Cell Proteases in Immunology and Biology

edited by
George H. Caughey
University of California, San Francisco, School of Medicine
San Francisco, California

Marcel Dekker, Inc.   New York • Basel • Hong Kong

Library of Congress Cataloging-in-Publication Data

Mast cell proteases in immunology and biology / edited by George H. Caughey.
p. cm. — (Clinical allergy and immunology ; 6)
Includes bibliographical references and index.
ISBN 0-8247-9484-2
1. Proteolytic enzymes. 2. Mast cells—Physiology. 3. Mast cells—Immunology. I. Caughey, George H. II. Series.
QP609.P78M37 1995
616.07'9—dc20

95-10261
CIP

The publisher offers discounts on this book when ordered in bulk quantities. For more information, write to Special Sales/Professional Marketing at the address below.

This book is printed on acid-free paper.

MARCEL DEKKER, INC.
270 Madison Avenue, New York, New York 10016

Current printing (last digit):
10 9 8 7 6 5 4 3 2 1

PRINTED IN THE UNITED STATES OF AMERICA

Preface

Recent advances in the understanding of mast cell proteases make this book both possible and necessary. Due to the expanding international cadre of investigators studying the nature, number, and biology of mast cell secretory proteases, it is increasingly difficult to keep up with the literature on the subject. This book aims to serve mast cell biologists, pathologists, immunologists, allergists, and others with a need to make sense of mast cell enzymes. It provides an overview of the field by 23 of its active investigators, with a focus on work that deciphers the biological roles of the enzymes. This volume does not try to be, and cannot hope to be, an encyclopedic record of work in this area. However, the book is intended to capture the complexity and diversity of these enzymes, which have attracted recent interest as pharmaceutical targets because of their proposed roles in allergic and nonallergic events.

Proteases have a long history in the timeline of mast cell research. Their presence was first detected in mast cells in the same year (1953) that mast cells were identified as the main tissue source of histamine. The first mast cell protease was purified in 1966 (a year before the purification and characterization of IgE) and was sequenced and crystallized in 1978 (a year before the chemical identification of leukotrienes). Nonetheless, recognition of the variety and importance of mast cell proteases was, until recently, slow to evolve. We now know that proteases are the major proteins of mast cell secretory granules, that their variable expression determines major differences in mast cell phenotype and function, and that they are likely to play a variety of roles in mammalian pathophysiology and maintenance of tissue homeostasis. Compared to other secreted mast cell products—especially biogenic amines, eicosanoids, and (most recently) cytokines—mast cell proteases have maintained a low profile, in part because information concerning their function has been late to emerge. This state of affairs is in a rapid flux. Research into mast cell proteases is quickly advancing from a phase characterized predominantly by enzyme discovery, localization, structure determination, and enumeration of subtypes and species differences into a new phase characterized by the emergence of an understanding of biological roles and regulation of mast cell protease expression, and by the development of useful antagonists.

Some of the actions of mast cell proteases suggested in this volume will surprise those who associate mast cell products mainly or exclusively with the classical events of immediate hypersensitivity, or who assume that the release of proteases inevitably worsens inflammation. Chapter 12, for example, proposes that mast cell proteases are involved in lipoprotein metabolism—a nonclassical role for mast cells. Chapter 5 and 13 summarize evidence that secretion of mast cell proteases is *anti*-inflammatory in such contexts as cutaneous neurogenic inflammation, in which mast cell proteases degrade inflammatory neuropeptides released from sensory nerves, and anaphylatoxin-generated peritonitis, in which mast cell chymase inactivates C3a. Other surprises include the stimulation of gland cell secretion by mast cell chymase and cathepsin G (Chapter 8), the apparent growth factor activity of tryptase for fibroblasts, and the possible involvement of mast cell proteases in psoriasis (Chapter 3).

The presence of granule-associated esteroproteases in mast cells was first suspected through the application of enzyme histochemical approaches introduced by George Gomori, 76 years after Paul Ehrlich's description of *mastzellen* in 1877. Gomori's histochemical studies and others that followed, as related by David Lagunoff in the first chapter, dissected mast cell esterolytic activity into trypsin- and chymotrypsin-like components and provided the first glimpses of the species variations that now vex and fascinate us. The enzyme histochemical studies also provided the basis for the "tryptase–chymase" rubric coined in 1963 by David Lagunoff and Earl Benditt and used throughout this volume. As is often the case in the general course of scientific inquiry, serendipity played a large role in the discovery of individual mast cell proteases. A relatively early example was the discovery of the mast cell origin of "group-specific protease," a serine protease isolated originally from rat small intestine by Nobuhiko Katunuma and colleagues and named for its ability to degrade the apo forms of a number of pyridoxal phosphate-requiring enzymes. We now know this enzyme as "rat mast cell protease II" or to be concise, "rat chymase II." In a more recent example, the purification and molecular cloning of the major angiotensin-converting enzyme of human ventricular myocardium by Ahsan Husain and colleagues revealed heart chymase. This enzyme is identical in most, if not all, respects to human skin mast cell chymase characterized and cloned previously by Norman Schechter, members of my laboratory, and others. The heart chymase story's unusual plot twist, in addition to its reinforcement of the long-hypothesized role for chymase as an important extravascular converting enzyme, is the finding that chymase may be a product not solely of mast cells but also of ventricular endothelium. The discovery of mast cell carboxypeptidase also benefited from serendipity, having been identified first in preparations of rat liver mitochondria found subsequently to be contaminated with chymase-containing mast cell secretory granules. These and many other twists and turns in the road to the discovery of mast cell proteases led to the state of the art reviewed here.

I suspected strongly that the past and present schemes used to label and classify mast cell proteases confuse those immersed in the field and overwhelm the uninitiated. To clarify matters, I provide a master guide to mast cell protease nomenclature and abbreviations, which follows the last chapter.

Finally, I thank the chapter authors as well as the many other contributors to the development of the mast cell protease field, all of whom made this book possible.

George H. Caughey

Contents

Contributors

George H. Caughey, M.D. Associate Professor, Cardiovascular Research Institute, University of California, San Francisco, School of Medicine, San Francisco, California

Sanford M. Goldstein Associate Clinical Professor, Department of Dermatology, University of California, San Francisco, San Francisco, California

Ilkka T. Harvima, M.D., Ph.D. Assistant Professor, Department of Dermatology, University of Kuopio, Kuopio, Finland

Rauno J. Harvima, M.D., Ph.D. Assistant Professor, Department of Dermatology, University of Kuopio, Kuopio, Finland

Maija Horsmanheimo, M.D., Ph.D. Professor, Department of Dermatology, University of Kuopio, Kuopio, Finland

Tony E. Hugli, Ph.D. Member, Department of Immunology, The Scripps Research Institute, La Jolla, California

John F. Huntley, Ph.D., M.Sc., M.I.Biol. Research Scientist, Division of Parasitology, Moredun Research Institute, Edinburgh, Scotland

Anne-Marie A. Irani, M.D. Associate Professor, Department of Pediatrics, Medical College of Virginia, Richmond, Virginia

Nobuhiko Katunuma, M.D., Ph.D. Institute for Health Science, Tokushima Bunri University, Tokushima, Japan

Hiroshi Kido, M.D., Ph.D. Professor, Department of Enzyme Chemistry, Institute for Enzyme Research, The University of Tokushima, Tokushima, Japan

Jorma O. Kokkonen, M.D., Ph.D. Wihuri Research Institute, Helsinki, Finland

Petri T. Kovanen, M.D., Ph.D. Scientific Director, Wihuri Research Institute, Helsinki, Finland

David Lagunoff, M.D. Chairman, Department of Pathology, Saint Louis University Health Sciences Center, St. Louis, Missouri

Ken A. Lindstedt, Ph.D. Wihuri Research Institute, Helsinki, Finland

Hugh R. P. Miller, B.V.M.S., Ph.D. Professor, Department of Veterinary Clinical Studies, The University of Edinburgh, Edinburgh, Scotland

Anita Naukkarinen Assistant Professor, Department of Pathology, University of Kuopio, Kuopio, Finland

George F. J. Newlands, F.I.M.L.S. Research Scientist, Division of Parasitology, Moredun Research Institute, Edinburgh, Scotland

Norman M. Schechter, Ph.D. Research Associate Professor, Department of Dermatology, University of Pennsylvania School of Medicine, Philadelphia, Pennsylvania

Lawrence B. Schwartz, M.D., Ph.D. Charles and Evelyn Thomas Professor of Medicine, Department of Internal Medicine, Virginia Commonwealth University, Richmond, Virginia

William E. Serafin, M.D. Assistant Professor, Departments of Medicine and Pharmacology, Vanderbilt University School of Medicine, Nashville, Tennessee

Christian P. Sommerhoff, M.D. Department of Clinical Chemistry and Clinical Biochemistry, Surgical Clinic City, University of Munich, Munich, Germany

Eric B. Springman, Ph.D. American Lung Association Research Training Fellow, Department of Pharmacology, Vanderbilt University School of Medicine, Nashville, Tennessee

Elizabeth K. Tam, M.D. Associate Professor, Departments of Medicine and Pharmacology, John A. Burns School of Medicine, University of Hawaii, Honolulu, Hawaii, and Assistant Clinical Professor, Department of Medicine, University of California, San Diego, San Diego, California

1

Mast Cell Proteases: A Historical Perspective

David Lagunoff
Saint Louis University Health Sciences Center, St. Louis, Missouri

The discovery as well as the initial characterization of chymase, the first of the mast cell proteases, was histochemical (1). The discovery belonged to the brilliant histochemist George Gomori, who effectively and widely applied his knowledge and intuition to a range of chemical reactions on cells and tissues. He found the mast cell enzyme in the course of devising a method for cholinesterase. Gomori synthesized the chloracetyl esters of α-naphthol and naphthol AS and observed that mast cells stained intensely in 1 or 2 min, neutrophils and their precursors stained after 10 min, and generalized cell staining parallel to that seen with non-chloro-substituted acetyl esters occurred only after 20 min.

Earl Benditt, a friend and colleague of Gomori, has written to me of seeing the reaction in Gomori's laboratory for the first time:

> George in the course of making substrates for various carboxylic esterases was trying out different naphtholics as the chromogenic component of his esters. George made the acetyl derivatives of the naphtholic compounds employed in the dye industry. The free naphthols formed by enzyme hydrolysis were capable of coupling with complex diazonium salts to form insoluble dyes. He would then test these substrates on various tissues. In this way he discovered that mast cells were dramatically revealed by the chloracetyl esters of α-naphthol or naphthol AS. One day I went by George's lab as he was showing Asboe Hanson, a visiting dermatologist from Denmark, a slide of skin with mast cells stained with the new substance. I can remember that my eyes almost popped out because I was

already interested in the cells based on studies of edema I was doing in the rat paw. This demonstration of George's was very exciting to me, in particular because John French had come to work with me at LaRabida, where I then had my lab. Ralph Lillie had published a paper reporting on the fact that incubation with malt diastase could remove the mucopolysaccharide from sections of cartilage. Lillie attributed the action of the diastase to its capacity to break down chondroitin sulphate. Ideas about mucopolysaccharides and their structure were then quite primitive. John and I dreamed up several possibilities as to why the metachromasia disappeared from the sections and showed that the malt extract removed dry weight and reduced the staining and that the same effect could be produced by trypsin or chymotrypsin. This stimulated my interest in the protease and led to the hypothesis that the chloracetoxy ester might be a chymotrypsin substrate.

Benditt and Arase (2) supported the hypothesis with the demonstration that nonchromogenic substrates of α-chymotrypsin competitively inhibited the mast cell enzyme responsible for hydrolysis of the chloracetyl naphthol AS. (Years later the editors of *Biochemistry* would force us to use the term *chloracetoxy benzosalicylanilide* in place of the traditional *chloracetyl naphthol AS,* the AS standing for "anilide sauer.") Additional support for the chymotrypsin hypothesis was provided when Benditt and Arase (3) established that the substrate specificity of the esterase activity of extracts of isolated rat peritoneal mast cells largely, although not exactly, paralleled the esterase activity of α-chymotrypsin.

I spent a significant part of medical school in Benditt's laboratory working on another mast cell issue, the synthesis of serotonin, and only came to work on the chymotrypsin-like enzyme after I had rejoined his laboratory in Seattle after an internship. The first thing I did was synthesize an histochemical substrate for the mast cell enzyme that looked more like a respectable α-chymotrypsin substrate, the phenylpropionoxy ester of benzosalicylanilide (4). Staining with this substrate was identical to that with the chloracetyl ester, only a bit slower to develop. The phenylpropionyl ester has the advantage of being considerably more stable in solution, but it has never become widely used, even though Sigma still carries it in its catalog as naphthol AS phenylpropionate. (In those days the idea of patenting an enzyme histochemical substrate would have been considered absurd, if it had been considered at all.) One interesting observation made with the phenylpropionyl ester was staining of a weakly metachromatic, poorly granulated cell in the rat intestinal mucosa; this was an early sighting of the mucosal mast cell. Detailed studies of substrate specificity led to the proposal that the affinity of chymotrypsin for the naphtholic substrates is determined by both the naphthol moiety and the acyl group, whereas the specificity of the histochemical reaction is largely a function of the rate of hydrolysis of the substrate, which is determined by the structure of the acyl moiety (5).

Surgical pathologists continue to use the esterolytic activity of the chymase to identify mast cells, at least the connective tissue type (MC_{TC}). The procedure carries the name of Leder (6), whose contributions were to the substitute naphthol-AS-D for naphthol AS and popularize the reaction for demonstrating mast cells in human tissue. In recent times, researchers have largely switched to the immunological localization of chymase.

The presence of chymotrypsin-like enzyme in the conspicuous granules of the mast cell was rather obvious, but it was necessary to go beyond the histochemical localization. By 1963 we had developed a usable if not entirely satisfactory method of isolating secretory granules from rat peritoneal mast cells (7). The method provided a preparation that could be used to test the putative localization of the enzyme in the granules, and to no one's surprise that is where we found it. The paper (8) containing the report of the definitive granule localization also included the first identification of the actual proteolytic activity of the granule enzyme as well as the demonstration that a high-salt concentration (0.5 *M* KCNS) was required to solubilize the enzyme and the proposal to name the enzyme mast cell chymase and its trypsin-like relative mast cell tryptase. Benditt and I wrote in 1963, 12 years after Gomori had provided the first evidence for the enzyme: "At this time, our knowledge of the enzyme, though far from complete, seems sufficiently adequate to grace the mast cell enzyme with a less cumbersome and more accurate epithet than 'chymotrypsin-like esterase of mast cells.' " The new names were not met with much enthusiasm by the few other laboratories interested in the enzymes, so Woodbury and I (9) opted for the more conservative rat mast cell protease when we published our finding that Katanuma's group-specific protease was a second chymotrypsin-like mast cell protease. Obviously, despite our recidivism, chymase and tryptase slowly took hold, and by 1990 there was no turning back; Schwartz (10) even introduced a set of rules for applying the terms.

Chiu and I (11) examined the distribution of chymase in a selected group of animals from different vertebrate classes and failed to find it in chicken, turtle, frog, or carp. Since chymase constitutes a major protein constituent of the granules in rat and a significant if lesser component of the human granules, identification of the protein that serves the same role in nonmammalian mast cells is of some considerable comparable interest. The mast cell carboxypeptidase is one possibility.

With Pritzl's help, I was able to obtain a fairly clean preparation of rat chymase (12); this procedure provided material for the first amino acid analysis, the first partial sequence in the hands of Woodbury, and the first antibody to the chymase. Meanwhile the first in a string of rediscoveries of the chymase had appeared. Pastan (13) identified an enzyme in the thyroid with properties suspiciously like those of the mast cell enzyme, and histochemical staining localized it to the mast cells in the thyroid. Chymase appeared subsequently in parotid gland homogenates (14) and in rat liver mitochondria (15). The latter presentation led to the discovery of carboxypeptidase in rat mast cells. The large amount of enzyme in mast cells and the widespread presence of these cells in the rat made the contamination of tissue preparations inevitable.

The most notorious masquerade and one with an important twist was uncovered by Woodbury. He had been involved in Neurath's laboratory in the sequencing of group-specific protease (GSP) isolated from rat small intestine by Katanuma and his associates. This enzyme, which had a sequence resembling chymotrypsin, had the interesting property of hydrolyzing the apo forms of several pyridoxal phosphate–requiring enzymes but not the holo forms—hence its name. Woodbury came to me with the proposal that we collaborate on localizing the group-specific enzyme. We prepared non-cross-reacting antibodies to GSP and peritoneal mast cell chymase. Connective tissue mast cells stained only with the antibody to chymase, whereas the antibody to GSP weakly stained the connective tissue mast cells but strongly stained a population of cells in intestinal epithelium and lamina propria that corresponded to the cells defined by Enerbäck (16) as atypical mast cells and demonstrated earlier by the histochemical reaction for chymase. Intestinal GSP (chymase II) was by sequence distinct from but closely related to chymase (17). Staining with the antibody to GSP in connective tissue mast cells turned out to be a cross reaction with chymase I. Gibson and Miller (18) found that the connective tissue mast cell staining capacity of antisera could be adsorbed with purified chymase I.

The valid claim to the title of second major endoprotease found in the mast cell belongs to tryptase, not chymase II. Reidel and Wunsch (19) originally synthesized a putative substrate for trypsin-like enzymes, β-naphthylarginyl amide. Glenner and Cohen (20) used this substrate and found a positive reaction in dog and human cells but not in rat mast cells. Glenner et al. (21), and Benditt and I with help from Ruth Watts (22), a more competent organic chemist than I, reported simultaneously on the ε-aminocaproate ester of naphthol AS, in some ways a superior substrate. With this substrate we found activity in rat mast cells but attributed it tentatively to hydrolysis by chymase. However, tryptase is present in rat mast cells; our most recent estimate puts the ratio of tryptase to chymase active sites in rat peritoneal mast cells at 0.05 (23). In contrast to mast cells, rat basophils were quite negative with the chymase substrate but stained with the tryptase substrate. What was very likely mast cell tryptase had been characterized about the same time as fibrinolytic activity in mast cell tumors from dog and human. In our comparative study, Chiu and I found tryptase activity in turtle but not in chicken, frog, or carp (11).

Haas and Heinrich (24) at the University of Freiburg were interested in histone-degrading proteases and demonstrated the surprising localization of two proteolytic enzymes in a preparation of inner mitochondrial membranes; one was a proteinase identical to Katanuma's group-specific proteinase, the other was a carboxypeptidase. When it became apparent that GSP was a mast cell enzyme, Haas et al. (15) quickly obtained evidence that the most likely source of the two proteolytic enzymes was contamination of their mitochondrial fraction with mast cell granules. Our antibody to chymase I gave a line of identity with their enzyme from liver mitochondria in double diffusion. Perhaps the most important aspect of the finding was the evidence for a third distinct proteolytic activity, carboxypeptidase, in the mast cell. Everitt (25) in Neurath's laboratory characterized the carboxypeptidase in terms of its substrate specificity and its extractability from purified rat peritoneal mast cell granules. Like chymase I, carboxypeptidase is firmly bound to rat mast cell granule heparin. It would be interesting to know the phylogenetic distribution of mast cell carboxypeptidase to test the possibility that it might be the protein that binds heparin in mast cells of birds and amphibians and a non-N-sulfated glycosaminoglycan in fish.

An enzyme that hydrolyzes leucyl naphthylamide has been histochemically localized in mast cells (26), but as far as I know, no one has bothered to characterize the enzyme biochemically. The evidence that a schistosomal leucine aminopeptidase can alter cytokine production by activating a T_H1-to-T_H2 switch (27) may stimulate renewed interest in the possibility of a mast cell leucine aminopeptidase.

A noteworthy feature of the mast cell proteolytic enzymes is their storage in granules apparently in their active forms rather than as inactive precursors as is the case in pancreatic zymogen granules, the other principal site in which this group of enzymes is gathered in high concentration. I know of no evidence for any precursor form of the enzymes in mast cells, and deliberate experiments have shown in the case of chymase I that the active sites are all accessible to the irreversible inhibitors diisopropyl fluorophosphate (DFP) and phenylmethylsulfonyl fluoride (PMSF) (28,29). Yurt and Austen (30) proposed that the activity of chymase was partially masked by binding to heparin proteoglycan, and suggested that within the granule, serotonin might also downregulate chymase activity. We have been unable to confirm Yurt and Austen's findings in terms of reduced activity of the heparin–chymase complex. Experiments by Johnson et al. (31) and Rickard and me (32) indicated a low pH value for granules in situ. Nuclear magnetic resonance data are also consistent with a low granule pH (34). Our experiments using the pseudo-first-order rate constant for inhibition of the chymase by PMSF in situ support the idea that low pH is a major determinant of the low activity of the enzyme in situ (32). Since serotonin would be expected to leak out of the granule as the pH increases, it could contribute to the control of the chymase in the rat, but of course, serotonin occurs only in the mast cells of a few rodent species.

Woodbury and co-workers (33) have proposed another control on chymase activity, this one affecting secreted granules. They obtained evidence that proteins larger than 17 kDa could not diffuse into the granule matrix and were thus inaccessible to the enzyme, or vice versa, depending on the molecular point of view. Experiments by Rickard and me indicate that their results depend on performing the experiments at low salt concentration. It has been known for some time that mast cell granules swell progressively as salt concentration in the suspending medium increases (34). Fernandez and co-workers (35) have neatly demonstrated this swelling phenomenon directly and have shown its reversibility. At 0.2 *M* univalent cation concentration molecules as large as α_1-antiprotease, 90 kDa, readily diffuse into the granule matrix and inactivate 90% of the chymase. Even thyroglobulin at 669 kDa is attacked at measurable rates by chymase bound to the granule matrix. Examination of the likely three-dimensional structure of chymase (36) indicates that the active site of the enzyme is about equidistant from the dipolar clusters of positive changes which can be presumed responsible for binding to heparin. When the granule matrix expands, the enzyme is apparently free to interact with protein substrates and inhibitors while still attached to heparin. Benditt has suggested to me the possibility that the heparin might even serve to direct the chymase to one or more of its substrates in the connective tissue. Everitt and Neurath (25) pointed out the complementary specificities of the mast cell carboxypeptidase and chymase, with the chymase leaving carboxy-terminal aromatic residues that are removable with carboxypeptidase activity.

Certainly, the greatest enigma presented by mast cell proteases is their biological function. These cells, situated largely in the connective tissue and often associated with small blood vessels and nerves, harbor an array of proteases comparable to or exceeding that of pancreatic zymogen granules. The chymase and the carboxypeptidase, but not the tryptase, are likely to be locally retained by their tight association with the glycosaminoglycan. Not only are several distinct proteolytic activities present, they are present in substantial amounts. The obvious classes of potential substrates are constitutive molecules of the connective tissue matrix, including basal lamina, soluble protein and peptide mediators of inflammation and their precursors, and surface molecules of local cells. The total of information available as to what these mast cell enzymes actually do as opposed to what they can do is not very impressive, but I expect that the still missing parts of the story will be supplied in fewer than the 42 years that have brought us from Gomori's discovery to where we are today.

REFERENCES

1. G. Gomori, Chloracetyl esters as histochemical substrates, *J. Histochem. Cytochem.* *1*:469 (1951).

2. E. P. Benditt and M. Arase, Enzyme kinetics in a histochemical system, *J. Histochem. Cytochem. 6*:431 (1958).
3. E. P. Benditt and M. Arase, An enzyme in mast cells with properties like chymotrypsin, *J. Exp. Med. 110*:451 (1959).
4. D. Lagunoff and E. P. Benditt, Histochemical examinations of chymotrypsin-like esterases, *Nature 192*:1198 (1961).
5. D. Lagunoff, Histochemistry of proteolytic enzymes, in *Methods and Achievements in Experimental Pathology,* Vol. 2 (E. Bajasz and G. Jasinin, eds.), S. Karger, Basel, 1967, p. 55.
6. L. D. Leder, Über die selektive fermentcytochemische Darstellung von neutrophilen myeloischen Zellen and Gewebsmastzellen in Paraffinschnitt, *Klin. Wochenschr. 42*:553 (1964).
7. D. Lagunoff, M. T. Phillips, O. A. Iseri, and E. P. Benditt, Isolation and preliminary characterization of rat mast cell granules, *Lab. Invest. 13*:1331 (1964).
8. D. Lagunoff and E. P. Benditt, Proteolytic enzymes of mast cells, *Ann. NY Acad. Sci. 103*:185 (1963).
9. R. G. Woodbury, M. Everitt, Y. Sanada, N. Katanuma, D. Lagunoff, and H. Neurath, A major serine protease in rat skeletal muscle: evidence for its mast cell origin, *Proc. Natl. Acad. Sci. USA 75*:5311 (1978).
10. L. B. Schwartz, Preface, in *Neutral Proteases of Mast Cells* (L. B. Schwartz, ed.), S. Karger, Basel, 1990, p. vii.
11. H. Chiu and D. Lagunoff, Histochemical comparison of vertebrate mast cells, *Histochem. J. 4*:135 (1972).
12. D. Lagunoff and P. Pritzl, Characterization of rat mast cell granule proteins, *Arch. Biochem. Biophys. 173*:554 (1976).
13. I. Pastan and S. Almqvist, Purification and properties of a mast cell protease, *J. Biol. Chem. 241*: 5090 (1966).
14. M. R. Rabinovitch, D. Lagunoff, and J. M. Iversen, The presence of mast cell granules in rat parotid secretory granule preparations, *J. Histochem. Cytochem. 28*:1351 (1980).
15. R. Haas, P. C. Heinrich, and D. Sasse, Proteolytic enzymes of rat liver mitochondria: evidence for a mast cell origin, *FEBS Lett. 103*:168 (1979).
16. L. Enerbäck, Mast cells in rat gastrointestinal mucosa. I. Effects of fixation, *Acta Pathol. Microbiol. Scand. 66*:289 (1966).
17. R. G. Woodbury, G. M. Gruzenski, and D. Lagunoff, Immunofluorescent localization of a serine protease in rat small intestine, *Proc. Natl. Acad. Sci. USA 75*:2785 (1978).
18. S. Gibson and H. Miller, Mast cell subsets in the rat distinguished immunohistochemically by their content of serine proteinases, *Immunology 58*:101 (1986).
19. A. Reidel and E. Wunsch, *N*-Benzol-arginin-β-naphthylamide als Substrat zur quantitaven kolorimetrischen Bestimmung von Trypsin, *Hoppe-Seyler's Z. Physiol. Chem. 316*:61 (1959).
20. G. G. Glenner and L. A. Cohen, Histochemical demonstration of a species-specific trypsin-like enzyme in mast cells, *Nature 185*:846 (1960).

21. G. G. Glenner, V. K. Hopsu, and L. A. Cohen, Histochemical demonstration of a trypsin-like esterase activity in mast cells, *J. Histochem. Cytochem. 10*:109 (1962).
22. D. Lagunoff, E. P. Benditt, and R. Watts, Histochemical study of esterases homospecific with trypsin, *J. Histochem. Cytochem. 10*:672 (1962).
23. D. Lagunoff, A. Rickard and C. Marquardt, Rat mast cell tryptase, *Arch. Biochem. Biophys. 291*:52 (1991).
24. R. Haas and P. C. Heinrich, A novel SH-type carboxypeptidase in the inner membrane of rat liver mitochondria, *Eur. J. Biochem. 96*:9 (1979).
25. M. T. Everitt and H. Neurath, Rat peritoneal mast cell carboxypeptidase localization, purification and enzymatic properties, *FEBS Lett. 110*:292 (1980).
26. O. Braun-Falco and K. Salfield, Leucine aminopeptidase activity in mast cells, *Nature 183*:51 (1959).
27. E. J. Pearce, reported in *J. N.I.H. Res. 5*:59 (1993).
28. B. A. V. Perera and J. L. Mongar, The role of a chymotrypsin-like enzyme in histamine release from mast cells, *Immunology 6*:472 (1963).
29. D. Lagunoff, Natural proteases of the mast cell, in *Biochemistry of the Acute Allergic Reactions* (K. F. Austen and E. L. Becker eds.), Alan R. Liss, New York, 1981.
30. R. Yurt and K. F. Austen, Preparative purification of the rat mast cell chymase, *J. Exp. Med. 146*:1405 (1977).
31. R. G. Johnson, S. E. Carty, B. J. Fingerhood, and A. Scarpa, The internal pH of mast cell granules, *FEBS Lett. 120*:75 (1980).
32. D. Lagunoff and A. Rickard, Evidence for control of mast cell granule protease in situ by low pH, *Exp. Cell Res. 144*:353 (1983).
33. H. Le Trong, H. Neurath, and R. G. Woodbury, Substrate specificity of the chymotrypsin-like protease in secretory granules isolated from rat mast cells, *Proc. Natl. Acad. Sci. USA 84*:364 (1987).
34. D. Lagunoff, Contributions of electron microscopy to the study of mast cells, *J. Invest. Dermatol. 58*:296 (1972).
35. C. Nanati and J. M. Fernandez, The secretory granule matrix: a fast-acting smart matrix, *Science 259*:963 (1993).
36. S. J. Remington, R. G. Woodbury, R. A. Reynolds, B. W. Matthews, and H. Neurath, The structure of rat mast cell protease II at 1.9 Å resolution, *Biochemistry 27*:8097 (1988).

2

Mast Cell Tryptase: Properties and Roles in Human Allergic Responses

Lawrence B. Schwartz
Virginia Commonwealth University, Richmond, Virginia

I. INTRODUCTION

Mast cells and basophils are the major effector cells in allergic diseases. When activated by either IgE-dependent or IgE-independent agonists, mast cells secrete an array of mediators that act on the vasculature, smooth muscle, connective tissue, mucous glands, and inflammatory cells to cause immediate-type hypersensitivity reactions. Although their immunopathological role in these reactions is acknowledged, their purpose in normal tissue homeostasis or in defense against microorganisms each remain less certain. Preformed mediators, stored in secretory granules and secreted with cell activation, include a biogenic amine, typically histamine; proteoglycans, either heparin, oversulfated chondroitin sulfates, or both; and a spectrum of neutral proteases.

Neutral proteases, accounting for the vast majority of the granule protein, serve as selective markers of mast cells and of different types of mast cells (1). Tryptase (EC 3.4.21.59) is the principal enzyme accounting for the trypsin-like activity first detected in human mast cells by histochemical techniques (2,3). Tryptase, along with chymase, carboxypeptidase, and a cathepsin G–like protease, are the dominant protein compounds in secretory granules of the MC_{TC} type of human mast cell, whereas only tryptase is detected in those of the MC_T type of mast cell. Whereas human mast cells are rich in serine-type proteases, human basophils exhibit relatively little protease activity (4). Because of the selective expression of these proteases in mast cells, one could hypothesize that these enzymes play an important role in the biology of this cell type, that such enzymes might serve as sensitive and specific markers of mast cells and of mast cell activation, and that regulation of the transcription of their corresponding genes is related to regulation of the differentiation of this cell type. In this chapter we focus on studies of human mast cell tryptase.

II. PHYSICAL AND ENZYMATIC PROPERTIES

Tryptase was first purified to apparent homogeneity from dispersed and enriched lung mast cells in 1981 (2), and later from crude lung cell dispersions (5), HMC-1 cells (6), and from lung (7), pituitary (8), and skin (9) tissue. The original technique involved sequential chromatography on Dowex-1X2, DEAE-Sephadex, and heparin-agarose, but required dispersed mast cells of at least 20% purity. Subsequently, hydrophobic chromatography was included to permit purification from cruder sources. Most recently, an immunoaffinity chromatography technique was developed with a mouse monoclonal IgG recognizing the active form of tryptase (10). Tryptase activity can be measured by cleavage of small synthetic ester [tosyl-Arg methyl ester (TAME)] or peptide (tosyl-Gly–Pro–Lys-*p*-nitroanilide) substrates with Arg or Lys in the C-terminal position.

Tryptase is a tetrameric endoprotease of 134,000 Da with subunits of 31,000 to 34,000 Da, each with an active enzymatic site (2) and common antigenic sites (11). A reduction in molecular weight of tryptase subunits of 2000 to 4000 after treatment with endoglycosidase indicates that carbohydrate is present on each subunit (8). Within mast cell granules and after secretion, tryptase associates with proteoglycan, presumably heparin, by ionic interactions. Because NaCl concentrations of at least 0.7 *M* are needed to dissociate tryptase from heparin, the mechanism for dissociation to occur in vivo, if one exists at all, is not clearly defined. However, antithrombin III, when present in large excess in vitro, can dissociate up to one-third of the tryptase bound to heparin. Chymase and carboxypeptidase appear to reside in a complex with proteoglycan that is distinct and separable from the macromolecular tryptase–proteoglycan complex (12), suggesting that tryptase is processed and packaged separately from these other two proteases.

Tryptase hydrolyzes proteins and peptides on the C-terminal side of the basic residues, lysine and arginine (13). Tripeptide substrates with basic amino acid residues in the S1 and S2 positions are preferred by tryptase (8). Such substrates also are cleaved by other trypsin-like enzymes and therefore are not specific. Although classified as a serine protease, the absence of inhibition by the classical inhibitors of serine esterases present in plasma, lung, and urine as well as by lima bean, soybean, and ovomucoid trypsin inhibitors clearly distinguishes tryptase from pancreatic trypsin and from most other serine esterases (14). Tryptase is inhibited by low-molecular-weight substances such as leupeptin, diisopropyl fluorophosphate, and phenylmethylsulfonyl fluoride. Divalent cations such as calcium (15), and benzamidine and its derivatives (16,17), are competitive inhibitors of tryptase. Interestingly, histamine shifts the substrate dose–response curve to the right and to a sigmoidal pattern, suggesting cooperative behavior (15).

Instead of the enzymatic activity of tryptase being regulated by the usual inhibitors of serine proteases, tryptase is uniquely stabilized in its active tetrameric form by heparin (5), to which it is ionically bound under physiological conditions (Fig. 1). When free in solution, tryptase subunits irreversibly dissociate from one another into inactive monomers, without any evidence for autodegradation. Substantial conformational changes occur during this process, as evidenced by circular dichroic spectral shifts and by distinct epitopes being detected on the active tetramer and inactive monomers (10). It also appears that inactivation proceeds as a multistep process, because limited recovery of lost enzyme activity reportedly occurs at an early stage of the inactivation process (18). Chondroitin sulfate E and polyglutamic acid stabilize tryptase somewhat

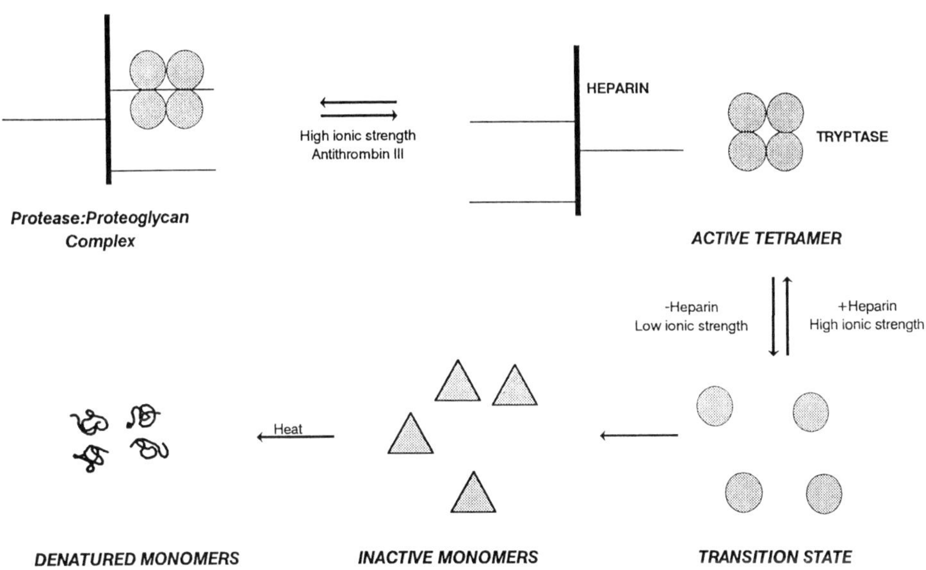

Figure 1 Regulation of tryptase activity. Tryptase is stabilized in its active tetrametric form by binding ionically to heparin proteoglycan. At high ionic strength or in the presence of excess antithrombin III, a portion of tryptase is solubilized from heparin. Free tryptase, particularly at a physiological salt concentration, dissociates spontaneously into monomers that associated with changes in conformation soon become irreversibly inactive. Heating tryptase will denature the protein to a random coil.

less well than heparin, and chondroitin monosulfates are little better than buffer at stabilizing tryptase under physiological conditions. Thus negative charge density rather than carbohydrate structure is the primary determinant for stabilizing tryptase activity (19). The physiological means of dissociating tryptase from heparin or chondroitin sulfate E in vivo is not totally clear, but the capacity of divalent cations (15) and of heparin-binding proteins such as antithrombin III (14) to destabilize heparin-bound tryptase at physiological concentrations in vitro may be relevant to the situation in vivo. In vitro, the relationship of heparin-stabilized tryptase activity to ionic strength is somewhat complex, because it reflects the destabilizing effect of dissociation of tryptase from heparin, the stabilizing effect of high ionic strength on free tetramer, and the direct effects of increasing ionic strength to raise the K_m values and increase the k_{cat} values (8,19).

III. BIOLOGICAL ACTIVITY

Even though the biological role of tryptase has not been convincingly demonstrated, several activities of potential biological interest have been examined in vitro (Table 1). Tryptase rapidly cleaves and inactivates fibrinogen as a coagulable substrate for thrombin (20); the lack of fibrin deposition and rapid resolution of urticaria–angioedema reactions may, in part, reflect this same activity in vivo. Tryptase degrades fibronectin directly (21). Tryptase activates latent collagenase derived from rheumatoid synovial cells, apparently by first activating prostromelysin (metalloproteinase III) (22), which in turn activates latent collagenase. The increased numbers of mast cells found in rheumatoid synovium (23) and in inflammatory cutaneous lesions of scleroderma (24), together with the increased tissue turnover that occurs in each condition, suggest a related role

Table 1 Potential Biological Activities of Human Tryptase

Enzymatic event	Potential significance
Fibrinogenolysis	Prevent clotting at local sites of mast cell activation
Augment histamine-mediated contractility of airway smooth muscle	Increase airway hyperreactivity to histamine and allergens
Degrade vasoactive intestinal peptide	Decrease endogenous bronchodilatory activity in lung
Activate prostomelysin	Facilitate activation of procollagenase and tissue remodeling
Degrade fibronectin	Tissue remodeling
Stimulate fibroblast proliferation	Enhance fibrogenesis or wound repair

for tryptase. Tryptase does not appear to affect complement anaphylatoxin or bradykinin generation (25); an earlier report of C3a generation from C3 could not be reproduced with immunoaffinity-purified tryptase (L. B. Schwartz, unpublished data). Bradykinin, lysylbradykinin, C3a, and C5a, if generated in vivo during immediate hypersensitivity events, is not due directly to tryptase activity. Tryptase does degrade calcitonin gene–related peptide and vasoactive intestinal peptide (26–28), two neuropeptides that relax smooth muscle. If human tryptase destroys vasoactive intestinal peptide in lung, bronchial tone might increase and increased airway hyperreactivity ensue. Tryptase also exhibits mitogenic activity on fibroblast lines in vitro (29), suggesting one mechanism by which mast cells might participate in fibrogenesis in diseases such as scleroderma.

IV. MOLECULAR BIOLOGY

Two tryptase cDNA molecules (α and β) have been cloned from a human lung mast cell cDNA library and sequenced (30,31), and three (I, II, and III) from a human skin mast cell library, as summarized in Table 2 (32). The α-tryptase cDNA codes for a 27,423-Da 244-amino acid catalytic protein with two putative carbohydrate binding sites (30). The β-tryptase cDNA codes for a 245-amino acid protein that is 90% identical to α-tryptase and 98 to 100% identical to tryptases I, II, and III. There are two N-linked carbohydrate binding sites

Table 2 Molecular Characteristics of Human Tryptase

Property	α-Tryptase	β-like Tryptases (I, II, III, β)
Number of amino acids		
Catalytic regions	244	245
Prepro regions	30	30
Homologies of catalytic regions		
α to β	91% identical	
β to I, II, III	98–100% identical	
Homologies of prepro regions		
α to β	87%	
β to I, II, III	100%	
Carbohydrate-binding regions	2	1–2
Calculated peptide molecular weights		
Catalytic region	27,423	27,449–27,585
Prepro region	3,048	3,089
Chromosome localization	16	16

in the α, I, and III types of tryptase, and one carbohydrate binding site in the other tryptase sequences, suggesting an explanation for a portion of the electrophoretic heterogeneity of natural tryptase in sodium dodecyl sulfate (SDS) polyacrylamide gels. The amino acid sequence for α-tryptase contains 24 amino acid differences and one amino acid deletion compared to the β-tryptase sequence. Interestingly, the amino acids in the region of β-tryptase where the deletion occurs is identical to the sequence reported for dog tryptase (33). Recently, an α-tryptase mRNA sequence from mono Mac 6 cells was cloned and sequenced and found to be identical to that described from the lung mast cell library except that the three guanine nucleotides accounting for the one amino acid deletion in α-tryptase were detected (34), raising the possibility that they had been missed earlier due to unresolved compressions.

The prepropeptide portion is 30 amino acids long in all cases; these regions are identical in the β, I, II, and III forms of tryptase, but differ by four amino acids from the α-tryptase prepropeptide. The activation peptide for tryptase has not been clearly defined, but its termination in Gly is unusual. The propeptide sequences for pancreatic serine proteases terminate in a basic residue; most leukocyte and mast cell granule proteases terminate in an acidic residue. Neither the prepropeptide nor the propeptide portions of tryptase have been detected in enzyme purified from human lung mast cells, consistent with tryptase being stored in an active form. The prepropeptide may be responsible for directing tryptase to secretory granules and/or the assembly of tryptase as an active tetramer, and probably is removed before or shortly after tryptase enters the secretory granules.

By applying PCR techniques to DNA taken from human:hamster somatic cell hybrids, at least two tryptase sequences were localized to the normal human haploid genome on chromosome 16 (31). One sequence corresponds to α-tryptase and tryptase III, the other to β-tryptase, tryptase I, and tryptase II. One tryptase gene has been sequenced that appears to match tryptase I cDNA (32). It contains six exons and five introns. The first intron falls between the 5′-noncoding and ATG-methionine start sites, which is rare, if not unique, among proteases. Positions of the four introns within the coding region are similar to those for trypsin and glandular kallikrein but different from those for proteases involved in blood coagulation and for mast cell chymases. The 5′ region has several consensus sites for regulating transcription, including TATA and CAAT boxes and a region showing homology to the enhancer elements for the rat chymase II gene (35) and for protease genes expressed in the pancreas. If the latter proves to be a mast cell–specific enhancer region, it would be of interest to examine its regulation during the differentiation of human mast cells.

Human tryptase has a very restricted substrate specificity for proteins, leaving most uncleaved, and being resistant to inhibition by biologic inhibitors of serine proteases. Two large insertions have been predicted to reside on either side of the substrate-binding cleft, possibly explaining this behavior (36).

V. CELLULAR MARKER OF MAST CELLS DURING DEVELOPMENT AND IN TISSUES

Neutral proteases or their subunit peptides each represent a single gene product and are often produced selectively and in great abundance in mast cells. These best reflect the heterogeneity or plasticity of mast cells in vivo and in vitro, particularly in humans, where histochemical heterogeneity is less apparent. Based on double-labeling techniques using immunohistochemistry for tryptase and histochemistry for tissue mast cells, tryptase was found in all mast cells and only in mast cells in human skin, lung, and bowel. In contrast, chymase, carboxypeptidase, and a cathepsin G–like protease were detected together in a subset of mast cells called MC_{TC} cells (37). Those mast cells with only tryptase were called MC_T cells. Substantial amounts of tryptase are present in MC_{TC} cells derived from foreskin (35 pg per cell) and in MC_T cells derived from lung (10 pg per cell), where it is located in secretory granules and is released in parallel with histamine during degranulation (38). These levels appear to account for perhaps 20% of the entire protein in the mast cell. Small amounts have been measured in human basophils (0.04 pg per cell). The basophil leukemia cell line called KU812 expressed only tryptase I mRNA, a β-like sequence, but whether this corresponds to normal basophils is not known (39). Transformed monocytic cell lines, U937 and Mono Mac 6, express β-tryptase and α-tryptase mRNAs, respectively, whereas normal monocytes express only trace amounts of tryptase mRNA and no detectable tryptase protein (34). Other cell types found in peripheral blood, such as eosinophils, neutrophils, and lymphocytes, have no detectable tryptase protein or mRNA content. Thus the enzyme is a discriminating marker of human mast cells.

Tryptase expression occurs at the time that granule formation begins during the development of human mast cells (40). Thus the enzyme serves as a useful differentiation marker for this cell type. Studies of *kit* ligand–mediated differentiation of mast cells from fetal liver cells (41) and from cord blood mononuclear cells (42) in vitro have utilized tryptase expression as a critical marker for identifying commitment to a mast cell lineage.

VI. ASSESSING THE BURDEN AND ACTIVATION OF MAST CELLS BY MEASURING TRYPTASE IN SERUM AND OTHER BIOLOGICAL FLUIDS

Several mediators of mast cells have been considered as potentially useful indicators of mast cell or basophil activation. Certain neutral proteases have greater potential utility than other mediators because of their abundant presence in and selective localization to mast cell secretory granules, and their persistence in the circulation. A sensitive and specific immunoassay for tryptase was developed to distinguish tryptase from other trypsin-like enzymes and to quantify tryptase in complex biological fluids (43,44) such as serum (45), bronchoalveolar lavage fluid (46), nasal lavage fluid (47), skin chamber fluid (48,49), and tears (50). This has enabled the measurement of tryptase release in vivo and thereby a precise assessment of mast cell activation. In serum, tryptase levels in normal subjects are undetectable (< 1 ng/mL), whereas elevated levels are detected in systemic mast cell disorders such as anaphylaxis and mastocytosis (Table 3). Other disorders resulting in various shock syndromes, such as toxic shock syndrome (L. B. Schwartz and D. Y. M. Leung, unpublished data) and the Mazzotti reaction (L. B. Schwartz and T. B. Nutman, unpublished data), seen with treatment of onchocerciasis with diethylcarbamazine (51), do not exhibit elevations of serum tryptase, suggesting that systemic mast cell activation is not the principal cell type involved. Thus tryptase levels provide a more precise assessment of mast cell activation in vivo than has previously been available.

Tryptase quantities in serum are elevated in systemic anaphylaxis (45). During beesting-induced anaphylaxis circulating levels are maximal at 60 to 120 min, and then decline with a half-life of 1.5 to 2.5 h (52). The magnitude of the tryptase level correlates with the clinical severity as measured by the drop in mean arterial pressure (53). Elevated levels of mast cell tryptase in postmortem sera reflect antemortem mast cell activation and suggest that tryptase levels also may be used as a practical indicator of fatal anaphylaxis (54). Elevated levels of tryptase in postmortem sera of victims of sudden infant death syndrome (SIDS) has implicated mast cell activation in the pathogenesis of this disorder (55), suggesting that anaphylaxis accounts for the pathogenesis of a substantial portion of SIDS victims.

Table 3 Levels of Tryptase in Serum or Plasma

Clinical condition	Tryptase (ng/mL)
Normal level	<1
Systemic anaphylaxis (mast cell-dependent)	
Severe within 4 h of hypotension	>5
Postmortem, fatal anaphylaxis	>10
Moderate within 2 to 4 h of signs	1–5
Severe after 4 h of hypotension	1–5
Systemic mastocytosis	
Nonacute, baseline	≥1 (ca. 50% of subjects)
Acute	>1 (nearly 100% of subjects)
Sudden infant death syndrome	>10 (ca. 40% of victims)
Septic shock, myocardial shock, vasovagal reactions, serum sickness, vasculitis, trauma, toxic shock syndrome, Mazzotti reaction	<1

Tryptase levels in serum may reflect the tissue burden of mast cells as well as their activation. In subjects with systemic mastocytosis, characterized by mast cell hyperplasia in skin, liver, spleen, and bone marrow, elevated levels of tryptase are found in about 50% of subjects at baseline (i.e., when not acutely symptomatic). Preliminary results with a more sensitive immunoassay for tryptase suggest that elevated levels of the protein will be detectable in nearly all subjects having this disease, making it a useful screening tool.

Ongoing mast cell activation in asthma, detected by elevated levels of tryptase and PGD_2 in bronchoalveolar lavage fluid, higher spontaneous release of histamine by mast cells obtained from the bronchoalveolar lavage fluid of asthmatics, and ultrastructural analysis of mast cells in pulmonary tissue, appears to be a characteristic of the inflammation of asthma (46,56–58). The early and late response to endobronchial challenge of atopic subjects without asthma at different doses of allergen showed findings analogous to those in the skin, with eosinophil and perhaps basophil recruitment and activation occurring during the late-phase response (59). Mast cell activation appears to be involved in immediate respiratory and vascular responses to oral aspirin challenge of sensitive subjects (60). On the other hand, mast cell activation may not be a feature of exercise-induced asthma (61).

VII. TRYPTASE FROM NONHUMAN SOURCES

Although the oligomeric structure and mast cell specificity of tryptases in different mammalian species appears to be conserved, many other properties differ, including affinity for heparin, susceptibility to inhibition by various biological inhibitors of serine-type proteases, and distribution among different types of mast cells. Analysis of amino acid sequences, where known, exhibit sufficient diversity as to suggest that the functional role of tryptases in different species may vary.

Rat mast cell tryptase has been purified from peritoneal mast cells (62) and skin (63). Unlike human tryptase, rat tryptase derived from peritoneal mast cells is inhibited by soybean trypsin inhibitor, α_1-antitrypsin, antipain, and aprotinin. In contrast, skin-derived tryptase was not found to be associated with an inhibitor and was not inhibited by soybean trypsin inhibitor and α_1-antitrypsin, but like tryptase from peritoneal mast cells, it was inhibited by antipain and aprotinin. In neither case did tryptase bind to heparin in physiological buffer, a property also deduced with subcellular fractions of rat peritoneal mast cells (64). Rat mast cell tryptase is a tetramer with four equally sized glycosylated subunits of 30,000 to 40,000 Da.

The distribution of tryptase among mast cells in rats is distinct from that in humans. Polyclonal antibodies prepared against skin-derived tryptase were used to show that although the enzyme was present only in Safranin-positive mast

structure and negative charge density of polysaccharides, *Biochem. J. 248*:821 (1987).
20. L. B. Schwartz, T. R. Bradford, B. H. Littman, and B. U. Wintroub, The fibrinogenolytic activity of purified tryptase from human lung mast cells, *J. Immunol. 135*:2762 (1985).
21. J. Lohi, I. Harvima, and J. Keski-Oja, Pericellular substrates of human mast cell tryptase: 72,000 dalton gelatinase and fibronectin, *J. Cell. Biochem. 50*:337 (1992).
22. B. L. Gruber, M. J. Marchese, K. Suzuki, L. B. Schwartz, Y. Okada, H. Nagase, and N. S. Ramamurthy, Synovial procollagenase activation by human mast cell tryptase dependence upon matrix metalloproteinase 3 activation, *J. Clin. Invest. 84*:1657 (1989).
23. D. G. Malone, R. L. Wilder, A. M. Saavedra-Delgado, and D. D. Metcalfe, Mast cell numbers in rheumatoid synovial tissues: correlations with quantitative measures of lymphatic infiltration and modulation by anti-inflammatory therapy, *Arthritis Rheum. 30*:130 (1987).
24. R. A. Hawkins, H. N. Claman, R. A. Clark, and J. C. Steigerwald, Increased dermal mast cell populations in progressive systemic sclerosis: a link in chronic fibrosis, *Ann. Intern. Med. 102*:182 (1985).
25. L. B. Schwartz, Tryptase from human mast cells: biochemistry, biology and clinical utility, *Monogr. Allergy 27*:90 (1990).
26. A. F. Walls, S. D. Brain, A. Desai, P. J. Jose, E. Hawkings, M. K. Church, and T. J. Williams, Human mast cell tryptase attenuates the vasodilator activity of calcitonin gene–related peptide, *Biochem. Pharmacol. 43*:1243 (1992).
27. G. H. Caughey, F. Leidig, N. F. Viro, and J. A. Nadel, Substance P and vasoactive intestinal peptide degradation by mast cell tryptase and chymase, *J. Pharmacol. Exp. Ther. 244*:133 (1988).
28. E. K. Tam and G. H. Caughey, Degradation of airway neuropeptides by human lung tryptase, *Am. J. Respir. Cell Mol. Biol. 3*:27 (1990).
29. T. Hartmann, S. J. Ruoss, W. W. Raymond, K. Seuwen, and G. H. Caughey, Human tryptase as a potent, cell-specific mitogen: role of signaling pathways in synergistic responses, *Am. J. Physiol. Lung Cell. Mol. Physiol. 262*:L528 (1992).
30. J. S. Miller, E. H. Westin, and L. B. Schwartz, Cloning and characterization of complementary DNA for human tryptase, *J. Clin. Invest. 84*:1188 (1989).
31. J. S. Miller, G. Moxley, and L. B. Schwartz, Cloning and characterization of a second complementary DNA for human tryptase, *J. Clin. Invest. 86*:864 (1990).
32. P. Vanderslice, S. M. Ballinger, E. K. Tam, S. M. Goldstein, C. S. Craik, and G. H. Caughey, Human mast cell tryptase: multiple cDNAs and genes reveal a multigene serine protease family, *Proc. Natl. Acad. Sci. USA 87*:3811 (1990).
33. P. Vanderslice, C. S. Craik, J. A. Nadel, and G. H. Caughey, Molecular cloning of dog mast cell tryptase and a related protease: structural evidence of a unique mode of serine protease activation, *Biochemistry 28*:4148 (1989).
34. R. Huang, M. Åbrink, A. E. Gobl, G. Nilsson, M. Aveskogh, L. G. Larsson, K. Nilsson, and L. Hellman, Expression of a mast cell tryptase in the human monocytic cell lines U-937 and Mono Mac 6, *Scand. J. Immunol. 38*:359 (1993).
35. J. Sarid, P. N. Benfey, and P. Leder, The mast cell–specific expression of a pro-

tease gene, RMCP II, is regulated by an enhancer element that binds specifically to mast cell trans-acting factors, *J. Biol. Chem. 264*:1022 (1989).
36. D. A. Johnson and G. J. Barton, Mast cell tryptases: examination of unusual characteristics by multiple sequence alignment and molecular modeling, *Protein Sci. 1*:370 (1992).
37. A. A. Irani and L. B. Schwartz, Neutral proteases as indicators of human mast cell heterogeneity, in *Neutral Proteases of Mast Cells* (L. B. Schwartz, ed.), S. Karger, Basel, 1990, p. 146.
38. L. B. Schwartz, R. A. Lewis, D. Seldin, and K. F. Austen, Acid hydrolases and tryptase from secretory granules of dispersed human lung mast cells, *J. Immunol. 126*:1290 (1981).
39. T. Blom and L. Hellman, Characterization of a tryptase mRNA expressed in the human basophil cell line KU812, *Scand. J. Immunol. 37*:203 (1993).
40. S. S. Craig, N. M. Schechter, and L. B. Schwartz, Ultrastructural analysis of maturing human T and TC mast cells in situ, *Lab. Invest. 60*:147 (1989).
41. A. A. Irani, G. Nilsson, U. Miettinen, S. C. Craig, L. K. Ashman, T. Ishizaka, K. M. Zsebo, and L. B. Schwartz, Recombinant human stem cell factor stimulates differentiation of mast cells from dispersed human fetal liver cells, *Blood 80*:3009 (1992).
42. H. Mitsui, T. Furitsu, A. M. Dvorak, A.-M. A. Irani, L. B. Schwartz, N. Inagaki, M. Takei, K. Ishizaka, K. M. Zsebo, S. Gillis, and T. Ishizaka, Development of human mast cells from umbilical cord blood cells by recombinant human and murine c-*kit* ligand, *Proc. Natl. Acad. Sci. USA 90*:735 (1993).
43. S. Wenzel, A. M. Irani, J. M. Sanders, T. R. Bradford, and L. B. Schwartz, Immunoassay of tryptase from human mast cells, *J. Immunol. Methods 86*:139 (1986).
44. I. Enander, P. Matsson, J. Nystrand, A.-S. Andersson, E. Eklund, T. R. Bradford, and L. B. Schwartz, A new radioimmunoassay for human mast cell tryptase using monoclonal antibodies, *J. Immunol. Methods 138*:39 (1991).
45. L. B. Schwartz, D. D. Metcalfe, J. S. Miller, H. Earl, and T. Sullivan, Tryptase levels as an indicator of mast-cell activation in systemic anaphylaxis and mastocytosis, *N. Engl. J. Med. 316*:1622 (1987).
46. S. E. Wenzel, A. A. Fowler, 3d, and L. B. Schwartz, Activation of pulmonary mast cells by bronchoalveolar allergen challenge: in vivo release of histamine and tryptase in atopic subjects with and without asthma, *Am. Rev. Respir. Dis. 137*:1002 (1988).
47. M. Castells and L. B. Schwartz, Tryptase levels in nasal-lavage fluid as an indicator of the immediate allergic response, *J. Allergy Clin. Immunol. 82*:348 (1988).
48. M. Shalit, L. B. Schwartz, N. Golzar, C. von Allman, M. Valenzano, P. Fleekop, P. C. Atkins, and B. Zweiman, Release of histamine and tryptase in vivo after prolonged cutaneous challenge with allergen in humans, *J. Immunol. 141*:821 (1988).
49. L. B. Schwartz, P. C. Atkins, T. R. Bradford, P. Fleekop, M. Shalit, and B. Zweiman, Release of tryptase together with histamine during the immediate cutaneous response to allergen, *J. Allergy Clin. Immunol. 80*:850 (1987).
50. S. I. Butrus, K. I. Ochsner, M. B. Abelson, and L. B. Schwartz, The level of tryptase in human tears: an indicator of activation of conjunctival mast cells, *Ophthalmology 97*:1678 (1990).

51. H. Francis, K. Awadzi, and E. A. Ottesen, The Mazzotti reaction following treatment of onchocerciasis with diethylcarbamazine: clinical severity as a function of infection intensity, *Am. J. Trop. Med. Hyg. 34*:529 (1985).
52. L. B. Schwartz, J. W. Yunginger, J. S. Miller, R. Bokhari, and D. Dull, The time course of appearance and disappearance of human mast cell tryptase in the circulation after anaphylaxis, *J. Clin. Invest. 83*:1551 (1989).
53. P. G. Van der Linden, C. E. Hack, J. Poortman, Y. C. Vivié-Kipp, A. Struyvenberg, and J. K. Van der Zwan, Insect-sting challenge in 138 patients: relation between clinical severity of anaphylaxis and mast cell activation, *J. Allergy Clin. Immunol. 90*:110 (1992).
54. J. W. Yunginger, D. R. Nelson, D. L. Squillace, R. T. Jones, K. E. Holley, B. A. Hyma, L. Biedrzycki, K. G. Sweeney, W. Q. Sturner, and L. B. Schwartz, Laboratory investigation of deaths due to anaphylaxis, *J. Forensic Sci. 36*:857 (1991).
55. M. S. Platt, J. W. Yunginger, A. Sekula-Perlamn, A. A. Irani, and L. B. Schwartz, Involvement of mast cells in sudden infant death syndrome, *J. Allergy Clin. Immunol. 94*:250 (1994).
56. M. C. Liu, E. R. Bleecker, L. M. Lichtenstein, A. Kagey-Sobotka, Y. Niv, T. L. McLemore, S. Permutt, D. Proud, and W. C. Hubbard, Evidence for elevated levels of histamine, prostaglandin D_2, and other bronchoconstricting prostaglandins in the airways of subjects with mild asthma, *Am. Rev. Respir. Dis. 142*:126 (1990).
57. P. H. Howarth, J. Wilson, R. Djukanovic, S. Wilson, K. Britten, A. Walls, W. R. Roche, and S. T. Holgate, Airway inflammation and atopic asthma: a comparative bronchoscopic investigation, *Int. Arch. Allergy Appl. Immunol. 94*:266 (1991).
58. N. N. Jarjour, W. J. Calhoun, L. B. Schwartz, and W. W. Busse, Elevated bronchoalveolar lavage fluid histamine levels in allergic asthmatics are associated with increased airway obstruction, *Am. Rev. Respir. Dis. 144*:83 (1991).
59. J. B. Sedgwick, W. J. Calhoun, G. J. Gleich, H. Kita, J. S. Abrams, L. B. Schwartz, B. Volovitz, M. Ben-Yaakov, and W. W. Busse, Immediate and late airway response of allergic rhinitis patients to segmental antigen challenge: characterization of eosinophil and mast cell mediators, *Am. Rev. Respir. Dis. 144*:1274 (1991).
60. J. V. Bosso, L. B. Schwartz, and D. D. Stevenson, Tryptase and histamine release during aspirin-induced respiratory reactions, *J. Allergy Clin. Immunol. 88*:830 (1991).
61. D. H. Broide, S. Eisman, J. W. Ramsdell, P. Ferguson, L. B. Schwartz, and S. I. Wasserman, Airway levels of mast cell–derived mediators in exercise-induced asthma, *Am. Rev. Respir. Dis. 141*:563 (1990).
62. H. Kido, Y. Yokogoshi, and N. Katunuma, Kunitz-type protease inhibitor found in rat mast cells: purification, properties, and amino acid sequence, *J. Biol. Chem. 263*:18104 (1988).
63. V. J. Braganza and W. H. Simmons, Tryptase from rat skin: purification and properties, *Biochemistry 30*:4997 (1991).
64. D. Lagunoff, A. Rickard, and C. Marquardt, Rat mast cell tryptase, *Arch. Biochem. Biophys. 291*:52 (1991).
65. Z. Chen, A. A. Irani, T. R. Bradford, S. S. Craig, G. Newlands, H. Miller, T.

Huff, W. H. Simmons, and L. B. Schwartz, Localization of rat tryptase to a subset of the connective tissue type of mast cell, *J. Histochem. Cytochem. 41*:961 (1993).
66. D. S. Reynolds, R. L. Stevens, W. S. Lane, M. H. Carr, K. F. Austen, and W. E. Serafin, Different mouse mast cell populations express various combinations of at least six distinct mast cell serine proteases, *Proc. Natl. Acad. Sci. USA 87*:3230 (1990).
67. W. Chu, D. A. Johnson, and P. R. Musich, Identification and molecular cloning of a mouse mast cell tryptase, *FASEB J. 4*:A2159 (1991).
68. G. H. Caughey, N. F. Viro, J. Ramachondran, S. C. Lazarus, D. B. Borson, and J. A. Nadel, Dog mastocytoma tryptase: affinity purification, characterization and amino-terminal sequence, *Arch. Biochem. Biophys. 258*:555 (1987).
69. G. H. Caughey, E. H. Zerweck, and P. Vanderslice, Structure, chromosomal assignment, and deduced amino acid sequence of a human gene for mast cell chymase, *J. Biol. Chem. 266*:12956 (1991).
70. K. Sekizawa, G. H. Caughey, S. C. Lazarus, W. M. Gold, and J. A. Nadel, Mast cell tryptase causes airway smooth muscle hyperresponsiveness in dogs, *J. Clin. Invest. 83*:175 (1989).
71. S. J. Ruoss, T. Hartmann, and G. H. Caughey, Mast cell tryptase is a mitogen for cultured fibroblasts, *J. Clin. Invest. 88*:493 (1991).
72. L. Fiorucci, F. Erba, and F. Ascoli, Bovine tryptase: purification and characterization, *Biol. Chem. Hoppe-Seyler 373*:483 (1992).

3

Skin Tryptase: Features and Expression in Human Dermatological Disorders

Ilkka T. Harvima, Rauno J. Harvima, Anita Naukkarinen, and Maija Horsmanheimo
University of Kuopio, Kuopio, Finland

I. INTRODUCTION

Human skin is rich in a variety of proteolytic enzymes (1). The first description of neutral protease activity in a high-salt extract of human skin hydrolyzing synthetic ester substrates of tryptic specificity was published in 1961 by Yamura and Cormia (2). Later, Seppä et al. (3) demonstrated a high-molecular-weight tryptic enzyme activity that hydrolyzed *N*-α-benzoyl-DL-arginine *p*-nitroanilide (pNA) at pH 7.8, in a high-salt extract of human skin. In the mid-1970s Fräki and Hopsu-Havu (4–7) published a series of reports on extraction and separation of several different proteolytic enzymes in human skin. The sequential low- and high-salt extraction of human skin showed the extractability of the neutral tryptic and chymotryptic enzyme activities in high-salt buffer (4). This finding facilitated the removal of acid proteases and of endogenous inhibitor of the chymotryptic activity, α_1-antitrypsin, and it improved markedly further separation of the neutral protease activities with chromatographic columns (5,6). The partially purified trypsin-like enzyme, skin tryptase, exhibited some characteristics of serine proteinases, and apparent M_r value of about 120,000 by gel filtration in 2 *M* KCl, and resistance to inhibition by plasma protease inhibitors (6). When injected intradermally into rabbit and human skin, this tryptase preparation first caused vasodilatation and erythema, and later long-lasting leukocyte and lymphocyte infiltration and slight induration at the injection site (7), indicating a marked role for tryptase in cutaneous inflammatory reactions.

The presence of trypsin-like enzyme activity in mast cells of human skin was observed in the early 1960s by using enzyme-histochemical staining techniques with *N*-benzoyl-DL-arginine-β-naphthylamide as the substrate (8,9). However, it took over 20 years before tryptase was purified to homogeneity from human pulmonary mast cells (10). A similar tryptase was purified from human skin (11–13). During the last few years, research on mast cell neutral proteinases has been expanding rapidly, due to realization of their potential biological effects.

II. PURIFICATION OF SKIN TRYPTASE

Skin tryptase has been purified from fresh, healthy-looking human skin obtained from mastectomies and leg amputations (11,12). Also, 1- to 4-day-old autopsy material is suitable for purification, since tryptase displays full enzyme activity on various tissue sections and in tissue high-salt extracts (14).

The extraction procedure begins with the removal of subcutaneous fat, followed by mechanical mincing of the skin. The minced tissue is combined with the low-salt buffer (10 m*M* sodium phosphate buffer, pH 6.5), and the cells are broken with five cycles of freezing and thawing. After centrifugation at 10,000 × *g*, the sediment is reextracted with the same low-salt buffer as a washing step. Tryptase and chymase can be extracted under high-salt conditions from the sediment with 1.5 *M* KCl containing 10 m*M* sodium phosphate buffer, pH 6.5. An equal volume of 3.5 *M* $(NH_4)_2SO_4$ is included in the high-salt extract and the precipitated protein is removed by filtration through Whatman No. 1 paper. The enzyme solution is then subjected to octyl-Sepharose CL-4B chromatography equilibrated with 1.75 *M* $(NH_4)_2SO_4$, 0.75 *M* KCl, and 5 m*M* sodium phosphate buffer, pH 6.5. The bound protein is detached with 2 *M* KCl, and fractions with tryptic activity are pooled and applied to a Sephacryl S-200 gel filtration column equilibrated with 2 *M* KCl buffered to pH 6.5 with 10 m*M* sodium phosphate. Up to this step, the partially purified tryptase has been kept in high-salt conditions, preventing its irreversible inactivation, which occurs rapidly in low-salt buffer (12). The next step exposes tryptase to low-salt conditions. After gel filtration, the tryptase preparation is dialyzed quickly against 0.3 *M* NaCl containing 10 m*M* Mes buffer, pH 6.1, and applied to a cellulose phosphate column. Alternatively, heparin-Sepharose column in 0.2 *M* NaCl can be used as the last chromatographic step. In both cases, tryptase can be eluted in a pure form with 2 *M* NaCl in buffer (12). The Mono S column (Pharmacia) attached to the high-performance liquid chromatographic (HPLC) system can be used for purification also (R. J. Harvima et al., unpublished results).

Both skin and lung tryptases have been purified successfully with the method described above (11,12). There are also other corresponding purification methods developed for lung (10,15) and pituitary (16) tryptases, and they can be used as well. The production of a specific mouse monoclonal B2 antibody against human lung tryptase has allowed the development of an immunoaffinity chromatographic material that enables rapid purification procedure with large quantities of active tetrametric tryptase (17).

III. STRUCTURE AND PROPERTIES OF SKIN TRYPTASE

Skin tryptase is a glycoprotein and serine proteinase functioning at a slightly alkaline pH. The native skin tryptase has an M_r value ranging between 120,000 and 140,000 Da, but it has a tetrameric substructure with two major electrophoretic bands at approximately 31,000 to 33,000 and 33,000 to 37,000 Da (10,12,16,17). The M_r value deduced from complementary DNA (cDNA) is 27,364 Da (244 amino acids) for α-tryptase and 27,458 Da (245 amino acids) for β-tryptase. The major difference between α- and β-tryptase cDNAs is the number of putative carbohydrate binding sites: α-tryptase has two but β-tryptase only one binding site (18–20).

Each of the four monomers of tryptase contains one active center (17–20). This properly agrees well with the finding that tryptase exhibits sigmoidal reaction kinetics in conditions exceeding physiological salt concentration (21) (Fig. 1). Furthermore, 10 m*M* histamine has been found to switch the kinetics from hyperbolic to sigmoidal (22). Hill equation slopes of nearly 2 suggest that strong cooperativity between at least two subunits or weaker cooperativity between three or four subunits occurs in this tetrameric enzyme (21,22). Thus this allosteric characteristic of tryptase should be considered carefully whenever tryptase activity is measured. Initial comparison of skin and lung tryptases showed differences in their preference for the hydrolysis of synthetic peptide substrates (23). However, the assay conditions were not equal—skin tryptase was mixed with 45 m*M* KCl, which affects the enzyme kinetics (21), and lung tryptase was mixed with 5 to 10 m*M* $CaCl_2$, which renders tryptase inactive (22), giving a probable explanation for the reported differences (23).

One striking feature of tryptase is its rapid and irreversible inactivation in low-salt conditions. Salt concentrations exceeding 1.4 *M* KCl or heparin glycosaminoglycan (GAG) or its fragments of M_r greater than 5700 Da are necessary for complete stabilization of tryptase activity (12,24). The tight binding of tryptase to negatively charged heparin GAG prevents the dissociation of tryptase into inactive monomers (24) without affecting the allosteric behavior of the enzyme (21). Another feature of tryptase is its resistance to inactivation by serum protease inhibitors (6,12,25). In fact, no specific physiological inhibitor has yet been found for the enzyme. In parallel with these findings, no inactive tryptase protein has been detected on histochemical sections of a chronic inflammatory skin disease where numerous protease inhibitors are likely to be present (26,27). These characteristics imply that tryptase bound to heparin proteoglycan

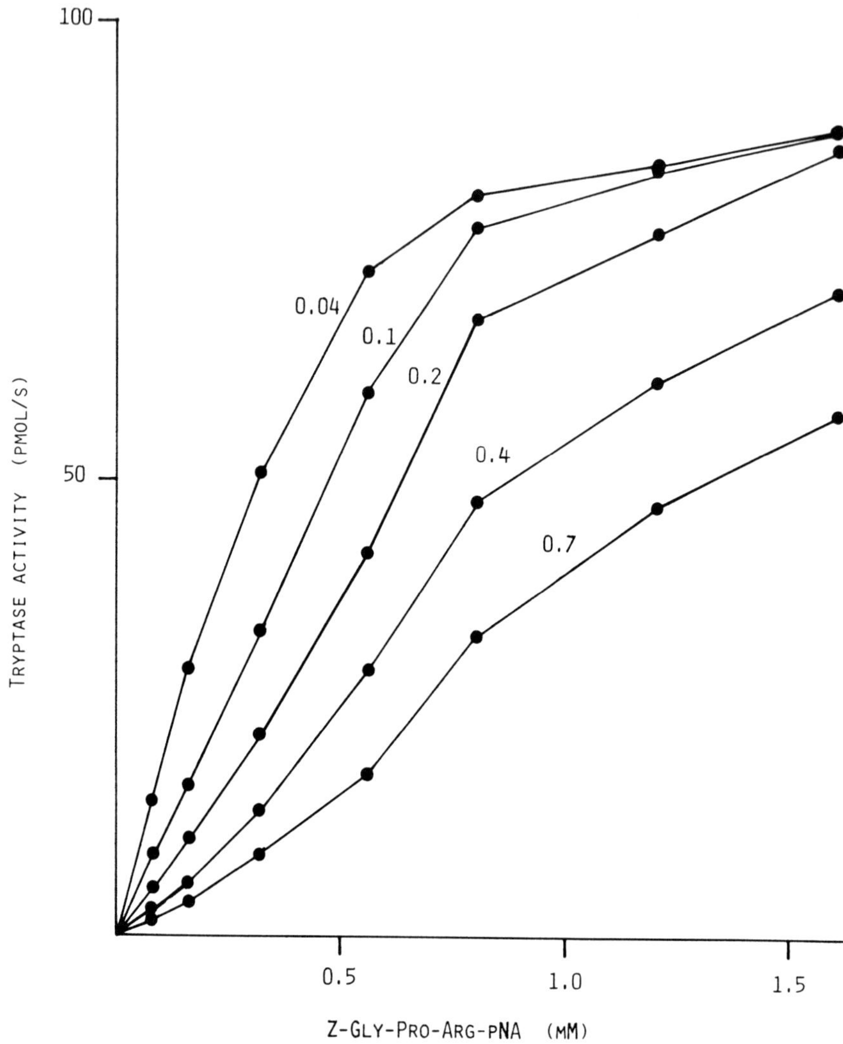

Figure 1 Effect of potassium chloride concentration on the hydrolysis of Z–Gly–Pro–Arg–pNA by human skin tryptase. Z–Gly–Pro–Arg–pNA was mixed with KCl in the final concentrations indicated in the figure, and the release of *p*-nitroaniline by tryptase was monitored continuously at a wavelength of 405 nm. The enzyme kinetic curve is distorted from a hyperbola into a sigmoid when increasing the KCl concentration from 0.1 *M* to 0.2 *M*. (From Ref. 21.)

has a prolonged action time following its release from the mast cell. Possible mechanisms for the catabolism of tryptase may include (a) cleavage of heparin GAG by heparinases to small fragments unable to protect tryptase from inactivation (24), (b) displacement of tryptase from heparin GAG by a protein with higher binding affinity (25), and (c) irreversible inactivation of tryptase by the calcium ion (22).

Mast cell tryptase is a heterogeneous glycoprotein. Fractionation of B2-immunoaffinity (17)-purified skin and lung tryptase with the Mono S column connected to a HPLC system results in two major protein and tryptase activity peaks, both of which comprise at least two overlapping peaks. Furthermore, variation in sodium dodecyl sulfate (SDS)-electrophoresis bands can be observed between fractionated tryptase preparations. Although at least four different cDNAs of tryptase have been cloned and sequenced (18–20) and two types of mast cells have been observed (27,28), this heterogeneity in Mono S peaks and electrophoretic bands could be explained by differences in glycosylation: only amino acid sequences indicative for β-tryptase have been detected so far, although differences in primary amino acid sequence have not been fully excluded (R. J. Harvima, I. T. Harvima, U. K. Miettinen, D. Dull, and L. B. Schwartz, unpublished results). Also of interest are the findings that human basophil cell line KU812 (29) and human HMC-1 mast cell line (Gunnar Nilsson, Department of Pathology, University of Uppsala, Sweden, personal communication) express only β-tryptase mRNA. Despite the variations in carbohydrate side chains of purified skin and lung tryptase, both native enzymes and their subfractions exhibit identical immunoreactivity with several monoclonal and polyclonal anti-tryptase antibodies, similar catalytic properties with different synthetic substrates (Table 1), and similar inactivation properties in low-salt buffer and by several protease inhibitors (12) (R. J. Harvima et al., unpublished results). Thus skin and lung tryptases are functionally and structurally closely related to each other.

Table 1 Comparison of the Catalytic Properties of Purified Skin Tryptase with Lung Tryptase

	Skin tryptase			Lung tryptase		
Substrate[a]	K_m (mM)	k_{cat} (s^{-1})	k_{cat}/K_m ($\times 10^4$)	K_m (mM)	k_{cat} (s^{-1})	k_{cat}/K_m ($\times 10^4$)
Z–Gly–Arg–AFC	1.53	104	6.80	1.77	107	6.05
Z–Gly–Pro–Arg–pNA	0.251	551	220	0.263	533	203
Z–Lys–Arg–AFC	0.392	45.7	11.7	0.409	45.7	11.2
Z–Arg–pNA	3.18	5.35	0.168	3.48	5.52	0.159

Source: Ref. 12.
[a]pNA, *p*-nitroanilide; AFC, 7-amide-4-trifluoromethylcoumarin; Z, carbobenzoxy.

IV. BIOLOGICAL EFFECTS OF SKIN TRYPTASE

The first description of the effects of partially purified skin tryptase injected into human and rabbit skin suggested a potential role for this enzyme in inflammatory skin conditions (7). Tryptase has been found to participate in several different regulatory systems of plasma. The enzyme has been reported to exert its activity into the complement system by generating active C3a anaphylatoxin from C3 (30). Tryptase also shows anticoagulant features by degrading the α and β chains of fibrinogen (31) and by destroying high-molecular-weight kininogen (32) without generating biologically active peptides from either molecule. However, tryptase seems to have no significant role in plasma proenzyme cascade systems since it has shown no efficacy in activating or inactivating various plasma proenzymes, including plasminogen, prothrombin, coagulation factor XII, protein C, complement C1s (31,33,34), and prokallikrein (35).

Other important functions of tryptase include its degradative effects on several proinflammatory neuropeptides, such as vasoactive intestinal peptide (VIP), peptide histidine–methionine and calcitonin gene–related peptide (CGRP), but it leaves untouched the mast cell degranulator substance P (SP) as well as neurokinins A and B (36,37). The biological activity of CGRP and VIP declines upon hydrolysis by tryptase (38,39). Hence tryptase could be one important mediator in modulating and/or attenuating neurogenic inflammation in skin induced by the release of neuropeptides from dermal sensory nerve endings (40,41).

The action of trypsin-like proteolytic enzymes is crucial for different types of virus infections (42–44). Several findings suggest that trypsin-like proteolytic cleavage of the peptide bond between Gly–Pro–Gly–Arg and Ala at the tip of the V3 loop of HIV-1 gp120 envelope glycoprotein is essential for HIV infection (45–49). Recent observations that human mast cell tryptase reacts very sensitively with synthetic V3 loop peptides (50) and that the enzyme cleaves specifically the critical peptide bond at the tip of the V3 loop (47) have led to the assumption that tryptase could activate the HIV-1 virus. This is especially exciting since persons with HIV infection sometimes develop aggressive psoriasis (51), and HIV transcripts have been localized to $FXIIIa^+$ and $CD4^+$ dendrocytes in the lesional psoriatic dermis, where these dendrocytes (52,53), together with tryptase-positive mast cells (26,27), are situated in the papillary dermis in increased numbers. A recent report has shown the failure of mast cell tryptase in a concentration of up to 10 μg/mL to augment the infection of permissive and nonpermissive cells by two HIV-1 virus strains (54). Nevertheless, this failure may be explained by the fact that tryptase may lose its activity (12,24) and/or ability to process proteins efficiently (30,31) during incubation with virions in low-salt buffer without any stabilizing compound. The other explanation could be that conformational changes in the virion's gp120 induced by binding to CD4 receptor are needed to distort V3 loop accessible to tryptase (47,48).

Like many trypsin-like enzymes (55), tryptases from human and dog mast cells have been reported to stimulate DNA synthesis and proliferation of fibroblasts. The signaling pathways leading to fibroblast activation are obscure but differ from those of thrombin (56,57). Furthermore, the extracellular matrix secreted by cultured human fibroblasts is susceptible to degradation by human skin tryptase. In particular, type IV collagenase/gelatinase is cleaved from the 72-kDa protein to 62-kDa protein by tryptase, which may indicate the activation of this type IV collagenolytic enzyme. Also, fibronectin, but not plasminogen activator inhibitor PAl-1, can be released from the matrix and cleaved into the fragments of 180, 160, and 27 kDa by skin tryptase (58). Tryptase is not capable of activating procollagenase directly from human skin fibroblasts, nor can it degrade collagen or proteoglycan (59). Tryptase can, however, activate prostromelysin to stromelysin (matrix metalloproteinase 3), which in turn can process latent collagenase to yield active collagenase (60). All these in vitro characteristics of tryptase suggest a significant role of tryptase in remodeling the extracellular matrix of skin in physiological and pathological processes.

V. LOCALIZATION OF TRYPTASE IN NORMAL AND DISEASED SKIN

A. Tryptase in Normal Skin

In skin, mast cells are typically located adjacent to blood vessels, hair follicles, dermal nerves, and various appendages, but they are most frequent in the upper dermis. Epidermis is normally devoid of mast cells, and tryptase is confined to mast cells exclusively (14,61,62). Mast cells can be subdivided into two major categories: MC_{TC} and MC_T cells, which differ from each other by their neutral proteinase composition. Originally, MC_{TC} and MC_T cells were named after their tryptase and chymase content: MC_T cells contain tryptase only, whereas MC_{TC} cells contain both tryptase and chymase in their secretory granules (28,63). Later, MC_{TC} cells were demonstrated to contain both carboxypeptidase and a cathepsin G–like proteinase as well (64,65). In addition, the ultrastructure of the secretory granules is different: MC_{TC} cells have granules with grating and lattice substructures, whereas granules in MC_T cells contain discrete scrolls (66), which, however, have occasionally been detected also in the peripheral regions of MC_{TC} granules (67). In normal human skin, MC_{TC} cells are the predominant cell type, accounting for 88 to 99.7% of all mast cells (28,63). Recently, a novel MC_C mast cell type, which contains chymase and carboxypeptidase but not tryptase, has been described at several body sites (68), although many investigators have failed to detect it (27,28,66,67). Nonetheless, MC_C cells, if truly present, can be found only on rare occasions in human skin, where 0.5% of the mast cells belong to this subpopulation (68).

Mast cells synthesize large quantities of tryptase, emphasizing its significance in effector functions of mast cells: dispersed skin MC_{TC} cells contain 35 pg per cell, but lung MC_T cells contain 10.8 pg of tryptase per cell (69). In mast cells, tryptase resides in secretory granules together with other neutral proteases (66). Although tryptase has been packed with chymase, carboxypeptidase, and a cathepsin G–like proteinase in the same granules, differences in their subgranular distribution have been observed. Tryptase is bound to proteoglycan complexes that are distinct from those that bind chymase and carboxypeptidase (70). At the ultrastructural level, tryptase is located in the electron-lucent areas of granules, whereas chymase and a cathepsin G–like proteinase favor the electron-dense regions (71). The physiological significance of this divergent protease distribution is not clear. Tryptase–proteoglycan complexes are smaller than chymase/carboxypeptidase–proteoglycan complexes (70), suggesting that tryptase–proteoglycan complexes more readily diffuse away from a mast cell upon degranulation.

The sensitive and specific tryptase substrate, Z–Gly–Pro–Arg-4-methoxy-2-naphthylamide (see also Table 1), applied to an enzyme-histochemical staining technique with either Fast Garnet GBC or Fast Black K as the chromogen, has produced an easy and quick method for staining cutaneous mast cells. This method can be extended to a double-staining technique, which comprises (a) identification of mast cells with an enzyme-histochemical stain for tryptase; (b) photographing the section; (c) dissolving the stain with Tween 20; (d) staining tryptase, chymase, α_1-antitrypsin, or α_1-antichymotrypsin immunohistochemically with its monoclonal or polyclonal antibodies; and (e) rephotographing the same section (26,27,62) (Fig. 2). By comparing the photographs, localization of the immunohistochemically stained substances in mast cells can be confirmed.

Tryptase activity measured with 0.23 m*M* Z–Gly–Pro–Arg–pNA at pH 7.6 (Table 1) is 584 $\pm$ 276 mU/g and 889 $\pm$ 431 mU/g (wet weight of tissue) in high-salt extract of human skin and lung, respectively. These activity levels are substantially higher than those detected in human pancreas (292 $\pm$ 133 mU/g), spleen (8.53 $\pm$ 8.39 mU/g), liver (27.2 $\pm$ 27.1 mU/g), and kidney (12.1 $\pm$ 11.9 mU/g) extracts. Furthermore, tryptase activity in these tissues parallels the histamine content but not cathepsin D and histamine-*N*-methyltransferase activities. Tryptase activity is confined to mast cells only as confirmed with the enzyme-histochemical staining method described above (14). Therefore, human skin and also lung are major sources for tryptase of several human tissues known to be rich in various proteolytic and metabolic enzymes. The divergent tissue distribution of tryptase versus cathepsin D and histamine-*N*-methyltransferase indicates that tryptase is one central constituent in skin biology.

B. Tryptase in Psoriasis and Lichen Planus

The role of mast cells in the pathophysiology of psoriasis has been emphasized by several investigators (27,41,72–74). One of the earliest morphological alterations in the developing psoriatic lesion is the appearance of degranulated mast cells together with endothelial swelling in postcapillary venules (72,73). In the chronic psoriatic lesion, increased numbers of mast cells are located immediately beneath the epidermis. Furthermore, mast cells with extruded secretory granules can be found in morphological contact with the lesional epidermis as well as in the epidermis (26,27,72,73) (Fig. 2a).

Practically all mast cells in the upper dermis of the psoriatic lesion contain immunoreactive chymase protein in addition to tryptase (Fig. 2), indicating that they are MC_{TC} cells which are increased in number (Fig. 3) (27). Although tryptase displays full enzyme activity in all mast cells of the psoriatic lesion (Fig. 2) (26,27), chymase has apparently become inactive in the upper part but not in the deeper part of the dermis (Fig. 3) (27,75). If healthy-looking psoriatic skin is slightly traumatized by the tape-stripping technique, a fresh psoriatic lesion develops on the area within 1 week, and an increase in mast cell number is observed in the lesion (27,74). The enzyme activity of chymase to-

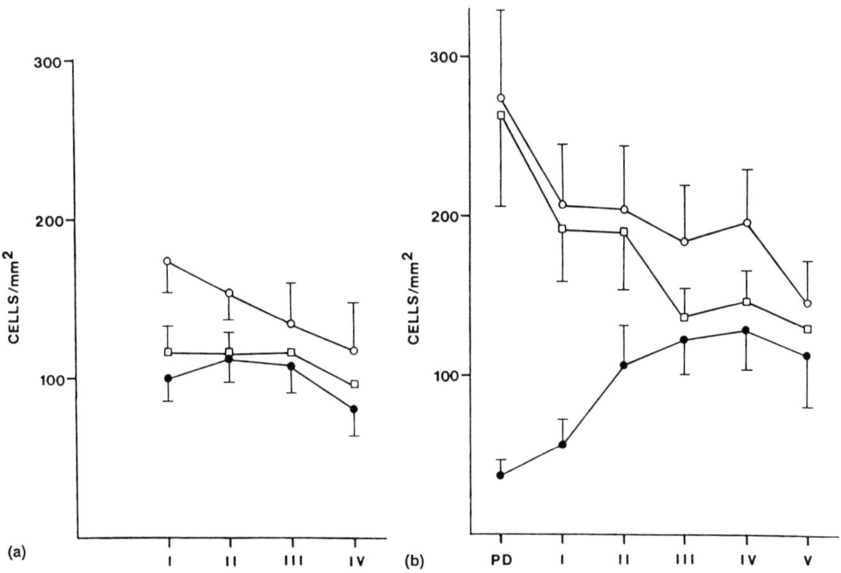

Figure 3 Distribution and density of tryptase- and chymase-containing mast cells in different levels of dermis in (a) nonlesional psoriatic skin and (b) lesional psoriatic skin. The abscissa represents the dermal levels I to V (the depth of each level is 0.2 mm) immediately below the papillary dermis (PD). ○—○, Tryptase-positive cells with Z-Gly–Pro–Arg–MNA; □—□, chymase-positive cells with monoclonal B7 antichymase antibody; ●—●, chymase-positive cells with Suc–Val–Pro–Phe–MNA as the substrate. Values are mean ± SEM (n = 13). (From Ref. 27.)

ward its specific substrate, Suc–Val–Pro–Phe–MNA, declines rapidly in mast cells of the upper dermis. As early as day 7 after tape stripping, 27 to 38% fewer mast cells with chymase activity are present in the young lesion. The first alterations in the distribution and number of tryptase-positive mast cells in the upper dermis occur relatively late, and they can be detected on day 14 after tape stripping as more frequent mast cell contacts with epidermis together with only a slight increase in mast cell count. A probable explanation for the apparent inactivation of chymase in the upper dermis of the psoriatic lesion is the localization of chymase inhibitors, α_1-antitrypsin, and α_1-antichymotrypsin (76) in most mast cells of lesional and nonlesional psoriatic skin (27).

In comparison to mast cells in psoriatic skin, tryptase-positive mast cells in cutaneous lichen planus show some differences. The major difference is that mast cells are clearly less frequent in the upper dermis of the lichen planus lesion than in that of psoriasis. In addition, the contacts between tryptase-positive mast cells and basal epidermal layer are approximately four to six times more frequent in the psoriatic lesion than in the lichen planus lesion. Furthermore, mast cells in the uppermost dermis of nonlesional lichen planus skin are more numerous than in nonlesional psoriatic or normal skin (77,78). Chymase activity in mast cells of the upper dermis of the lichen planus lesion is markedly diminished, as is also the case in psoriasis (77). Hence the relatively high count of tryptase-positive mast cells in nonlesional lichen planus skin supports the hypothesis that mast cells are involved in the initial phases of the developing disease. These cells are not likely to play any significant role in the damage of epidermal basal cells characteristic of lichen planus, but they could participate in the chronic inflammatory reaction of dermis, where tryptase-positive mast cells are numerous (77).

Development of a simultaneous double-staining method for tryptase-positive mast cells and neurofilament- or neuropeptide-positive sensory nerves has allowed the morphometric quantitation of apparent mast cell–nerve contacts in cutaneous psoriasis and lichen planus (41,78) (Fig. 4). In general, neurofilament-positive sensory nerves and their contacts with mast cells are increased in the psoriatic lesion, but this increase is not due solely to extensive inflammatory reaction in psoriasis since no such increase was observed in lichen planus (41,78). Sensory nerves stained immunohistochemically for neuropeptides SP and VIP are longer and more frequent in the chronic psoriatic lesion than in corresponding nonlesional skin. In fact, both of these neuropeptides can be observed in the developing psoriatic lesion as early as day 7 after tape stripping (41,79), with simultaneous decline of chymase activity (27). Because tryptase degrades VIP but not SP, and chymase cleaves both VIP and SP (36,37), neither active tryptase nor inactive chymase is able to control the SP-mediated neurogenic inflammation in the developing psoriatic lesion, but tryptase could

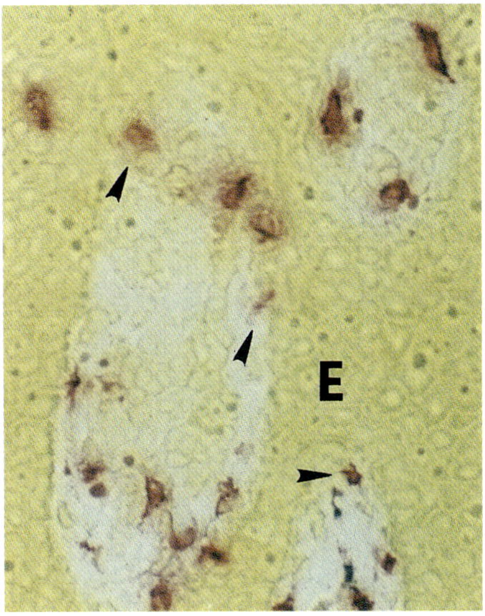

(a)

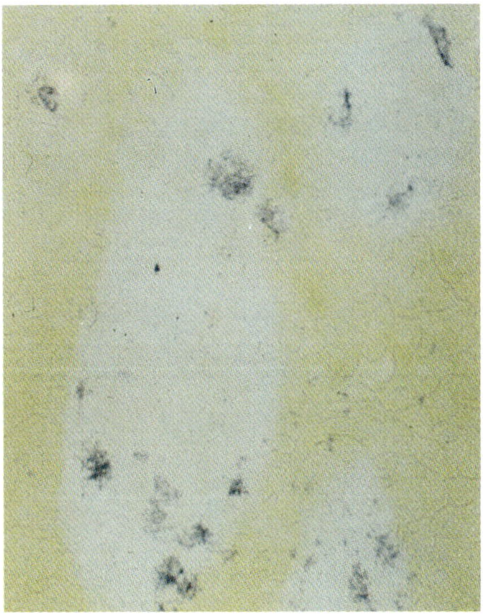

(b)

Figure 2 Section of lesional psoriatic skin stained with (a) Z–Gly–Pro–Arg–MNA as the substrate and Fast Garnet GBC as the chromogen for 15 min. This method stains tryptase activity in mast cells exclusively (bright red cells). After photographing, the tryptase stain was removed with 15% Tween 20 overnight, followed by staining the same section with (b) monoclonal B7 antichymase antibody. Most of the tryptase-positive cells exhibit chymase immunoreactivity indicating MC_{TC} cells. Note the chymase-negative MC_T cells (arrowheads); E, epidermis; magnification × 280. (From Ref. 27.)

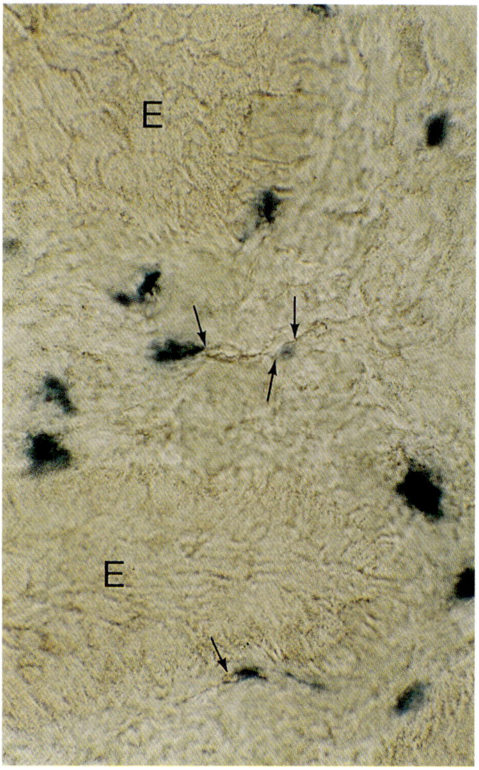

Figure 4 Delicate dermal nerve fibers positive for substance P (arrows) are seen in contact with tryptase-positive mast cells in a section of lesional psoriatic skin. Tryptase was stained with Z–Gly–Pro–Arg–MNA as the substrate and Fast Black K as the chromogen, which gives dark blue mast cells. E, epidermis; magnification × 570. (From Ref. 41.)

exert its activity against VIP. Our preliminary results suggest that psychic stress could also induce alterations in neuropeptides and mast cells of the psoriatic lesion (80).

C. Tryptase in Cutaneous Mastocytosis

Skin is a frequently affected organ in mastocytosis, and cutaneous lesions show a variety of clinical manifestations. Typically, lesional dermis is infiltrated by large numbers of mast cells that appear morphologically atypical but that function normally in response to immunological and nonimmunological secretory stimuli (82,83).

Mast cells in cutaneous mastocytosis exhibit high levels of active tryptase when detected enzyme-histochemically using different tryptic substrates or immunohistochemically with polyclonal antitryptase antibodies (62). Similarly, elevated levels of tryptase immunoreactivity have been detected in suction-blister fluid of urticaria pigmentosa (83). Both trypsin-like and chymotrypsin-like enzyme activities have been measured in significantly increased levels in high-salt extracts of urticaria pigmentosa lesions compared with those of nonlesional, normal, and hemangioma skin. This suggests that the mast cells that infiltrate the skin lesions in mastocytosis are of MC_{TC} type. Interestingly, chymotryptic activity increases as much as 15-fold, but tryptase activity increases only 3-fold in lesional skin compared to control groups (84), implying that mast cells in cutaneous mastocytosis are switched to synthesize large amounts of chymotryptic enzyme(s) over tryptase. Using immunohistochemical stains for tryptase and chymase, mast cells in cutaneous mastocytosis were shown subsequently to be of the MC_{TC} type exclusively (85,86).

D. Tryptase in Cutaneous Hypersensitivity Conditions

The crucial role for mast cells in cutaneous immediate-type allergic reactions is well recognized, but their function in late-phase reactions or in the rash in chronic atopic dermatitis has not been clear. Mast cells release numerous mediators upon activation, but tryptase determinations in biological fluids specifically reflect mast cell involvement in allergic inflammation (87). Although basophils contain significant amounts of histamine, only minute amounts of tryptase can be measured in them (0.04 pg per cell) (88).

Tryptase is released together with histamine during the immediate-type allergic reaction in skin. However, the time course needed for the appearance of tryptase in skin chamber fluid is slower than that needed for histamine. This probably reflects the slower diffusion rate of large tryptase–proteoglycan complexes from the site of mast cell activation (89). The release kinetics of tryptase and histamine are again different if the allergen exposure on skin is prolonged to 9 h. Both tryptase and histamine are released during the first hour of exposure, but thereafter tryptase levels in skin chamber fluid decline progressively to baseline by the fifth hour, while histamine levels fall down but remain constantly above baseline for the entire 9-h exposure period. The reason for this continuous low secretion of histamine is not clear; probably basophils are recruited to the site of allergic inflammation, or there could be variations in histamine and tryptase release from mast cells or synthesis in them (90,91). Anaphylactic release of tryptase together with histamine into the circulation has been detected after challenging sensitized patients to yellow jacket or honeybee stings under experimental conditions. The magnitude of this release was related to clinical severity of the reactions, possibly indicating a crucial role for mast cells and their major secretory protein, tryptase, in systemic anaphylaxis originated from skin challenge (92).

In acute vesicular atopic eczema, mast cells are degranulated but not increased in number. Upon chronic house dust mite exposure on mildly abraded skin of sensitized atopic dermatitis patients, mast cells have been shown to increase in number by 6 days, replacing the earlier basophil, eosinophil, and mononuclear cell infiltrates (93,94). Although mast cells are generally increased in number in lichenified lesions of atopic dermatitis, variations between different clinical patient groups have been reported; patients with concomitant atopic respiratory disease have significantly greater mast cell numbers in the upper dermis of the lesional skin than do patients without respiratory involvement (95). The mast cell type in the acute erythematous skin of four atopic dermatitis patients has been identified as MC_{TC} cells, although in three cases a substantial proportion (2 to 17%) of the mast cells belonged to MC_T cells. Of special interest is the occasional presence of MC_T cells in the epidermis among the basal cells (96). Using the double-staining method (27) (see Fig. 2), tryptase-positive mast cells in lesional skin of atopic dermatitis were shown to contain significantly higher levels of interleukin-4 (IL-4) immunoreactivity with respect to staining intensity and the percentage of IL-4–positive mast cells than those in nonlesional skin. These characteristics link mast cells to T_H2-lymphocyte maturation and proliferation in atopic dermatitis (97,98).

In chronic idiopathic urticaria, mast cells have been found in increased numbers together with T lymphocytes and monocytes, suggesting crosstalking mechanisms between these cells as one etiopathogenic factor (99). Substances such as tumor necrosis factor α released from monocytes can induce mast cell activation and tryptase and histamine release (99,100). Histamine, in turn, has several effects on modulating the immune system (101). Interestingly, significantly elevated levels of tryptase have been measured in suction-blister fluids of involved and even noninvolved skin of patients with chronic idiopathic urticaria, indicating that mast cells are activated readily in this hypersensitivity disorder (83).

E. Tryptase in Cutaneous Connective Tissue Disorders

Several lines of evidence suggest an important role for mast cells in connective tissue reactions. Mast cells have an intimate relationship with fibroblasts and endothelial cells, which are capable of ingesting and degrading mast cell granules (102). In fact, several mast cell mediators have been reported to induce the growth and/or collagen synthesis of fibroblasts. These mediators include histamine (103), interleukin-4 (104), and tryptase (56,57). On the other hand, factors released from fibroblasts seem to be necessary for the growth of MC_{TC} mast cells (105). Mast cells by themselves may produce basement membrane components, type IV collagen, laminin, and heparan sulfate, thus participating in early overproduction of basement membrane components in fibrosis or in areas of injury (106). As presented above, several proteins synthesized and secreted by fibroblasts, such as fibronectin, type IV collagenase/gelatinase (58), and prostromelysin (60), serve as the substrate for mast cell tryptase. Hence tryptase has a wide variety of effects on fibroblasts and extracellular matrix remodeling.

The assumption that mast cells are involved in sclerotic skin processes is based on the findings that increased mast cell numbers have been counted in the sclerotic skin of porphyria cutanea tarda and scleroderma (107–109). Although mast cells have been found in increased numbers in the early phase of diffuse scleroderma, these cells are less frequent in the late phase of skin sclerosis (107,108). One disadvantage of the earlier studies is the use of formaldehyde fixation for skin specimens (107–109). This fixative is not optimal for cutaneous mast cells, and a part of these cells can escape detection (110,111). Recently, mast cells and their subtypes in diffuse scleroderma have been analyzed for tryptase and chymase immunoreactivity (112). In that study, mast cells were rather decreased in number, which, however, tended to increase with disease duration, exceeding normal values. Furthermore, in 12 of 22 lesional skin specimens, MC_T cells were observed, and in 5 specimens the MC_T cell was of the exclusive mast cell type. Eight skin specimens exhibited apparent mast cell fragmentation, which seemed to coincide with severe clinical itching (112). Thus significant alterations in mast cells and their subtypes have been observed, which encourages further studies on the role of MC_T cell and tryptase in scleroderma.

Participation of mast cells in tissue-repair and wound-healing processes has been suggested in several studies that report rapid increase in mast cell numbers at the site of skin injury (74,113). Occasionally, this may lead to prominent, persisting mast cell infiltrations in hypertrophic scars and keloids (113,114). These findings may be due in part to the stimulatory effects of mast cells on fibroblast growth, collagen synthesis (57,103,104), and angiogenesis (115). The involvement of mast cells in chronic ulcers has been suggested as well. Mast cells seem to have a particular predilection for epidermis in venous stasis ulcers rather than in decubitus and arteriosclerotic ulcers (116). Whether mast cell tryptase can influence the growth of keratinocytes is not known but would be very exciting to investigate in the epithelialization of wounds as well as in psoriasis (26,27). In contrast to tissue repair processes in skin injuries, mast cell involvement in solar skin degeneration has been described previously. In chronic sun damage of skin, called *chronic heliodermatitis,* mast cells are frequently present in inflamed areas, and many of them are in close apposition to fibroblasts (117).

VI. CONCLUSIONS

Increasing experimental data indicate that mast cells have many different functions in skin physiology and pathology. Of special interest is the major secretory protein of skin mast cells, tryptase, which shows properties similar to those of tryptases found in other human tissues. The biological role of skin tryptase is still an open question, but its abundant presence in active form in normal skin and in different types of skin diseases suggests that tryptase functions in promoting and modulating inflammation and in remodeling extracellular matrix to make space for inflammatory and malignant cells. In addition, tryptase could participate in tissue-repair mechanisms after skin injuries.

ACKNOWLEDGMENTS

The authors wish to thank the Medical Council of the Academy of Finland for financial support.

REFERENCES

1. J. E. Fräki, G. S. Lazarus, and V. K. Hopsu-Havu, Protein catabolism in the skin, in *Biochemistry and Physiology of the Skin* (L. A. Goldsmith, ed.), Oxford University Press, New York, 1983, p. 338.
2. T. Yamura and F. E. Cormia, Studies on human skin protease, *J. Invest. Dermatol. 37*:121 (1961).
3. H. E. J. Seppä, C. T. Jansén, and V. K. Hopsu-Havu, Proteolytic enzymes in the skin. V. Extraction and column chromatographic separation of the human skin proteinases, *Acta Dermato-Venereol. (Stockh.) 51*:35 (1971).
4. J. E. Fräki and V. K. Hopsu-Havu, Human skin proteases: differential extraction of proteases and of endogenous protease inhibitors, *Arch. Dermatol. Forsch. 242*:329 (1972).
5. J. E. Fräki and V. K. Hopsu-Havu, Human skin proteases: fractionation and characterization, *Arch. Dermatol. Forsch. 243*:52 (1972).
6. J. E. Fräki and V. K. Hopsu-Havu, Human skin proteases: separation and characterization of two alkaline proteases, one splitting trypsin and the other chymotrypsin substrates, *Arch. Dermatol. Res. 253*:261 (1975).

7. J. E. Fräki, Human skin proteases: effect of separated proteases on vascular permeability and leukocyte emigration in skin, *Acta Dermato-Venereol. (Stockh.) 57*:393 (1977).
8. G. G. Glenner and L. A. Cohen, Histochemical demonstration of a species-specific trypsin-like enzyme in mast cells, *Nature 185*:846 (1960).
9. V. K. Hopsu and G. G. Glenner, A histochemical enzyme kinetic system applied to the trypsin-like amidase and esterase activity in human mast cells, *J. Cell Biol. 17*:503 (1963).
10. L. B. Schwartz, R. A. Lewis, and K. F. Austen, Tryptase from human pulmonary mast cells: purification and characterization, *J. Biol. Chem. 256*:11939 (1981).
11. J. E. Fräki, N. M. Schechter, I. T. Harvima, and G. S. Lazarus, Human skin tryptase, in *The Biological Role of Proteinases and Their Inhibitors in Skin* (H. Ogawa, G. S. Lazarus, and V. K. Hopsu-Havu, eds.), University of Tokyo Press, Tokyo, 1985, p. 127.
12. I. T. Harvima, N. M. Schechter, R. J. Harvima, and J. E. Fräki, Human skin tryptase: purification, partial characterization and comparison with human lung tryptase, *Biochim. Biophys. Acta 957*:71 (1988).
13. I. Harvima, Human skin tryptase: purification, characterization, and localization in mast cells, doctoral thesis, publications of the University of Kuopio, series Medicine, original reports 6/1989.
14. I. T. Harvima, K. Karkola, R. J. Harvima, A. Naukkarinen, H. Neittaanmäki, M. Horsmanheimo, and J. E. Fräki, Biochemical and histochemical evaluation of tryptase in various human tissues, *Arch. Dermatol. Res. 281*:231 (1989).
15. T. J. Smith, M. W. Hougland, and D. A. Johnson, Human lung tryptase: purification and characterization, *J. Biol. Chem. 259*:11046 (1984).
16. J. A. Cromlish, N. G. Seidah, M. Marcinkiewicz, J. Hamelin, D. A. Johnson, and M. Chrétien, Human pituitary tryptase: molecular forms, NH_2-terminal sequence, immunocytochemical localization, and specificity with prohormone and fluorogenic substrates, *J. Biol. Chem. 262*:1363 (1987).
17. L. B. Schwartz, T. R. Bradford, D. C. Lee, and J. F. Chlebowski, Immunologic and physicochemical evidence for conformational changes occurring on conversion of human mast cell tryptase from active tetramer to inactive monomer, *J. Immunol. 144*:2304 (1990).
18. J. S. Miller, E. H. Westin, and L. B. Schwartz, Cloning and characterization of complementary DNA for human tryptase, *J. Clin. Invest. 84*:1188 (1989).
19. J. S. Miller, G. Moxley, and L. B. Schwartz, Cloning and characterization of a second complementary DNA for human tryptase, *J. Clin. Invest. 86*:864 (1990).
20. P. Vanderslice, S. M. Ballinger, E. K. Tam, S. M. Goldstein, C. S. Craik, and G. H. Caughey, Human mast cell tryptase: multiple cDNAs and genes reveal a multigene serine protease family, *Proc. Natl. Acad. Sci. USA 87*:3811 (1990).
21. I. T. Harvima, R. J. Harvima, T. O. Eloranta, and J. E. Fräki, The allosteric effect of salt on human mast cell tryptase, *Biochim. Biophys. Acta 956*:133 (1988).
22. S. C. Alter and L. B. Schwartz, Effect of histamine and divalent cations on the activity and stability of tryptase from human mast cells, *Biochim. Biophys. Acta 991*:426 (1989).

23. T. Tanaka, B. J. McRae, K. Cho, R. Cook, J. E. Fräki, D. A. Johnson, and J. C. Powers, Mammalian tissue trypsin-like enzymes: comparative reactivities of human skin tryptase, human lung tryptase and bovine trypsin with peptide 4-nitroanilide and thioester substrates, *J. Biol. Chem. 258*:13552 (1983).
24. S. C. Alter, D. D. Metcalfe, T. R. Bradford, and L. B. Schwartz, Regulation of human mast cell tryptase: effects of enzyme concentration, ionic strength and the structure and negative charge density of polysaccharides, *Biochem. J. 248*:821 (1987).
25. S. C. Alter, J. A. Kramps, A. Janoff, and L. B. Schwartz, Interactions of human mast cell tryptase with biological protease inhibitors, *Arch. Biochem. Biophys. 276*:26 (1990).
26. I. T. Harvima, A. Naukkarinen, R. J. Harvima, and M. Horsmanheimo, Enzyme- and immunohistochemical localization of mast cell tryptase in psoriatic skin, *Arch. Dermatol. Res. 281*:387 (1989).
27. I. T. Harvima, A. Naukkarinen, K. Paukkonen, R. J. Harvima, M.-L. Aalto, L. B. Schwartz, and M. Horsmanheimo, Mast cell tryptase and chymase in developing and mature psoriatic lesions, *Arch. Dermatol. Res. 285*:184 (1993).
28. A. A. Irani, T. R. Bradford, C. L. Kepley, N. M. Schechter, and L. B. Schwartz, Detection of MC_T and MC_{TC} types of human mast cells by immunohistochemistry using new monoclonal anti-tryptase and anti-chymase antibodies, *J. Histochem. Cytochem. 37*:1509 (1989).
29. T. Blom and L. Hellman, Characterization of a tryptase mRNA expressed in the human basophil cell line KU812, *Scand. J. Immunol. 37*:203 (1993).
30. L. B. Schwartz, M. S. Kawahara, T. E. Hugli, D. Vik, D. T. Fearon, and K. F. Austen, Generation of C3a anaphylatoxin from human C3 by human mast cell tryptase, *J. Immunol. 130*:1891 (1983).
31. L. B. Schwartz, T. R. Bradford, B. H. Littman, and B. U. Wintroub, The fibrinogenolytic activity of purified tryptase from human lung mast cells, *J. Immunol. 135*:2762 (1985).
32. M. Maier, J. Spragg, and L. B. Schwartz, Inactivation of human high molecular weight kininogen by human mast cell tryptase, *J. Immunol. 130*:2352 (1983).
33. I. T. Harvima, R. J. Harvima, I. M. Penttilä, T. O. Eloranta, M. Horsmanheimo, and J. E. Fräki, Effect of human mast cell tryptase on human plasma proenzymes, *Int. Arch. Allergy Appl. Immunol. 90*:104 (1989).
34. S. C. Alter, B. Lämmle, J. H. Griffin, and L. B. Schwartz, Tryptase from human mast cells does not activate purified human Hageman factor, *Clin. Allergy 18*:151 (1988).
35. L. B. Schwartz, T. R. Bradford, and J. H. Griffin, The effect of tryptase from human mast cells on human prekallikrein, *Biochem. Biophys. Res. Commun. 129*:76 (1985).
36. G. H. Caughey, F. Leidig, N. F. Viro, and J. A. Nadel, Substance P and vasoactive intestinal peptide degradation by mast cell tryptase and chymase, *J. Pharmacol. Exp. Ther. 244*:133 (1988).
37. E. K. Tam and G. H. Caughey, Degradation of airway neuropeptides by human lung tryptase, *Am. J. Respir. Cell Mol. Biol. 3*:27 (1990).
38. A. F. Walls, S. D. Brain, A. Desai, P. J. Jose, E. Hawkins, M. K. Church, and

T. J. Williams, Human mast cell tryptase attenuates the vasodilator activity of calcitonin gene-related peptide, *Biochem. Pharmacol. 43*:1243 (1992).
39. G. M. Franconi, P. D. Graf, S. C. Lazarus, J. A. Nadel, and G. H. Caughey, Mast cell tryptase and chymase reverse airway smooth muscle relaxation induced by vasoactive intestinal peptide in the ferret, *J. Pharmacol. Exp. Ther. 248*:947 (1989).
40. J. Wallengren and R. Hakanson, Effects of substance P, neurokinin A and calcitonin gene-related peptide in human skin and their involvement in sensory nerve-mediated responses, *Eur. J. Pharmacol. 143*:267 (1987).
41. A. Naukkarinen, I. Harvima, K. Paukkonen, M.-L. Aalto, and M. Horsmanheimo, Immunohistochemical analysis of sensory nerves and neuropeptides, and their contacts with mast cells in developing and mature psoriatic lesions, *Arch. Dermatol. Res. 285*:341 (1993).
42. G. Appleyard and M. Tisdale, Inhibition of the growth of human coronavirus 229E by leupeptin, *J. Gen. Virol. 66*:363 (1985).
43. K. B. Andersen and H. Skov, Retrovirus-induced cell fusion is enhanced by protease treatment, *J. Gen. Virol. 70*:1921 (1989).
44. K. Sakai, Y. Kawaguchi, Y. Kishino, and H. Kido, Electron immunohistochemical localization in rat bronchiolar epithelial cells of tryptase clara, which determines the pneumotropism and pathogenicity of Sendai virus and influenza virus, *J. Histochem. Cytochem. 41*:89 (1993).
45. A. Koito, T. Hattori, T. Murakami, S. Matsushita, Y. Maeda, T. Yamamoto, and K. Takatsuki, A neutralizing epitope of human immunodeficiency virus type 1 has homologous amino acid sequences with the active site of inter-α-trypsin inhibitor, *Int. Immunol. 1*:613 (1989).
46. H. Kido, A. Fukutomi, and N. Katunuma, A novel membrane-bound serine esterase in human $T4^+$ lymphocytes immunologically reactive with antibody inhibiting syncytia induced by HIV-1: purification and characterization, *J. Biol. Chem. 265*:21979 (1990).
47. G. J. Clements, M. J. Price-Jones, P. E. Stephens, C. Sutton, T. F. Schulz, P. R. Clapham, J. A. McKeating, M. O. McClure, S. Thomson, M. Marsh, J. Kay, R. A. Weiss, and J. P. Moore, The V3 loops of the HIV-1 and HIV-2 surface glycoproteins contain proteolytic cleavage sites: a possible function in viral fusion? *AIDS Res. Hum. Retrovir. 7*:3 (1991).
48. Q. J. Sattentau and J. P. Moore, Conformational changes induced in the human immunodeficiency virus envelope glycoprotein by soluble CD4 binding, *J. Exp. Med. 174*:407 (1991).
49. A. Werner and J. A. Levy, Human immunodeficiency virus type 1 envelope gp120 is cleaved after incubation with recombinant soluble CD4, *J. Virol. 67*:2566 (1993).
50. I. T. Harvima, R. J. Harvima, G. Nilsson, L. Ivanoff, and L. B. Schwartz, Separation and partial characterization of proteinases with substrate specificity for basic amino acids from human MOLT-4 T lymphocytes: identification of those inhibited by variable-loop-V3 peptides of HIV-1 (human immunodeficiency virus-1) envelope glycoprotein, *Biochem. J. 292*:711 (1993).

51. M. Duvic, T. M. Johnson, R. P. Rapini, T. Freese, G. Brewton, and A. Rios, Acquired immunodeficiency syndrome–associated psoriasis and Reiter's syndrome, *Arch. Dermatol. 123*:1622 (1987).
52. S. E. Mahoney, M. Duvic, B. J. Nickoloff, M. Minshall, L. C. Smith, C. E. M. Griffiths, S. W. Paddock, and D. E. Lewis, Human immunodeficiency virus (HIV) transcripts identified in HIV-related psoriasis and Kaposi's sarcoma lesions, *J. Clin. Invest. 88*:174 (1991).
53. F. van Neer, V. Zemelman, R. Cerio, J. Langtry, and R. C. D. Staughton, The role of factor XIIIa–positive dermal dendrocytes in HIV-1 positive psoriatics, *Br. J. Dermatol. 128*:29 (1993).
54. P. R. A. Meylan, E. K. Tam, R. S. Kornbluth, and D. D. Richman, HIV infectivity is not augmented by treatment with trypsin, factor Xa or human mast-cell tryptase, *AIDS 6*:128 (1992).
55. B. R. Zetter, L. B. Chen, and J. M. Buchanan, Effects of protease treatment on growth, morphology, adhesion, and cell surface proteins of secondary chick embryo fibroblasts, *Cell 7*:407 (1976).
56. S. J. Ruoss, T. Hartmann, and G. H. Caughey, Mast cell tryptase is a mitogen for cultured fibroblasts, *J. Clin. Invest. 88*:493 (1991).
57. T. Hartmann, S. J. Ruoss, W. W. Raymond, K. Seuwen, and G. H. Caughey, Human tryptase as a potent, cell-specific mitogen: role of signaling pathways in synergistic responses, *Am. J. Physiol. 262*:L528 (1992).
58. J. Lohi, I. Harvima, and J. Keski-Oja, Pericellular substrates of human mast cell tryptase: 72,000 dalton gelatinase and fibronectin, *J. Cell. Biochem. 50*:337 (1992).
59. D. A. Johnson and T. E. Cawston, Human lung mast cell tryptase fails to activate procollagenase or degrade proteoglycan, *Biochem. Biophys. Res. Commun. 132*:453 (1985).
60. B. L. Gruber, M. J. Marchese, K. Suzuki, L. B. Schwartz, Y. Okada, H. Nagase, and N. S. Ramamurthy, Synovial procollagenase activation by human mast cell tryptase dependence upon matrix metalloproteinase 3 activation, *J. Clin. Invest. 84*:1657 (1989).
61. R. A. J. Eady, T. Cowen, T. F. Marshall, V. Plummer, and M. W. Greaves, Mast cell population density, blood vessel density and histamine content in normal human skin, *Br. J. Dermatol. 100*:623 (1979).
62. I. T. Harvima, A. Naukkarinen, R. J. Harvima, and J. E. Fräki, Immunoperoxidase and enzyme-histochemical demonstration of human skin tryptase in cutaneous mast cells in normal and mastocytoma cells, *Arch. Dermatol. Res. 280*:363 (1988).
63. A. A. Irani, N. M. Schechter, S. S. Craig, G. DeBlois, and L. B. Schwartz, Two types of human mast cells that have distinct neutral protease compositions, *Proc. Natl. Acad. Sci. USA 83*:4464 (1986).
64. A. A. Irani, S. M. Goldstein, B. U. Wintroub, T. Bradford, and L. B. Schwartz, Human mast cell carboxypeptidase: selective localization to MC_{TC} cells, *J. Immunol. 147*:247 (1991).
65. N. M. Schechter, A. A. Irani, J. L. Sprows, J. Abernethy, B. Wintroub, and L.

B. Schwartz, Identification of a cathepsin G-like proteinase in the MC_{TC} type of human mast cells, *J. Immunol. 145*:2652 (1990).
66. S. S. Craig, N. M. Schechter, and L. B. Schwartz, Ultrastructural analysis of human T and TC mast cells identified by immunoelectron microscopy, *Lab. Invest. 58*:682 (1988).
67. S. S. Craig and L. B. Schwartz, Human MC_{TC} type of mast cell granule: the uncommon occurrence of discrete scrolls associated with focal absence of chymase, *Lab. Invest. 63*:581 (1990).
68. N. Weidner and K. F. Austen, Heterogeneity of mast cells at multiple body sites: fluorescent determination of avidin binding and immunofluorescent determination of chymase, tryptase, and carboxypeptidase content, *Pathol. Res. Pract. 189*:156 (1993).
69. L. B. Schwartz, A. A. Irani, K. Roller, M. Castells, and N. M. Schechter, Quantitation of histamine, tryptase, and chymase in dispersed human T and TC mast cells, *J. Immunol. 138*:2611 (1987).
70. S. M. Goldstein, J. Leong, L. B. Schwartz, and D. Cooke, Protease composition of exocytosed human skin mast cell protease–proteoglycan complexes: tryptase resides in a complex distinct from chymase and carboxypeptidase, *J. Immunol. 148*:2475 (1992).
71. D. Whitaker-Menezes, N. M. Schechter, and G. F. Murphy, Localization of cathepsin G to cutaneous mast cell granules: evidence for regional segregation of mast cell proteinases, *Clin. Res. 41*:473A (1993) (abstract).
72. I. Brody, Mast cell degranulation in the evolution of acute eruptive guttate psoriasis vulgaris, *J. Invest. Dermatol. 82*:460 (1984).
73. C. Schubert and E. Christophers, Mast cells and macrophages in early relapsing psoriasis, *Arch. Dermatol. Res. 277*:352 (1985).
74. B. Toruniowa and S. Jablonska, Mast cells in the initial stages of psoriasis, *Arch. Dermatol. Res. 280*:189 (1988).
75. I. T. Harvima, A. Naukkarinen, R. J. Harvima, M.-L. Aalto, H. Neittaanmäki, and M. Horsmanheimo, Quantitative enzyme-histochemical analysis of tryptase- and chymase-containing mast cells in psoriatic skin, *Arch. Dermatol. Res. 282*:428 (1990).
76. N. M. Schechter, J. L. Sprows, O. L. Schoenberger, G. S. Lazarus, B. S. Cooperman, and H. Rubin, Reaction of human skin chymotrypsin-like proteinase chymase with plasma proteinase inhibitors, *J. Biol. Chem. 264*:21308 (1989).
77. I. T. Harvima, A. Naukkarinen, R. J. Harvima, M.-L. Aalto, H. Neittaanmäki, and M. Horsmanheimo, Quantitation of tryptase- and chymase-containing mast cells in cutaneous lichen planus, *Acta Dermato-Venereol. (Stockh.) 71*:394 (1991).
78. A. Naukkarinen, I. T. Harvima, M.-L. Aalto, R. J. Harvima, and M. Horsmanheimo, Quantitative analysis of contact sites between mast cells and sensory nerves in cutaneous psoriasis and lichen planus based on a histochemical double staining technique, *Arch. Dermatol. Res. 283*:433 (1991).
79. A. Naukkarinen, B. J. Nickoloff, and E. M. Farber, Quantification of cutaneous sensory nerves and their substance P content in psoriasis, *J. Invest. Dermatol. 92*:126 (1989).

80. I. T. Harvima, H. Viinamäki, A. Naukkarinen, K. Paukkonen, H. Neittaanmäki, R. J. Harvima, and M. Horsmanheimo, Association of cutaneous mast cells and sensory nerves with psychic stress in psoriasis, *Psychother. Psychosom. 60*:168 (1993).
81. N. A. Soter, The skin in mastocytosis, *J. Invest. Dermatol. 96*:32S (1991).
82. M. D. Tharp, B. Chaker, M. J. Glass, R. Burton, and L. L. Seelig, In vitro functional reactivities of cutaneous mast cells from patients with mastocytosis, *J. Invest. Dermatol. 89*:264 (1987).
83. B. Deleuran, M. Kristensen, C. Grønhøj Larsen, P. Matsson, I. Enander, A.-S. Andersson, and K. Thestrup-Pedersen, Increased tryptase levels in suction-blister fluid from patients with urticaria, *Br. J. Dermatol. 125*:14 (1991).
84. J. E. Fräki, N. M. Schechter, and G. S. Lazarus, Serine proteinases in human cutaneous mastocytosis, *Arch. Dermatol. Res. 278*:363 (1986).
85. A. A. Irani, M. M. Garriga, D. D. Metcalfe, and L. B. Schwartz, Mast cells in cutaneous mastocytosis: accumulation of the MC_{TC} type, *Clin. Exp. Allergy 20*:53 (1990).
86. N. Weidner, R. F. Horan, and K. F. Austen, Mast-cell phenotype in indolent forms of mastocytosis: ultrastructural features, fluorescence detection of avidin binding, and immunofluorescent determination of chymase, tryptase, and carboxypeptidase, *Am. J. Pathol. 140*:847 (1992).
87. I. T. Harvima and L. B. Schwartz, Mast cell products in the monitoring of allergic inflammation in *Progress in Allergy and Clinical Immunology, Vol. 2,* (Miyamoto T, Okuda M, eds.). Hogrefe & Huber Publishers, Seattle, pp. 184–190, 1992.
88. M. Castells, A. A. Irani, and L. B. Schwartz, Evaluation of human peripheral blood leukocytes for mast cell tryptase, *J. Immunol. 138*:2184 (1987).
89. L. B. Schwartz, P. C. Atkins, T. R. Bradford, P. Fleekop, M. Shalit, and B. Zweiman, Release of tryptase together with histamine during the immediate cutaneous response to allergen, *J. Allergy Clin. Immunol. 80*:850 (1987).
90. M. Shalit, L. B. Schwartz, N. Golzar, C. von Allmen, M. Valenzano, P. Fleekop, P. C. Atkins, and B. Zweiman, Release of histamine and tryptase in vivo after prolonged cutaneous challenge with allergen in humans, *J. Immunol. 141*:821 (1988).
91. M. Shalit, L. B. Schwartz, C. von Allmen, P. C. Atkins, R. M. Lavker, and B. Zweiman, Release of histamine and tryptase during continuous and interrupted cutaneous challenge with allergen in humans, *J. Allergy Clin. Immunol. 86*:117 (1990).
92. P.-W. G. van der Linden, C. E. Hack, J. Poortman, Y. C. Vivie-Kipp, A. Struyvenberg, and J. K. van der Zwan, Insect-sting challenge in 138 patients: relation betwen clinical severity of anaphylaxis and mast cell activation, *J. Allergy Clin. Immunol. 90*:110 (1992).
93. M. C. Mihm, N. A. Soter, H. F. Dvorak, and K. F. Austen, The structure of normal skin and the morphology of atopic eczema, *J. Invest. Dermatol. 67*:305 (1976).
94. E. B. Mitchell, J. Crow, G. Williams, and T. A. E. Platts-Mills, Increase in skin mast cells following chronic house dust mite exposure, *Br. J. Dermatol. 114*:65 (1986).

95. H. Sugiura, Y. Hirota, and M. Uehara, Heterogeneous distribution of mast cells in lichenified lesions of atopic dermatitis, *Acta Dermato-Venereol. (Stockh.) Suppl. 144*:115 (1989).
96. A. A. Irani, H. A. Sampson, and L. B. Schwartz, Mast cells in atopic dermatitis, *Allergy 44 (Suppl. 9)*:31 (1989).
97. L. Horsmanheimo, I. T. Harvima, A. Järvikallio, R. J. Harvima, A. Naukkarinen, and M. Horsmanheimo, Mast cells are one major source of interleukin-4 in atopic dermatitis. *Br. J. Dermatol. 131*: 348–353, 1994.
98. F. C. van Reijsen, C. A. F. M. Bruijnzeel-Koomen, F. S. Kalthoff, E. Maggi, S. Romagnani, J. K. T. Westland, and G. C. Mudde, Skin-derived aeroallergen-specific T-cell clones of Th2 phenotype in patients with atopic dermatitis, *J. Allergy Clin. Immunol. 90*:184 (1992).
99. J. Elias, E. Boss, and A. P. Kaplan, Studies of the cellular infiltrate of chronic idiopathic urticaria: prominence of T-lymphocytes, monocytes, and mast cells, *J. Allergy Clin. Immunol. 78*:914 (1986).
100. F. J. van Overveld, Ph. G. Jorens, M. Rampart, W. De Backer, and P. A. Vermeire, Tumor necrosis factor stimulates human skin mast cells to release histamine and tryptase, *Clin. Exp. Allergy 21*:711 (1991).
101. D. J. Beer and R. E. Rocklin, Histamine modulation of lymphocyte biology: membrane receptors, signal transduction, and functions, *Crit. Rev. Immunol. 7*:55 (1987).
102. F. M. Atkins, M. M. Friedman, P. V. Subba Rao, D. D. Metcalfe, Interactions between mast cells, fibroblasts and connective tissue components, *Int. Arch. Allergy Appl. Immunol. 77*:96 (1985).
103. A. Hatamochi, H. Ueki, C. Mauch, and T. Krieg, Effect of histamine on collagen and collagen m-RNA production in human skin fibroblasts, *J. Dermatol. Sci. 2*:407 (1991).
104. P. Gillery, C. Fertin, J. F. Nicolas, F. Chastang, B. Kalis, J. Banchereau, and F. X. Maquart, Interleukin-4 stimulates collagen gene expression in human fibroblast monolayer cultures: potential role in fibrosis, *FEBS Lett. 302*:231 (1992).
105. T. Ishizaka, T. Furitsu, and N. Inagaki, In vitro development and functions of human mast cells, *Int. Arch. Allergy Clin. Immunol. 94*:116 (1991).
106. H. L. Thompson, P. D. Burbelo, G. Gabriel, Y. Yamada, and D. D. Metcalfe, Murine mast cells synthesize basement membrane components: a potential role in early fibosis, *J. Clin. Invest. 87*:619 (1991).
107. K. Nishioka, Y. Kobayashi, I. Katayama, and C. Takijiri, Mast cell numbers in diffuse scleroderma, *Arch. Dermatol. 123*:205 (1987).
108. H. N. Claman, On scleroderma: mast cells, endothelial cells, and fibroblasts, *JAMA 262*:1206 (1989).
109. W. Torinuki, K. Kudoh, and H. Tagami, Increased mast cell numbers in the sclerotic skin of porphyria cutanea tarda, *Dermatologica 178*:75 (1989).
110. J. H. Olafsson, G. Roupe, and L. Enerbäck, Dermal mast cells in mastocytosis: fixation, distribution and quantitation, *Acta Dermato-Venereol. (Stockh.) 66*:16 (1986).
111. A. C. Markey, L. J. Churchill, and D. M. MacDonald, Human cutaneous mast

cells: a study of fixative and staining reactions in normal skin, *Br. J. Dermatol. 120*:625 (1989).
112. A. A. Irani, B. L. Gruber, L. D. Kaufman, M. B. Kahaleh, and L. B. Schwartz, Mast cell changes in scleroderma: presence of MC_T cells in the skin and evidence of mast cell activation, *Arthritis Rheum. 35*:933 (1992).
113. J. D. Reich, A. L. Cazzaniga, P. M. Mertz, F. A. Kerdel, and W. H. Eaglstein, The effect of electrical stimulation on the number of mast cells in healing wounds, *J. Am. Acad. Dermatol. 25*:40 (1991).
114. S. S. Craig, G. DeBlois, L. B. Schwartz, Mast cells in human keloid, small intestine, and lung by an immunoperoxidase technique using a murine monoclonal antibody against tryptase, *Am. J. Pathol. 124*:427 (1986).
115. R. M. Marks, W. R. Roche, M. Czerniecki, R. Penny, and D. S. Nelson, Mast cell granules cause proliferation of human microvascular endothelial cells, *Lab. Invest. 55*:289 (1986).
116. L. L. Bolton and W. Montagna, Mast cells in human ulcers, *Am. J. Dermatopathol. 15*:133 (1993).
117. R. M. Lavker and A. M. Kligman, Chronic heliodermatitis: a morphologic evaluation of chronic actinic dermal damage with emphasis on the role of mast cells, *J. Invest. Dermatol. 90*:325 (1988).

4

Chymotrypsin-like Proteinases of Human Skin Mast Cells

Norman M. Schechter
University of Pennsylvania School of Medicine, Philadelphia, Pennsylvania

I. INTRODUCTION

Mast cells are inflammatory cells that reside in connective and mucosal tissue (1). Like other inflammatory cells produced from stem cells in the bone marrow, they express and store relatively large concentrations of serine class proteolytic enzymes (Table 1). Serine proteinases are enzymatically active at neutral pH, suggesting that their release from inflammatory cells may affect extracellular environments.

The subset of human mast cells found in skin, MC_{TC}, contains at least two serine proteinases, termed skin chymase and skin tryptase according to the nomenclature for mast cell proteinases suggested originally by Lagunoff and Benditt (2). These proteinases are endopeptidases with general substrate specificities paralleling those of bovine pancreatic chymotrypsin and trypsin, respectively. There also is evidence that cutaneous mast cells contain a second chymotrypsin-like proteinase similar to cathepsin G, a major proteinase of neutrophils and monocytes (Table 1). In this chapter we review the work identifying human skin chymase as a mast cell constituent, describe the biochemical properties, regulation, and possible biological substrates of this proteinase, and then review the evidence for the presence of cathepsin G in mast cells.

II. IDENTIFICATION OF CHYMASE AND TRYPTASE IN SKIN

Histochemical studies examining the staining of skin tissue with chromogenic substrates hydrolyzed by bovine pancreatic chymotrypsin and trypsin were the

Table 1 Serine Proteinases Stored Within Human Inflammatory Cells

Cell type	Proteinase	Specificity[a]	μg/10⁶ cells
Neutrophil[b]	Cathespin G	Chymotrypsin-like	1
	Neutrophil elastase	Elastase-like	1
	Proteinase 3	Elastase-like	>0.4
	Azurocidin	Catalytically inactive[e]	>0.2
Monocyte[c]	Cathepsin G	Chymotrypsin-like	0.05
	Neutrophil elastase	Elastase-like	0.05
Lymphocyte[d]			
Cytotoxic and natural killer	Granzyme A	Trypsin-like	ND[e]
	Granzyme B	Asp-ase	
Mast cell[f]			
MC_{TC}	Tryptase	Trypsin-like	35
	Chymase	Chymotrypsin-like	4.5
	Cathepsin G	Chymotrypsin-like	ND[e]
MC_T	Tryptase	Trypsin-like	11

[a]Trypsin-like proteinases cleave proteins at positively charged residues (Lys–X and Arg–X bonds); chymotrypsin-like proteinases cleave at large hydrophobic residues (Trp–X, Try–X, Phe–X, Leu–X, and Met–X bonds); and elastase-like proteinases cleave at small hydrophobic residues (Ala–X and Val–X bonds).
[b]References 3 to 6. Azurocidin is structurally similar to neutrophil elastase, but the active-site Ser residue is replaced by a Gly residue, making the protein catalytically inactive.
[c]Reference 3.
[d]References 7 to 10.
[e]ND, not determined.
[f]References 11 and 12.

first to demonstrate that cells within the dermis of human skin contain chymotrypsin and trypsin-like proteinases (13–15). These cells were identified as mast cells because they contained numerous cytoplasmic granules exhibiting metachromatic staining with basic histochemical dyes. The in situ hydrolysis of chromogenic substrates implied that serine proteinases are stored within mast cells in a catalytically active form.

Solubilization of a trypsin- and a chymotrypsin-like proteinase from human skin was first achieved by Fräki and Hopsu-Havu (16). They showed that solubilization of both proteinases requires extraction of skin with a high-salt solution and that it was necessary to precede the high-salt extraction by a low-salt extraction to obtain the chymotrypsin-like proteinase (17). The purpose of the low-salt extraction was to remove proteinase inhibitors capable of inhibiting the chymotrypsin- but not the trypsin-like proteinase (16,17). Analysis of the extract by gel filtration chromatography showed that the two proteinases have widely different molecular weights (17). The trypsin-like proteinase, now identified as skin tryptase, eluted at an apparent molecular weight of approximately 120,000, while the chymotrypsin-like proteinase, now identified as skin chymase, eluted at a molecular weight of 30,000. Using the solubilization procedure of Fräki and Hopsu-Havu (16) as a starting point, purification procedures for each human skin proteinase have been developed (18–21).

III. PURIFICATION AND PROPERTIES OF HUMAN SKIN CHYMASE

In this laboratory, chymase extracted from approximately 500 g of human skin is purified through the use of several affinity chromatography steps that take advantage of the proteinase's substrate specificity and heparin-binding properties. After solubilization from skin, the high-salt extract is dialyzed from 2.0 to 0.4 *M* NaCl (pH 6.8) and chromatographed on a large heparin-Sepharose column (400 mL) equilibrated with the same solution (22–24). At this reduced ionic strength, tryptase quantitatively binds to the column, whereas only a fraction (≤25%) of chymase typically binds. Chymase eluting in 0.4 *M* NaCl is pooled, dialyzed extensively against pH 6.8 water, and precipitated from solution by the addition of protamine chloride. The precipitated material is sedimented by centrifugation, solubilized in 2 *M* NaCl, and chromatographed on a phenylbutylamine-Affigel resin (25). Binding of chymase to phenylbutylamine-Affigel is presumably through interaction of the attached ligand with the proteinase's substrate-binding pocket (25), although adsorption through nonspecific hydrophobic interactions cannot be eliminated. Chymase loaded onto the column in a solution of 2 *M* NaCl is eluted in a sharp peak by stepwise addition of 0.2 *M* D-Trp-OMe dissolved in 0.4 *M* NaCl. If washed too long with 2 *M* NaCl after loading, chymase will slowly elute. This can be prevented by increasing the size of the column. Following elution from phenylbutylamine-Affigel, chymase is chromatographed on a second heparin-Sepharose column of smaller size. In this chromatography, the proteinase binds quantitatively to the heparin resin and is eluted with a salt gradient. We find that the chymase elutes between 0.7 and 0.8 *M* NaCl.

The change in affinity for heparin-Sepharose after fractionation by phenylbutylamine-Affigel implies the presence of a component in the high-salt extract that binds strongly to chymase when the ionic strength of the extract is reduced, thereby blocking its interaction with the first heparin-Sepharose column. Chromatography on the phenylbutylamine resin apparently separates chymase from this component. In some extracts, the fraction of chymase binding to first heparin-Sepharose column is greater than 50%. When this occurs, chymase and tryptase are eluted from the column together in a step containing 2 *M* NaCl. The proteinase mixture is then fractionated on a soybean trypsin inhibitor

(SBTI)–Sepharose column (20). Because SBTI is a proteinase inhibitor that interacts with chymase but not tryptase, this step results in the separation of the two proteinases. Chymase bound to the SBTI column is eluted with a solution of 1 to 10 m*M* HCl. Recoveries from this fractionation step are low, averaging 15 to 20% of the enzyme loaded onto the column.

Chymase purified by either procedure just discussed has identical catalytic, inhibition, and immunological properties. SDS-PAGE analysis of chymase demonstrates a broad band centered around 30 kDa (20,21,24,26). Treatment of the purified enzyme with endoglycosidase F sharpens the banding pattern and slightly reduces the size of the protein, consistent with the presence of N-linked carbohydrate (27). The substrate specificity of skin chymase was shown to be that of a typical chymotrypsin-like enzyme using a library of peptide 4-nitroanilide substrates (28). Like rat mast cell chymase I (RMCP-I), human chymase binds to heparin, interacting at salt concentrations as high as 0.4 *M* NaCl (29). Interaction with heparin does not significantly change the catalytic properties of the enzyme for peptide substrates, but may affect its interaction with protein substrates through charge repulsion and/or steric hindrance (29). The amino acid sequence for the first 35 residues of skin chymase was determined by NH_2-terminal sequencing of the isolated protein (27). Except for uncertainties in two residue assignments which may be due to protein sequencing artifacts, the sequence corresponded to that deduced from the human chymase gene identified by Caughey et al. (30).

The human chymase gene was identified using cDNA for dog chymase as a probe (30). Human and dog chymases demonstrate substantial immunological cross-reactivity, suggesting a close structural relationship between both proteinases (24). Alignment of the deduced amino acid sequence for human chymase with that of dog chymase revealed 83% identity (30), confirming the structural relationship. Similar cross-reaction was not observed with rat chymase I (31), which shows 60% identity by alignment with human chymase (30).

IV. LOCALIZATION OF SKIN CHYMASE TO A SUBSET OF HUMAN MAST CELLS

Immunocytochemical staining of skin sections with polyclonal and monoclonal antibodies raised against skin chymase revealed that the proteinase is an exclusive constituent of mast cells (20,32). Consistent with immunocytochemical results, high-salt extracts of isolated adult foreskin mast cells were shown to contain hydrolytic activity consistent with that of purified chymase and to demonstrate a single band of 30 kDa on Western blots (12,27). Ultrastructural localization by immunoelectronmicroscopy identified chymase as a mast cell secretory granule constituent present in virtually all granules of skin mast cells

(20,29,33), and double-immunolabeling experiments demonstrated that chymase and tryptase are stored in the same secretory granules (33).

Contemporary with the identification of skin mast cell proteinases, a trypsin-like proteinase with size and inhibitor properties similar to those of skin tryptase was isolated from human lung tissue and preparations of isolated human lung mast cells (34,35). Antibodies to this proteinase confirmed its location to lung mast cells and the enzyme was given the name *lung tryptase* (34,35). An interesting aspect of these studies was the absence of a companion chymotrypsin-like proteinase. This observation implied a difference in the proteinase composition of lung and skin mast cells, and led to a series of immunocytochemical investigations examining the presence of chymase and tryptase in mast cells of different human tissues (11,32,33).

Using antibodies to human chymase and tryptase, two major subsets of human mast cells distinguishable by proteinase immunoreactivity were demonstrated. The predominant mast cell type found in skin and other connective tissues (e.g., gut submucosa) contains both chymase and tryptase, and the predominant mast cell type found in lung alveolar tissue and gut mucosa contains only tryptase. The mast cell subset having immunoreactivity for both proteinases was labeled MC_{TC} to denote the presence of chymase and tryptase, and correspondingly, the subset only showing immunoreactivity for tryptase was labeled MC_T.

The demonstration of two mast cell subsets differing in the presence of a chymotrypsin-like proteinase has also been demonstrated using chromogenic substrates to stain tissue sections for active chymotrypsin and trypsin-like proteinases (36). The lack of chymotrypsin-like hydrolytic activity in a population of mast cells demonstrated by these histochemical studies is further evidence that a significant proportion of human mast cells do not express a proteinase with chymotrypsin-like substrate specificity.

Because the biological function of chymase has not been established definitely, the significance of its restriction to a subset of human mast cells is unclear. A proteinase phylogenetic tree based on the amino acid sequence homologies between serine proteinases of inflammatory cells shows that an ancestral chymase gene probably existed prior to the divergence of rodent–human evolutionary lines from a common ancestor (9). Mice and rats have mast cells in tissue locations identical to those in humans. In all these mast cells, chymase is the predominant proteinase (37–39). Thus the absence of a proteinase with chymotrypsin-like activity from a set of human mast cells appears to be a development of human evolution. The need for chymase in one type of human mast cell at the same time that evolution was eliminating its need from another type of mast cell indicates that this proteinase may be essential to the biological function of MC_{TC} mast cells.

V. CONCENTRATION OF CHYMASE AND TRYPTASE IN MC_{TC} MAST CELLS

The specific activity of human skin chymase for the hydrolysis of the substrate Suc–Ala–Ala–Pro–Phe–*p*-nitroanilide is 2.7 μmol/nmol per minute under assay conditions of 1 m*M* substrate; 1.8 *M* NaCl; 0.45 *M* Tris-HCl, pH 8.0; and 9% dimethyl sulfoxide (ε410 nm *p*-nitroaniline = 8.800 M^{-1} cm^{-1}). Using this value and assuming that the recovery of chymase hydrolytic activity in extracts of isolated adult foreskin mast cells is a quantitative measure of stored proteinase, the amount of chymase in MC_{TC} mast cells is approximately 4.5 μg per 10^6 mast cells (12), as shown in Table 1. Using a similar manner of evaluation, the amount of tryptase in MC_{TC} mast cells is approximately 35 μg per 10^6 cells (12). This value makes tryptase the major serine proteinase stored within human MC_{TC} mast cells. As shown in Table 1, the amounts of chymase and tryptase in mast cells are significantly higher than the amounts of other serine proteinases stored in other inflammatory cells. The relatively high quantities of proteinases produced by human mast cells further indicates that both chymase and tryptase may have important physiological functions.

The amount of chymase in human MC_{TC} mast cells and its ratio by weight to tryptase differs from that of similar proteinases found in rat peritoneal mast cells. In rat mast cells, chymase is the major proteinase, present at a level of approximately 30 to 50 μg per 10^6 mast cells (40,41). Although the amount of rat tryptase in peritoneal mast cells has not been determined accurately, it appears to be a small fraction (in the range of 5%) of the chymase value (42,43). The large differences in chymase and tryptase content in human and rat mast cells represent another important evolutionary distinction between these mast cells.

VI. EXTRACELLULAR PROTEOLYTIC POTENTIAL OF HUMAN SKIN MAST CELLS

Subcellular localization within secretory granules suggests that chymase is released from mast cells by inflammatory stimuli capable of producing mast cell degranulation and that its biological function is displayed in the extracellular environment. Although extracellular human chymase has not been demonstrated in vivo, co-release of chymase and histamine by degranulating agents was demonstrated in preparations of isolated mast cells (44).

The extracellular tissue concentrations possibly attained by chymase due to mast cell degranulation can be estimated assuming the amount of chymase in a mast cell, the volume of a mast cell, and the concentration of mast cells in skin. If the average MC_{TC} mast cell is a sphere 20 μm in diameter (45) containing 4.5 pg of chymase, the internal chymase concentration of a mast cell is calculated as 45 μ*M*. This concentration would be an estimate of the upper limit for chymase in the vicinity of a mast cell immediately after degranulation. Taking 5 million cells in 1 cm^3 (gram of skin) as the average number of mast cells in skin (46,47), complete release and diffusion of chymase into the interstitial space would result in a enzyme concentration of approximately 1 μ*M*. Mast cells probably do not degranulate completely in response to stimuli, and the rate of chymase inhibition by physiological inhibitors has not been definitively established (see below); thus the values just described are probably overestimates. Nevertheless, their magnitudes suggest that mast cells have considerable proteolytic potential with respect to chymase and an even greater potential with respect to tryptase.

MC_{TC} mast cells are found throughout connective tissues. In skin, they are situated in close proximity to blood and lymphatic vessels, along the dermal–epidermal junction, and surrounding hair follicles and glands. In these locations, proteolytic enzymes released from mast cells could significantly affect the microenvironments of fibroblast, epithelial, and endothelial cells. One of the most potent inflammatory mediators stored within mast cell granules is histamine (48–50). Histamine released from mast cells interacts with receptors on postcapillary venules, producing local vasodilation and vascular permeability (48–50). The rapid action of histamine suggests that secreted mast cell proteinases not only have the potential to interact with proteins of tissues but that they probably come into contact with a flood of incoming plasma proteins and inflammatory cells. Several in vitro studies, which are described next, have established the reaction properties of human chymase with important biological mediators and structures. The ability of these reactions to affect human physiology significantly, however, remains to be demonstrated.

VII. POTENTIAL BIOLOGICAL SUBSTRATES FOR HUMAN SKIN CHYMASE

A. Conversion of Angiotensin I to Angiotensin II

Angiotensin I is a 10-residue peptide (Asp^1–Arg–Val–Tyr–Ile–His–Pro–Phe–His–Leu^{10}) derived from of the NH_2 terminus of angiotensinogen by renin, an aspartic proteinase secreted into the blood by the kidney. Cleavage of angiotensin I at the Phe^8–His^9 bond produces angiotensin II, a potent mediator of vasoconstriction that is central to the regulation of blood pressure (51). The physiological converter of angiotensin I to angiotensin II is a metalloproteinase with dipeptidyl carboxypeptidase activity, termed *angiotensin converting enzyme* (ACE). ACE is secreted into plasma by the lung.

The presence of the Pro–Phe sequence prior to the COOH-terminal His–Leu in angiotensin I suggests that chymotrypsin-like enzymes also may be capable of producing angiotensin II (28). In a study comparing the hydrolysis of angiotensin I by several chymotrypsin-like proteinases, skin chymase was shown to be the most efficient of the group in converting angiotensin I to angiotensin II (52,53). Chymase was approximately 10 times more efficient than neutrophil cathepsin G in producing angiotensin II, and 5 times more efficient than bovine chymotrypsin. The kinetic parameters for the catalysis of angiotensin I by chymase were somewhat better than those observed for ACE, suggesting that the chymase–angiotensin I reaction may be physiologically relevant. Chymase did not cleave angiotensin $I_{des\ Leu}$, a potent inhibitor of ACE (44).

As reviewed in this book, a carboxypeptidase capable of removing the COOH-terminal Leu residue from angiotensin I also is found in the secretory granules of MC_{TC} mast cells (54–57). This capability appears to provide MC_{TC} mast cells with roughly equivalent amounts of two enzymes having opposing functions with respect to vasoconstriction. Because lysates of mast cells presumably containing both proteinases demonstrated efficient conversion of angiotensin I to angiotensin II (44), the balance of angiotensin reactivity appears to be in favor of chymase.

In contrast to human skin chymase, rat chymase I, the closest rat analog to human skin chymase, degrades angiotensin I, preferring to cleave the Try^4–Ile^5 bond instead of the Phe^8–His^9 bond (53,58). The difference in hydrolysis of angiotensin I by human and rat chymases indicates that angiotensin II production is not a property of chymases that has been conserved in evolution.

B. Activation of Interleukin-1β

Interleukin-1β (IL-1β) is a cytokine with pro-inflammatory properties similar to interleukin-α (59,60). It is synthesized as a 31-kDa precursor which must be processed to a COOH-terminal fragment of 17 kDa to be biologically active (61). Although many cells, including monocyte macrophages, fibroblasts, and keratinocytes, produce and secrete the 31-kDa precursor, the only cell with the capacity to form active IL-1β is the monocyte macrophage (60). The converting enzyme producing this activation is a cysteine proteinase with a specificity for aspartic acid residues (62,63).

Studies on the susceptibility of precursor IL-1β to proteolysis suggest that activation may be a process within the capability of many proteinases (64,65). Because mast cells reside in tissues near cells producing IL-1β, activation of this cytokine by secreted mast cell proteinases could have a significant effect on local inflammatory reactions. In an in vitro study to evaluate the interaction of human skin chymase with precursor IL-1β, almost immediate formation of a stable 17- to 18-kDa product was observed in reactions performed at high cytokine/enzyme molar ratios (26). Formation of this product occurred through a series of cleavages that successively reduced the molecular weight of precursor IL-1β to that of the final species. The biological activity of the final product measured using an interleukin 4–producing T-helper-cell assay system was comparable to that of monocyte-derived IL-1β, suggesting that chymase could have an important proinflammatory function.

The NH_2 terminus of the chymase-generated product is Val^{114}, which is three residues from Ala^{117}, the NH_2 terminus of the monocyte-derived product formed by cleavage of an Asp–Ala bond. Identification of Val^{114} as the NH_2 terminus shows that the final step in activation by chymase is hydrolysis of the Try^{113}–Val^{114} bond.

C. Separation of the Dermal–Epidermal Junction

The dermal–epidermal junction is a basement membrane that mediates the attachment of the epidermis to the dermis (66,67). Damage to this structure has been associated with a number of severe skin blistering diseases (68). Because basement membranes are complex structures formed of protein and proteoglycans (66,67,69), they are a likely target for proteolytic enzymes (66,70). In skin, high numbers of mast cells are found along the dermal–epidermal junction in association with the rich network of blood vessels that run just below the epidermis. Although gross morphological changes to the dermal–epidermal junction and basement membranes surrounding blood vessels do not appear to occur in immediate hypersensitivity reactions, there is evidence implicating mast cells in bullous phemphigoid, a skin-blistering disease exhibiting splitting of the dermal–epidermal junction (71).

Evidence that chymase could affect the structure of basement membranes was obtained from organ culture expriments where sections of normal human skin were incubated in media containing purified chymase. At concentrations of approximately 0.1 μM proteinase, morphologic detachment of the epidermis from the dermal–epidermal junction was observed after 8 h of incubation (72). The pattern of separation had many features in common with that of bullous phemphigoid blisters. Immunoreacive laminin, type IV collagen, and other basement membrane structural proteins were detected on the dermal side of the split, while immunoreactive bullous phemphigoid antigen, a plasma membrane marker found only on the dermal face of epidermal basal cells, was detected on the epidermal side of the split. The demonstration of these antigens suggested that the major structural elements of basement membranes were not the targets of chymase, although their limited degradation could not be ruled out. Integrins, membrane receptors involved in the attachment of cells to matrix structures, also did not appear to be targets for chymase (73). Other basement membranes in skin (e.g., those around blood vessels) did not show structural damage, implying that the putative chymase substrate(s) was a protein(s) specific to the dermal–epidermal junction.

Identical studies replacing chymase with bovine trypsin or neutrophil cathepsin G showed similar structural damage, suggesting a general sensitivity of skin to proteolysis at the epidermal basal cell–dermal–epidermal junction interface. In contrast to the morphologically defined split observed with the serine proteinases just described, incubations of skin with neutrophil elastase resulted in apparent obliteration of the dermal–epidermal junction and complete disappearance of laminin and type IV collagen immunoreactivity over affected areas in treated skin sections. Thus proteinases released from mast cells potentially affect the integrity of the dermal–epidermal junction, but not to the same extent as proteinases released from neutrophils.

The three examples just discussed show that human skin chymase is a true endopeptidase, having the potential to interact with a diverse range of biologically important proteins and peptides. Not discussed was the ability of human skin chymase to degrade certain neuropeptides (74). Potential roles for mast cell chymase in addition to those discussed in this section are presented throughout this book in chapters dealing with human heart, rodent, and dog mast cell proteinases.

VIII. NATURAL INHIBITORS OF HUMAN SKIN CHYMASE

Physiological regulation of proteinases is usually the function of other proteins, termed *proteinase inhibitors*. Like proteinases, families of proteinase inhibitors have evolved exhibiting different inhibition mechanisms (75,76). Generally, low-molecular-weight inhibitors (below 25,000), such as those of the Kunitz, Kazal, Bowman-Birk, potato, and mucus inhibitor families, bind to the active site of the proteinases, forming thermodynamically stable complexes; dissociation constants (K_{diss}) $\leq 10^{-9}$ M are considered very good inhibitors (75,76). Large proteinase inhibitors belonging to the serpin (serine proteinase inhibitor) family interact at the active site, forming 1:1 complexes which have been characterized as "kinetically trapped" because they break down slowly in a catalytic-like manner, releasing free enzyme and a cleaved inhibitor that is inactive (77). Serpin–proteinase complexes demonstrate virtually no reversibility; therefore, effectiveness is usually measured by the magnitude of the second-order rate constant for association (k_{ass}) with a target proteinase (78). Macroglobulin inhibitors bind to proteinases through a conformational change which essentially "cages" the enzyme through interactions that do not involve the active site. Consequently, proteinase–macroglobulin complexes have the ability to hydroyze small substrates that can penetrate to the active site of the bound proteinase (76,79).

A. Low-Molecular-Weight Inhibitors

The susceptibility of skin chymase to small proteinase inhibitors from plants and animals is shown in Table 2. Except for elafin, all are moderate-to-good inhibitors of bovine pancreatic chymotrypsin (21). Chymase shows no inhibition by several of these inhibitors, indicating that its substrate-binding pocket is more restricted to the access of protein proteinase inhibitors than that of chymotrypsin. Bovine pancreatic trypsin inhibitor (80) and trypstatin (81) are found in bovine and rat mast cells, respectively. At present, similar analogs have not been found in human mast cells. Trypstatin is a moderately effective inhibitor of rat chymase (81). The last four inhibitors in Table 2 are found in human tissues and fluids. Even though human chymase has not been tested with several of these inhibitors, inhibition profiles developed from studies with other proteinases are a basis for speculation on their effectiveness toward human chymase.

Proteinase nexin 2 is an isoform of Alheimer's precursor protein containing a Kunitz inhibitor domain (82). The inhibitor is produced primarily by cells in

Table 2 Interaction of Skin Chymase with Low-Molecular-Weight Proteinase Inhibitors from Plants and Animals

Inhibitor	Family	Inhibition[a]
Soybean trypsin inhibitor	Kunitz	Good
Bovine pancreatic inhibitor	Kunitz	No inhibition
Trypstatin	Kunitz	ND
Chicken ovoinhibitor	Kazal	No inhibition
Turkey ovomucoid	Kazal	No inhibition
Lima bean trypsin inhibitor	Bowman-Birk	Moderate
Eglin C	Potato inhibitor I	Good
Potato inhibitor I	Potato inhibitor I	Moderate
Chymostatin	Fungal peptide	Moderate
Proteinase nexin II	Kunitz	No inhibition
SLPI (HUSI-I, MPI)[b]	Mucus	ND
Elafin (SKALPs)[c]	Mucus	ND
Inter-α-proteinase inhibitor	Kunitz	No inhibition

[a]Good inhibitors have K_{diss} values ≤ n*M*, and moderately effective inhibitors have K_{diss} values between 1 and 100 n*M*. No effect on hydrolytic activity at a concentration of 1 m*M* is considered no inhibition. ND, not determined. Data for interaction of human chymase with several inhibitors are found in Refs. 21, 23, and 51. Interactions of human chymase with potato inhibitor I and eglin C are unpublished observations from the author's laboratory using commercially available inhibitors.
[b]SLPI, secretory leukocyte proteinase inhibitor; other names for this inhibitor are HUSI-I (human leukoproteinase inhibitor) and MPI (mucous proteinase inhibitor).
[c]SKALP (skin-derived antileukoproteinase) and elafin are identical.

the brain but has also been identified in platelets (83). Although reported to be moderately effective against human skin chymase (84), this finding was not confirmed in a subsequent study by another laboratory (85), nor was it confirmed in the author's laboratory with a comparable inhibitor obtained from Athena Neurosciences. In addition, proteinase nexin 2 does not inhibit cathepsin G or rat chymase (81,86), making its inhibition of chymase unlikely. SLPI and elafin are proteinase inhibitors found in bronchial and parotid secretions (87,88). They are highly effective inhibitors of human neutrophil elastase (87,89,90). Elafin also is produced by psoriatic epidermal cells, apparently in response to local inflammation (89,90). Because mucus secretions may result from mast cell degranulation and because mast cells are found in close proximity to the epidermal layer of skin, inhibition of human chymase by these inhibitors may have physiological importance. SLPI was shown to be moderately effective against a chymase isolated from porcine lung and neutrophil cathepsin G, suggesting that it also should react with human chymase (87,91). Elafin, on the other hand, did not inhibit bovine chymotrypsin or cathepsin G (89,90), indicating a highly focused specificity of interaction. Inter-α-trypsin inhibitor is found in human plasma (76). Its removal from plasma was not required to deplete plasma of chymase inhibitory activity, implying that it is not an effective chymase inhibitor (23). The insensitivity of chymase to low-molecular-weight inhibitors so far identified in human tissues leaves open the demonstration of a tissue inhibitor that is focused toward chymase.

B. Plasma Inhibitors

Plasma contains numerous inhibitors with differing specificities that belong to the serpin family of proteins (76). These inhibitors regulate circulating plasma serine proteinases, which have important roles in inflammation: thrombin, C1-esterase, plasmin, and plasminogen activator (76,92). In addition, serpins rapidly inactivate serine proteinases stored in neutrophils (93,94), implying that they are responsible for the regulation of inflammatory cell–derived proteinases as well. Plasma also contains α_2-macroglobulin (α_2-M), a nonserpin inhibitor capable of inhibiting a wide range of proteinases from several different protease families.

Titration of human skin chymase with whole plasma led to an 85% loss in hydrolytic activity toward a model peptide substrate, demonstrating that plasma inhibiors are effective against chymase (23). The 15% residual chymase hydrolytic activity is believed due to inhibition by α_2-M, because it is not sensitive to the addition of lima bean trypsin inhibitor, a protein proteinase inhibitor that cannot penetrate α_2-M proteinase complexes. Using antibodies to deplete plamsa of specific inhibitors, α_1-proteinase inhibitor (α_1-PI) and α_1-antichymotrypsin (ACT) were identified as the serpins capable of inhibiting chymase (23). Plasma depleted of α_1-PI, ACT, and α_2-M retained chymotrypsin inhibitory activity but not chymase inhibitory activity, suggesting that chymase is sensitive to a more limited range of plasma inhibitors than is chymotrypsin.

In Table 3 the reaction properties for chymase inhibition by α_1-PI and ACT are compared to those for two other human chymotrypsin-like proteinases, neutrophil cathepsin G and pancreatic chymotrypsin (93,95,96). Of the three proteinases, chymase has the lowest k_{ass} values and demonstrates an unusually high stoichiometry of inhibition; that is, 4 mol of ACT and 5 mol of α_1-PI are required to fully inactivate 1 mol of chymase. The high stoichiometry of inhibition is due to a concurrent hydrolytic reaction, resulting in the inactivation of each inhibitor by chymase at rates three to four times faster than the rate of chymase inhibition (22). The stoichiometry of inhibition values obtained for each inhibitor does not vary in titrations performed at different chymase concentrations, indicating that they are an invariant reaction property, possibly like partition ratios in the case of enzyme reactions with suicide substrates (96). Thus inhibition of chymase by ACT and α_1-PI can be viewed as a balance between a substrate (hydrolytic) and an inhibition reaction. The reaction mechanism that produces the high level of stoichiometry of inhibition (SI) and its significance with respect to serpin–proteinase interactions is currently under investigation (22).

Despite the high SI and relatively low k_{ass} values, chymase–ACT and chymase–α_1-PI complexes are stable for at least 24 h (22,23). Thus both serpins would be effective inhibitors of extracellular chymase as long as they were present in large molar excess (greater than or equal to the stoichiometry of inhibition) to chymase. Based on the combined concentration of α_1-PI and ACT (52 μM and 6 μM, respectively) in plasma, chymase would be inhibited with a half-life of about 1 s if secreted in plasma (22). Because mast cells reside in

Table 3 Interaction of Serpin with Human Chymotrypsin-like Proteinases

	α_1-Antichymotrypsin		α_1-Proteinase inhibitor	
Proteinase	k_{ass} (M^{-1} s^{-1})	SI	k_{ass} (M^{-1} s^{-1})	SI
Chymase[a]	2×10^4	4	7×10^3	5
Cathepsin G[b]	5×10^7	1	7×10^4	1
Chymotrypsin[c]	1×10^5	1	1×10^6	1

[a]k_{ass} values were determined under pseudo-first-order conditions where the [Inhibitor]/[Chymase] concentration ratio was at least 10-fold higher than the stoichiometry of inhibition (SI) measured for each inhibitor (21,22).
[b]Data from Refs. 93 and 94.
[c]Data from Ref. 95.

tissues and are not found in the circulation, the rate of chymase inhibition by plasma inhibitors is difficult to evaluate accurately. Increased vascular permeability associated with histamine release might allow for the rapid influx of plasma inhibitors to inactivate chymase shortly after degranulation. In addition, immunocytochemical evidence indicates that human skin mast cells contain α_1-PI and ACT capable of inhibiting chymase after degranulation (97,98).

IX. HUMAN MAST CELLS CONTAIN A SECOND CHYMOTRYPSIN-LIKE PROTEINASE

A. Biochemical Properties of Chymase Versus Neutrophil Cathepsin G

As shown in Table 1, cathepsin G is a chymotrypsin-like proteinase stored within human neutrophils and monocytes (3, 99–101). Its genetic locus is on chromosome 14 (102) near that of chymase (30,103), and its amino acid sequence demonstrates approximately 50% homology with that of chymase (30). Although probably related through a common ancestral gene (9), the catalytic and inhibitory properties of each enzyme for model substrates and inhibitors are significantly different (21,23,51,104,105). Most notable are the unusually slow catalytic efficiencies for the hydrolysis of many model substrates by cathepsin G. In addition to differences involving active site structure, both proteinases display virtually no immunological cross-reactivity (27). Like chymase, the biological function of cathepsin G is not clearly defined. The differences in catalytic and structural properties may suggest that both proteinases evolved to different selective pressures. This is supported by the fact that the amino acid sequence (83%) and immunological properties of human chymase are more similar to dog chymase than to cathepsin G, its human relative (24,30).

B. Identification of Cathepsin G in Human Mast Cells

Evidence from radioimmunoassays has suggested that human lung mast cells contain antigens cross-reactive with human neutrophil cathepsin G and elastase (106). Immunocytochemical staining of tissue sections from different organs (skin, gut, and lung) confirmed the presence of a cathepsin G antigen in mast cells but failed to demonstrate the presence of neutrophil elastase (27). Cathepsin G immunoreactivity was specific for MC_{TC} mast cells, indicating the presence of a second chymotrypsin-like proteinases within these cells (27).

Mast cell cathepsin G has been further characterized using enriched populations MC_{TC} mast cells isolated from adult human foreskins. Western blots of cell extracts identified a single immunoreactive band migrating with an apparent molecular weight of 30,000. The molecular weight of this protein was slightly higher than that of neutrophil cathepsin G, indicating that the mast cell and neutrophil enzymes are not structurally identical (27). Although differing substantially in their ability to hydrolyze model substrates, neutrophil cathepsin G and chymase hydrolyze casein, a protein substrate, at nearly equal rates (27). Using casein as a substrate and specific inhibitors of cathepsin G to distinguish cathepsin G hydrolytic activity from that of chymase, approximately 30% of the chymotrypsin-like activity in mast cell extracts appeared to be due to cathepsin G, indicating that the antigen was an active proteinase (27). The nearly equal proportion of chymase and cathepsin G casein hydrolysis displayed by extracts also suggests that mast cells contain approximately equal amounts of each proteinase. mRNA for chymase and cathepsin G in mast cells has been demonstrated by PCR amplification of cDNA derived from a human leukemic cell line with primers specific for each proteinase (103).

So far, cathepsin G has not been purified from skin to complete its biochemical characterization. Examination of purified chymase by Western blotting and NH_2-terminal analysis has failed to detect cathepsin G in these preparations (27), suggesting that the proteinase is lost during purification.

X. CONCLUSIONS

Mast cells were first described by Paul Ehrlich in 1878 as cells engorged with material as if from overfeeding (107). Over 100 years later we are continuing to define the biological products stuffed into these cells and are continually being amazed by the assortment of biologically potent molecules that we are finding. Despite the presence of an arsenal of inflammatory mediators that include proteinases, cytokines, leukotrienes, prostaglandins, heparin, and histamine (108–110), and numerous physiological arguments for a more general inflammatory role (111), the involvement of mast cells in biological processes other than allergic reactions is not clear. Because most allergens, themselves, are not life threatening and do not appear to require the action of mast cell mediators for disposition, the advantage of mast cells to host physiology presents a riddle [as described by Kaliner (107) and Galli (1)] that remains to be solved. In this chapter the presence and potential functions of two chymotrypsin-like proteinases stored within human mast cell secretory granules was described. Why human cutaneous mast cells contain large amounts of proteinases and why they contain two enzymes with apparent overlapping proteolytic ability adds to the mystery of these cells. Providing a complete understanding of mast cells and their proteolytic enzymes in health and disease is an important goal for the future.

REFERENCES

1. S. J. Galli, New insights into "the riddle of the mast cells": microenvironment regulation of mast cell development and phenotypic heterogeneity, *Lab. Invest. 62*:5 (1990).

2. D. Lagunoff and E. P. Benditt, Proteolytic enzymes of mast cells, *Ann. NY Acad. Sci. 103*:185 (1963).
3. E. J. Campbell, E. K. Silverman, and M. A. Campbell, Elastase and cathepsin G of human monocytes, quantification of cellular content, release in response to stimuli, and heterogeneity in elastase-mediated proteolytic activity, *J. Immunol. 143*:2961 (1989).
4. D. Campanelli, P. A. Detmers, C. F. Nathan, and J. E. Gabay, Azurocidin and a homologous serine protease from neutrophils: differential antimicrobial and proteolytic properties, *J. Clin. Invest. 85*:904 (1990).
5. D. Campanelli, M. Melchoir, Y. Fu, M. Nakata, H. Shuman, C. Nathan, and J. E. Gabay, Cloning of cDNA for proteinase 3: a serine protease, antibiotic, and autoantigen from human neutrophils, *J. Exp. Med. 172*:1709 (1990).
6. N. V. Rao, N. G. Wehner, B. C. Marshall, W. R. Gray, B. H. Gray, and J. R. Hoidal, Characterization of proteinase-3 (PR-3), a neutrophil serine protease: structural and functional properties, *J. Biol. Chem. 266*:9540 (1991).
7. A. Hameed, D. M. Lowrey, M. Lichtenheld, and E. R. Podack, Characterization of three serine esterases isolated from human IL-2 activated killer cells, *J. Immunol. 141*:3142 (1988).
8. O. Krahenbuhl, C. Rey, D. Jenne, A. Lanzavecchia, P. Groscurth, S. Carrel, and J. Tschopp, Characterization of granzymes A and B isolated from granules of cloned human cytotoxic T lymphocytes, *J. Immunol. 141*:3471 (1988).
9. D. F. Jenne, D. Masson, M. Zimmer, A. Haefliger, W.-H. Li, and J. Tschopp, Isolation and complete structure of the lymphocyte serine protease granzyme G, a novel member of the granzyme multigene family in murine cytolytic T lymphocytes: evolutionary origin of lymphocytes proteases, *Biochemistry 28*:7953 (1989).
10. S. Odake, C.-M. Kam, L. Narasimhan, M. Poe, J. T. Blake, O. Krahenbuhl, J. Tschopp, and J. C. Powers, Human and murine cytotoxic T lymphocytes serine proteases: subsite mapping with peptide thioester substrates and inhibition of enzyme activity and cytolysis by isocumarins, *Biochemistry 30*:2217 (1991).
11. A. A. Irani, N. M. Schechter, S. S. Craig, G. DeBlois, and L. B. Schwartz, Two types of human mast cells that have distinct neutral protease compositions, *Proc. Natl. Acad. Sci. USA 83*:4464 (1986).
12. L. B. Schwartz, A.-M. A. Irani, K. Roller, M. C. Castells, and N. M. Schechter, Quantitation of histamine, tryptase, and chymase in dispersed human T and TC mast cells, *J. Immunol. 138*:2611 (1987).
13. G. G. Glenner and L. A. Cohen, Histochemical demonstration of a species specific trypsin-like proteinase in mast cells, *Nature 185*:846 (1960).
14. V. K. Hopsu and G. G. Glenner, Further observations on histochemical esterase and amidase activities with similarities to trypsin, *J. Histochem. Cytochem. 11*:520 (1963).
15. D. Lagunoff, Histochemistry of proteolytic enzymes, *Methods Achiev. Exp. Pathol. 2*:55 (1967).
16. J. E. Fräki and V. K. Hopsu-Havu, Human skin proteinases: separation and characterization of two alkaline proteases, one splitting trypsin and the other chymotrypsin substrates, *Arch. Dermatol. Res. 253*:261 (1975).

17. J. E. Fräki and V. K. Hopsu-Havu, Human skin proteases: partial purification and characterization of a protease inhibitor, *Arch. Dermatol. Forsch. 243*:153 (1972).
18. I. T. Harvima, N. M. Schechter, R. J. Harvima, and J. E. Fräki, Human skin tryptase: purification, partial characterization and comparison with human lung tryptase, *Biochim. Biophys. Acta 957*:71 (1988).
19. N. M. Schechter, G. Y. Eng, and D. R. McCaslin, Human skin tryptase: kinetic characterization of its spontaneous inactivation, *Biochemistry 32*:2617 (1992).
20. N. M. Schechter, J. K. Choi, D. A. Slavin, D. T. Deresienski, S. Sayama, G. Dong, R. M. Lavker, D. Proud, and G. S. Lazarus, Identification of a chymotrypsin-like proteinase in human mast cells, *J. Biol. Chem. 137*:962 (1986).
21. N. M. Schechter, J. E. Fräki, J. C. Geesin, and G. S. Lazarus, Human skin chymotryptic proteinase: isolation and relation to cathepsin G and rat mast cell proteinase 1, *J. Biol. Chem. 258*:2973 (1983).
22. N. M. Schechter, L. M. Jordan, A. M. James, B. S. Cooperman, Z.-M. Wang, and H. Rubin, Reaction of human chymase with reactive site variants of α1-antichymotrypsin: modulation of inhibitor vs. substrate properties, *J. Biol. Chem. 268*:23626 (1993).
23. N. M. Schechter, J. L. Sprows, O. L. Schoenberger, G. S. Lazarus, B. S. Cooperman, and H. Rubin, Reaction of human skin chymotrypsin-like chymase with plasma proteinase inhibitors, *J. Biol. Chem. 264*:21308 (1989).
24. N. M. Schechter, D. Slavin, R. D. Fetter, G. L. Lazarus, and J. E. Fräki, Purification and identification of two serine proteinases from dog mast cells biochemically and immunologically similar to human proteinases tryptase and chymase, *Arch. Biochem. Biophys. 262*:232 (1988).
25. K. J. Stevenson and A. Landman, The isolation of chymotrypsin-like enzymes by affinity chromatography using Sepharose-4-phenylbutylamine, *Can. J. Biochem. 49*:119 (1971).
26. H. Mizutani, N. M. Schechter, G. S. Lazarus, R. A. Black, and T. S. Kupper, Rapid and specific conversion of precursor interleukin 1β (IL-1β) to an active IL-1 species by human mast cell chymase, *J. Exp. Med. 174*:821 (1991).
27. N. M. Schechter, A.-M. A. Irani, J. L. Sprows, J. Abernethy, B. Wintroub, and L. B. Schwartz, Identification of a cathepsin G–like proteinase in the MC_{TC} type of human mast cell, *J. Immunol. 145*:2652 (1990).
28. J. C. Powers, T. Tanaka, J. W. Harper, Y. Minematsu, L. Barker, D. Lincoln, K. V. Crumley, J. E. Fräki, N. M. Schechter, G. L. Lazarus, K. Nakajima, K. Nakashino, H. Neurath, and R. G. Woodbury, Mammalian chymotrypsin-like enzymes: comparative reactivities of rat mast cell proteases, human and dog chymases, and human cathepsin G with peptide 4-nitroanilide substrates and with peptide chloromethyl ketone and sulfonyl fluoride inhibitors, *Biochemistry 24*:2048 (1985).
29. Sayama, R. V. Iozzo, G. S. Lazarus, and N. M. Schechter, Human skin chymotrypsin-like proteinase chymase: subcellular localization to mast cell granules and interaction with heparin and other glycosaminoglycans, *J. Biol. Chem. 262*:6808 (1987).
30. G. H. Caughey, E. H. Zerweck, and P. Vanderslice, Structure, chromosomal as-

signment, and deduced amino acid sequence of a human gene for mast cell chymase, *J. Biol. Chem. 266*:12956 (1991).
31. N. M. Schechter, Human chymase, *Monogr. Allergy 27*:114 (1990).
32. A.-M. A. Irani, T. R. Bradford, C. L. Kepley, N. M. Schechter, and L. B. Schwartz, Detection of MC_T and MC_{TC} types of human mast cells by immunohistochemistry using new monoclonal anti-tryptase and anti-chymase antibodies, *J. Histochem. Cytochem. 37*:1509 (1989).
33. S. S. Craig, N. M. Schechter, and L. B. Schwartz, Ultrastructural analysis of human T and TC mast cells identified by immunoelectron microscopy, *Lab. Invest. 58*:682 (1988).
34. L. B. Schwartz, R. A. Lewis, and K. F. Austen, Tryptase from human pulmonary mast cells: purification and characterization, *J. Biol. Chem. 256*:11939 (1981).
35. T. J. Smith, M. W. Hougland, and D. A. Johnson, Human lung tryptase: purification and characterization, *J. Biol. Chem. 259*:11046 (1984).
36. I. A. R. Osman, J. R. Garrett, and R. E. Smith, Enzyme histochemical discrimination between tryptase and chymase in mast cells of human gut, *J. Histochem. Cytochem. 37*:415 (1989).
37. S. Gibson, A. Mackeller, G. F. J. Newlands, and H. R. P. Miller, Phenotypic expression of mast cell granule proteinases: distribution of mast cell proteinases I and II in the rat digestive system, *Immunology 62*:621 (1987).
38. S. Gibson and H. R. P. Miller, Mast cell subsets in the rat distinguished immunohistochemically by their content of serine proteinases, *Immunology 58*:101 (1986).
39. D. S. Reynolds, W. E. Stevens, W. S. Lane, M. H. Carr, K. F. Austen, and W. E. Serafin, Different mouse mast cell populations express various combinations of at least six distinct mast cell serine proteinases, *Proc. Natl. Acad. Sci. USA 87*:3230 (1990).
40. L. B. Schwartz, C. Reidel, J. P. Caulfield, S. I. Wasserman, and K. F. Austen, Cell association of complexes of chymase, heparin proteoglycan, and protein after degranulation of rat mast cells, *J. Immunol. 126*:2071 (1981).
41. D. Lagunoff and P. Pritzl, Characterization of rat mast cell granule proteins, *Arch. Biochem. Biophys. 173*:554 (1976).
42. V. J. Braganza and W. H. Simmons, Tryptase from rat skin: purification and properties, *Biochemistry 30*:4997 (1991).
43. D. Lagunoff, A. Rickard, and C. Marquardt, Rat mast cell tryptase, *Arch. Biochem. Biophys. 291*:52 (1991).
44. B. U. Wintroub, C. E. Kaempfer, N. M. Schechter, and D. Proud, A human lung mast cell chymotrypsin-like enzyme: identification and partial characterization, *J. Clin. Invest. 77*:196 (1986).
45. B. Goldberg and M. Rabinovich, Connective tissue, in *Histology: Cell and Tissue Biology* (L. Weiss, ed.), Elsevier Biomedical, New York, 1983, p. 170.
46. G. R. Mikhail and A. Miller-Milinska, Mast cell population in human skin, *J. Invest. Dermatol. 43*:249 (1964).
47. A. A.-M. Irani, M. M. Garriga, D. D. Metcalfe, and L. B. Schwartz, Mast cells

in cutaneous mastocytosis: accumulation of the MC_{TC} type, *Clin. Exp. Allergy 20*:53 (1990).

48. P. Kubes, Polymorphonuclear leukocyte–endothelium interactions: a role for proinflammatory and anti-inflammatory molecules, *Can. J. Physiol. Pharmacol. 71*:88 (1993).
49. G. L. Larsen and P. M. Henson, Mediators of inflammation, *Annu. Rev. Immunol. 1*:335 (1983).
50. H. F. Dvorak, D. R. Senger, A. M. Dvorak, V. S. Harvey, and J. McDonagh, Regulation of extravascular coagulation by microvascular permeability, *Science 227*:1059 (1985).
51. D. J. Campbell, Circulating and tissue angiotensin systems, *J. Clin. Invest. 79*:1 (1987).
52. C. F. Reilly, D. A. Tweksbury, N. M. Schechter, and J. Travis, Rapid conversion of angiotensin I to angiotensin II by neutrophil and mast cell proteinases, *J. Biol. Chem. 257*:8619 (1982).
53. B. U. Wintroub, N. M. Schechter, G. S. Lazarus, C. E. Kaempfer, and L. B. Schwartz, Angiotensin I conversion by human and rat chymotrypic proteinases, *J. Invest. Dermatol. 83*:336 (1984).
54. S. M. Goldstein, C. E. Kaempfer, D. Proud, L. B. Schwartz, A.-M. Irani, and B. U. Wintroub, Detection and partial characterization of a human mast cell carboxypeptidase, *J. Immunol. 139*:2724 (1987).
55. S. M. Goldstein, C. E. Kaempfer, J. T. Kealey, and B. U. Wintroub, Human mast cell carboxypeptidase: purification and characterization, *J. Clin. Invest. 83*:1630 (1989).
56. A.-M. A. Irani, S. M. Goldstein, B. U. Wintroub, T. Bradford, and L. B. Schwartz, Human mast cell carboxypeptidase: selective localization to MC_{TC} cells, *J. Immunol. 147*:247 (1991).
57. D. S. Reynolds, D. S. Gurley, R. L. Stevens, D. J. Sugarbaker, K. F. Austen, and W. E. Serafin, Cloning of cDNAs that encode human mast cell carboxypeptidase A, and comparison of the protein with mast cell carboxypeptidase A and rat pancreatic carboxypeptidases, *Proc. Natl. Acad. Sci. USA 86*:9480 (1989).
58. H. L. Trong, H. Neurath, and R. G. Woodbury, Substrate specificity of the chymotrypsin-like protease in secretory granules isolated from rat mast cells, *Proc. Natl. Acad. Sci. USA 84*:364 (1987).
59. S. B. Mizel, The interleukins, *FASEB J. 3*:2379 (1989).
60. T. S. Kupper, Immune and inflammatory processes in cutaneous tissues: mechanisms and speculations, *J. Clin. Invest. 86*:1783 (1990).
61. C. J. March, B. Mosley, A. Larsen, D. P. Cerretti, G. Braedt, V. Price, S. Gillis, C. S. Henny, S. R. Kronheim, K. Grabstein, P. J. Conlon, T. P. Hopp, and D. Cosman, Cloning, sequence and expression of two distinct human interleukin-1 complementary DNAs, *Nature 315*:641 (1985).
62. S. M. Molineaux, F. J. Casano, A. M. Rolando, E. P. Peterson, G. Limjuco, J. Chin, P. R. Griffin, J. R. Calaycay, G. J. F. Ding, T. T. Yamin, O. C. Palyha, S. Luell, D. Fletcher, D. K. Miller, A. D. Howard, N. A. Thornberry, and M. J. Kostura, Interleukin-1β (IL-1β) processing in murine macrophages requires a

structurally conserved homologue of human IL-1β converting enzyme, *Proc. Natl. Acad. Sci. USA 90*:1809 (1993).
63. D. K. Miller, J. M. Ayala, L. A. Egger, S. M. Raju, T. T. Yamin, G. J. F. Ding, E. P. Gaffney, A. D. Howard, O. C. Palyha, A. M. Rolando, J. P. Salley, N. A. Thornberry, J. R. Weidner, J. H. Williams, K. T. Chapman, J. Jackson, M. J. Kostura, G. Limjuco, S. M. Molineaux, R. A. Mumford, and J. R. Calaycay, Purification and characterization of active human interleukin-1β-converting enzyme from THP-1 monocytic cells, *J. Biol. Chem. 268*:18062 (1993).
64. H. R. Mizutani, R. Black, and T. S. Kupper, Human keratinocytes produce but do not process pro-interleukin-1β (IL-1β): different strategies of IL-1 production and processing in monocytes and keratinocytes, *J. Clin. Invest. 87*:1066 (1991).
65. D. J. Hazuda, J. Strickler, F. Kuppers, P. L. Simon, and P. R. Young, Processing of precursor interleukin 1β and inflammatory disease, *J. Biol. Chem. 265*:6318 (1990).
66. N. M. Schechter, Structure of the dermal–epidermal junction and potential mechanisms for its degradation: the possible role of inflammatory cells, *Immunol. Ser. 46*:477 (1989).
67. R. A. Briggaman, Biochemical composition of the epidermal–dermal junction and other basement membranes, *J. Invest. Dermatol. 78*:1 (1982).
68. J. Uitto and A. M. Christiano, Molecular genetics of the cutaneous basement membrane zone: perspectives of epidermolysis bullosa and other blistering skin diseases, *J. Clin. Invest. 90*:687 (1992).
69. P. D. Yurchenco and J. C. Schittny, Molecular architecture of basement membranes, *FASEB J. 4*:1577 (1990).
70. J. F. J. Woessner, Matrix metalloproteinases and their inhibitors in connective tissue remodeling, *FASEB J. 5*:2145 (1991).
71. S. M. Goldstein, S. I. Wasserman, and B. U. Wintroub, Mast cell and eosinophil mediated damage in bullous phemigoid, *Immunol. Ser. 46*:527 (1989).
72. R. A. Briggaman, N. M. Schechter, J. Fräki and G. S. Lazarus, Degradation of the epidermal–dermal junction by proteolytic enzymes from human skin and human polymorphonuclear leukocytes, *J. Exp. Med. 160*:1027 (1984).
73. K. Fujii, G. S. Lazarus, and N. M. Schechter, Modification of integrin-mediated cell attachment to substrata by serine proteinases in the presence and absence of divalent ions, *Exp. Cell Res. 208*:94 (1993).
74. S. M. Goldstein, J. Leong, and N. W. Bunnett, Human mast cells hydrolyze neurotensin, kinetensin, and leu^5-enkephalin, *Peptides 12*:995 (1991).
75. M. J. Laskowski and I. Cato, Protein inhibitors of proteinases, *Annu. Rev. Biochem. 49*:593 (1980).
76. J. Travis and G. S. Salvesen, Human plasma proteinase inhibitors, *Annu. Rev. Biochem. 52*:655 (1983).
77. B. S. Cooperman, E. Nickbarg, E. Stravida, E. Rescorla, N. M. Schechter, and H. Rubin, The kinetic mechanism of antichymotrypsin interaction with chymotrypsin: the effects of ionic strength, human neutrophil elastase, and site-specific mutagenesis of the P1 position of antichymotrypsin, *J. Biol. Chem. 268*:23616 (1993).

78. J. Bieth, In vivo significance of kinetic constants of protein proteinase inhibitors, *Biochem. Med. 32*:387 (1984).
79. W. Borth, α2-Macroglobulin, a multifunctional binding protein with targeting characteristics, *FASEB J. 6*:3345 (1992).
80. H. Fritz, H. Kruck, J. Russe, and H. G. Liebich, Immunofluorescent studies indicate that the basic trypsin-kallikrein-inhibitor of bovine organs (Trasylol) originates from mast cells, *Hoppe-Seyler's Z. Physiol. Chem. 360*:437 (1979).
81. H. Kido, A. Fukutomi, J. Schilling, Y. Wang, B. Cordell, and N. Katunuma, Protease-specificity of Kunitz inhibitor domain of Alzheimer's disease amyloid protein precursor, *Biochem. Biophys. Res. Commun. 167*:716 (1990).
82. W. E. Van Nostrand, S. M. Wagner, M. Suzuki, B. H. Choi, J. S. Farrow, J. W. Gettes, C. W. Cotman, and D. D. Cunningham, Protease nexin-II, a potent anti-chymotrypsin, shows identity to amyloid β-protein precursor, *Nature 341*:546 (1989).
83. W. E. Van Nostrand, A. H. Schmaier, J. S. Farrow, and D. D. Cunningham, Protease nexin-II (amyloid β-protein precursor): a platelet a-granule protein, *Nature 248*:745 (1990).
84. W. E. Van Nostrand, S. L. Wagner, J. S. Farrow, and D. D. Cunningham, Immunopurification and protease inhibitory properties of protease nexin-2/amyloid β-protein precursor, *J. Biol. Chem. 265*:9591 (1990).
85. S. Sinha, J. Knops, F. Esch, E. D. Moyer, and T. Oltersdorf, Conversion of the Alzheimer's precursor protein (APP) Kunitz domain into a potent human neutrophil elastase inhibitor, *J. Biol. Chem. 266*:21011 (1991).
86. S. Sinha, H. F. Dovey, P. Seubert, P. J. Ward, R. W. Blacher, M. Blaber, R. A. Bradshaw, M. Ariei, W. C. Mobley, and I. Lieberburg, The protease inhibitory properties of the Alzheimer's β-amyloid precursor protein, *J. Biol. Chem. 265*:8983 (1990).
87. C. Boudier and J. G. Bieth, The proteinase: mucus proteinase inhibitor binding stoichiometry, *J. Biol. Chem. 267*:4370 (1992).
88. J.-M. Sallenave, M. D. Marsden, and A. P. Ryle, Isolation of elafin and elastase-specific inhibitor (ESI) from bronchial secretions, *Biol. Chem. Hoppe-Seyler 373*:27 (1992).
89. J. Schalkwijk, A. Chang, P. Jannsen, G. J. de Jongh, and P. D. Mier, Skin-derived antileukoproteinase (SKALPs): characterization of two new elastase inhibitors from psoriatic epidermis, *Br. J. Dermatol. 122*:631 (1990).
90. O. Wiedow, J.-M. Schroder, H. Gregory, J. A. Young, and E. Christophers, Elafin: an elastase-specific inhibitor of human skin: purification, characterization, and complete amino acid sequence, *J. Biol. Chem. 265*:14791 (1990).
91. E. Fink, R. Nettelbeck, and H. Fritz, Inhibition of mast cell chymase by eglin c and antileukoproteinase (HUSI-I): indications for potential biological functions of these inhibitors, *Biol. Chem. Hoppe-Seyler 373*:567 (1986).
92. R. Carrell, P. A. Pemberton, and D. R. Boswell, The serpins: evolution and adaption of a family or protease inhibitors, *Cold Spring Harbor Symp. Quant. Biol. 52*:527 (1987).
93. K. Beatty, J. Bieth, and J. Travis, Kinetics of association of serine proteinases with

native and oxidized α1-proteinase inhibitor and α1-antichymotrypsin, *J. Biol. Chem. 255*:3931 (1980).

94. S. Jallat, D. Caryallo, L. H. Tessier, D. Roecklin, C. Roitsch, F. Ogushi, R. G. Crystal, and M. Courtney, Altered specificities of genetically engineered α1-antitrypsin variants, *Protein Eng. 1*:29 (1986).
95. D. Marko-Vercaigne, J. Carrere, M.-P. Ducourouble, M. Davril, A. Laine, M. Amouric, C. Figarella, and A. Hayem, In vivo and in vitro inhibition of human pancreatic chymotrypsin A by serum inhibitors, *Biol. Chem. Hoppe-Seyler 368*:37 (1987).
96. H. Rubin, Z.-M. Wang, E. B. Nickbarg, S. McLarney, N. Naidoo, O. L. Schoenberger, J. Johnson, and B. S. Cooperman, Cloning, expression, purification and bioactivity of recombinant, native, and variant α1-antichymotrypsin, *J. Biol. Chem. 265*:1199 (1990).
97. P. Ruck, H.-P. Horny, and E. Kaiserling, Immunoreactivity of human tissue mast cells: nonspecific binding of primary antibodies against regulatory peptides by ionic linkage, *J. Histochem. Cytochem. 38*:859 (1990).
98. I. T. Harvima, A. Naukkarinen, K. Paukkonen, R. J. Harvima, M.-L. Aalto, L. B. Schwartz, and M. Horsemanheimo, Mast cell tryptase and chymase in developing and mature psoriatic lesions, *Arch. Dermatol. Res. 285*:184 (1993).
99. W. Schmidt and K. Havemann, Isolation of elastase-like and chymotrypsin-like neutral proteases from human granulocytes, *Hoppe-Seyler's Z. Physiol. Chem. 355*:1077 (1974).
100. G. Feinstein and A. Janoff, A rapid method for purification of human granulocyte cationic neutral proteases: purification and characterization of human granulocyte chymotrypsin-like enzyme, *Biochim. Biophys. Acta 403*:477 (1975).
101. P. M. Starkey and A. J. Barrett, Neutral proteinases of human spleen: purification and criteria for homogeneity of elastase and cathepsin G, *Biochem. J. 155*:255 (1976).
102. P. A. Hohn, N. C. Popescu, R. D. Hanson, G. Salvesen, and T. J. Ley, Genomic organization and chromosomal localization of the human cathepsin G gene, *J. Biol. Chem. 264*:13412 (1989).
103. G. H. Caughey, T. H. Schaumberg, E. H. Zerweck, J. H. Butterfield, R. D. Hanson, G. A. Silverman, and T. J. Ley, The human mast cell chymase gene (CMA1): mapping to the cathepsin G/granzyme gene cluster and lineage-restricted expression, *Genomics 15*:614 (1993).
104. T. Tanaka, Y. Minematsu, C. F. Reilly, J. Travis, and J. C. Powers, Human leukocyte cathepsin G: subsite mapping with 4-nitroanilides, chemical modification, and effect of possible cofactors, *Biochemistry 24*:2040 (1985).
105. K. Nakajima, J. C. Powers, B. M. Ashe, and M. Zimmerman, Mapping the extended substrate binding site of cathepsin G and human leukocyte elastase: studies with peptide substrates related to the α1-proteinase inhibitor reactive site, *J. Biol. Chem. 254*:4027 (1979).
106. H. L. Meir, L. W. Heck, E. S. Schulman, and D. W. MacGlashan, Purified human mast cells and basophils release human elastase and cathepsin G by an IgE-mediated mechanism, *Int. Arch. Allergy Appl. Immunol. 13*:179 (1985).

107. M. A. Kaliner, The mast cell: a fascinating riddle, *N. Engl. J. Med. 301*:498 (1979).
108. R. P. Schleimer, D. W. MacGlashen, S. P. Peters, R. Naclerio, D. Proud, N. F. Adkinson, and L. M. Lichtenstein, Inflammatory mediators and mechanisms of release from purified human basophils and mast cells, *J. Allergy Clin. Immunol. 74*:473 (1984).
109. W. E. Serafin and K. F. Austen, Mediators of immediate hypersensitivity reactions, *N. Engl. J. Med. 317*:30 (1987).
110. S. J. Galli, New concepts about the mast cells, *N. Engl. J. Med. 328*:257 (1993).
111. S. J. Galli, New approaches for the analysis of mast cell maturation, heterogeneity, and function, *Fed. Proc. 46*:1906 (1992).

5

Mast Cell Neuropeptidases

Elizabeth K. Tam
John A. Burns School of Medicine, University of Hawaii, Honolulu, Hawaii, and University of California at San Diego, San Diego, California

I. INTRODUCTION

The anatomical association between mast cells and the nervous system was first described in 1890 (1), and experimental evidence of bidirectional communication between elements of the immune and nueroendrocrine systems has been growing since at least 1926 (2). The last decade has been particularly dynamic, however, as immunohistochemical, molecular, and cell culture techniques provided new approaches to the exploration of mast cell biology and the nervous system.

Two developments in the fields of mast cell and neurobiology are particularly important to this review of mast cell neuropeptidases. In mast cell biology, recent appreciation of the functional and compositional heterogeneity of mast cells emphasized the potential for differential mediator release at particular anatomical sites. In neurobiology, elucidation of a peptidergic, "nonadrenergic, noncholinergic" nervous system suggested a means by which neuropeptides may modulate mast cell function and by which mast cell proteases may modify the activity of neuropeptides. The context of mast cell protease interactions with peptide neurotransmitters will be reviewed by summarizing, first, important advances in our understanding of the peptidergic nervous system; second, the anatomical and physiological evidence for communication between the peptidergic nervous system and mast cells; and finally, the specific biochemical evidence that the secretory proteases of mast cells hydrolyze neuropeptides.

II. THE PEPTIDERGIC NERVOUS SYSTEM

A. Nonadrenergic, Noncholinergic System

Although mast cells have been associated with nearly all elements and levels of the nervous system, in this chapter we focus on their interaction with peripheral nerves of the autonomic nervous system. The reader is referred to several recent reviews for more detailed discussion of mast cells in the central nervous system (3–6) and to reviews of the diffuse neuroendocrine system (7–9), another source of potential interactions with mast cells that will not be addressed specifically here.

Until recently, the motor pathways of the autonomic nervous system were thought to be comprised of the parasympathetic and sympathetic pathways, with acetylcholine and norepinephrine as their respective classical neurotransmitters. Only in the last decade has the existence of a third, nonadrenergic, noncholinergic system been suspected. It was observed that certain target organ effects of nerve stimulation—for instance, bronchomotor changes in airways or vascular permeability and edema in the skin—persisted despite cholinergic or adrenergic receptor blockade (10,11). The list of candidate neurotransmitters mediating these atropine- and propranolol-resistant effects includes other biogenic amines, amino acids, purines, prostaglandins, nitric oxide, and other molecules, but immunohistochemical identification in nerves of a number of peptides, and pharmacological confirmation of their potent target effects focused much attention on neuropeptides as important transmitters in the nonadrenergic, noncholinergic system.

B. Colocalization of Neuropeptide and Classical Transmitters

Although pharmacologically distinct from acetylcholine (ACh) and norepinephrine (NE), it is now clear that neuropeptides coexist with their more classical counterparts, and often with many other neuropeptides, in the same neurons that subserve autonomic motor and sensory functions (12–14). There is thus no separate "hardwiring" for this system, and understanding the projections of the peptidergic nervous system requires a review of classic autonomic circuits. The general associations of neuropeptides with ACh, NE, or other peptides in autonomic nerves is summarized here; biological effects of individual peptides are discussed later in the chapter.

1. *Parasympathetic Motor Efferents*

ACh is the classical transmitter between pre- and postganglionic neurons in both the parasympathetic and sympathetic nervous pathways. In the relatively simple parasympathetic system, long cholinergic preganglionic fibers synapse with the soma of postganglionic neurons—also cholinergic—in ganglia distributed in or near the target organ (Fig. 1). Parasympathetic fibers in the enteric plexus have

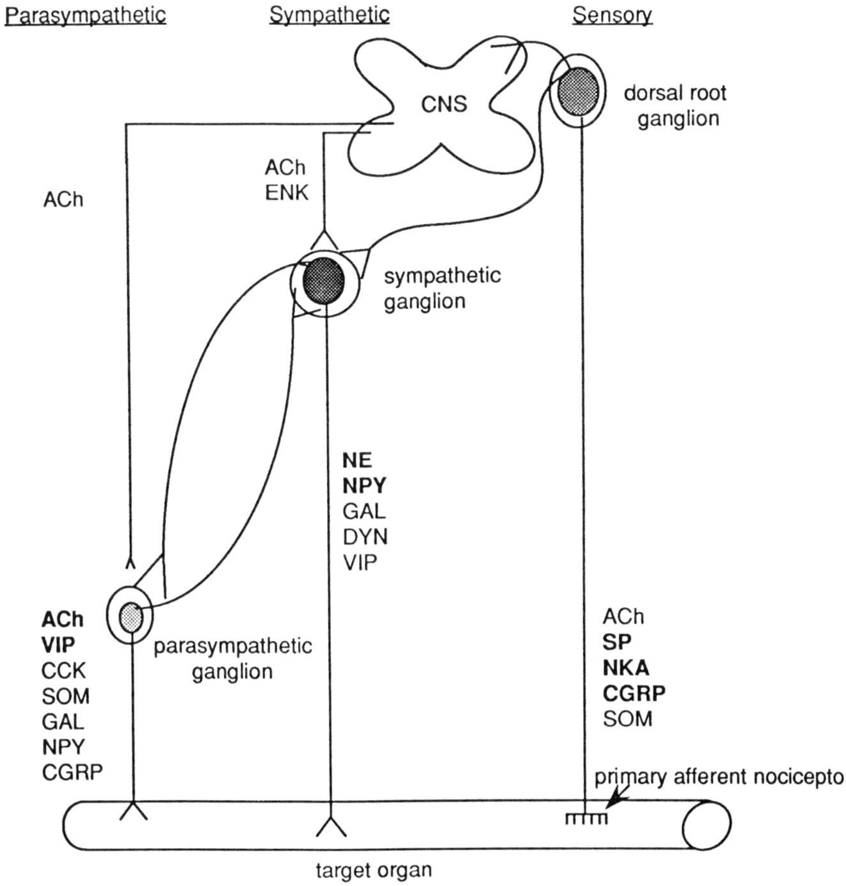

Figure 1 Schematic representation of the parasympathetic, sympathetic, and sensory pathways of the autonomic nervous system. Several neuropeptides may coexist with acetylcholine (Ach) or norepinephrine (NE) in a single neuron, and the pattern of neuropeptide content may vary according to species, development, the tissue that is innervated, and tissue or nerve injury. Transmitters so far most frequently associated with each pathway are shown in bold print. CCK, cholecystokinin; CGRP, calcitonin gene-related peptide; DYN, dynorphin; ENK, enkephalin; GAL, galanin; NPY, neuropeptide Y; SOM, somatostatin; SP, substance P; VIP, vasoactive intestinal peptide.

been characterized extensively in several mammalian species. These are reviewed briefly here, with the guarded assumption that some of the features can be generalized to other tissues or other species. As many as six peptides (depending on species) have been detected in cholinergic excitatory motor neurons of the myenteric plexus (15). In the submucous plexus (projections from which, we shall see, are often associated with mucosal mast cells) parasympathetic fibers contain not only ACh but may also contain vasoactive intestinal peptide (VIP), cholecystokinin (CCK), somatostatin (SOM), galanin (GAL), calcitonin gene-related peptide (CGRP), and neuropeptide Y (NPY). These fibers project to and ramify within the mucosal epithelium, where they characteristically form close associations—but not true synaptic contacts—with target cells. Rather, sites of transmitter release appear as vesicle-containing varicosities along the length of the axon (16,17). The combination of peptides colocalizing in parasympathetic nerves varies from tissue to tissue. In studies to date, VIP is the peptide most frequently associated with cholinergic postganglionic efferents.

2. *Sympathetic Motor Efferents*

Co-localization, heterogeneity, and plasticity of peptide content are important features of the sympathetic nervous system as well. Short, cholinergic, preganglionic fibers from the spinal cord synapse with sympathetic neurons in paravertebral and prevertebral ganglia (Fig. 1). These preganglionic sympathetic fibers have been reported to contain a number of peptides in different combinations—including enkephalin (ENK), neurotensin (NT), substance P (SP), and SOM—co-localizing with ACh (13). Sympathetic neurons also receive synaptic connections from other sympathetic neurons, from SP-containing sensory neurons via the dorsal root ganglia, and in the gastrointestinal tract, from parasympathetic neurons which may contain VIP, bombesin, and CCK (reviewed in Ref. 16). Three kinds of postganglionic sympathetic neurons project from paravertebral sympathetic ganglia. One kind, containing NE and NPY, innervates blood vessels in target organs; another, containing NE alone, innervates the parenchyma of target tissues; and one containing ACh (yes, ACh) and VIP innervates sweat glands (16). A fourth type of postganglionic sympathetic neuron is seen in prevertebral ganglia. This contains NE and SOM and probably innervates the gut. Peripherally, for instance in the submucous plexis, adrenergic nerves have been described which contain VIP, dynorphin (DYN), and/or GAL, whose axons project to submucosal arterioles (reviewed in Refs. 17 and 18).

To add to this complexity, cultured sympathetic ganglia cells are known to change the expression of both their classical and peptide neurotransmitters with alterations in culture conditions (13); it is likely, then, that the array of transmitters in sympathetic and perhaps other nerves in vivo is developmentally regulated and dynamic, subject to changes in the local milieu.

3. *Parasympathetic and Sympathetic Nerves to the Respiratory Tract*

In the respiratory tract, autonomic motor nerves are most abundant in the airways and proximal vessels of the pulmonary circulation; innervation of the distal airways, smaller vessels of the pulmonary circulation, and lung parenchyma are relatively sparse (19–22). Parasympathetic nerves are abundant in the airways of mammalian species. Some of these cholinergic motor fibers also contain the bronchodilatory peptides VIP and peptide histidine-methionine (PHM) in human airways or peptide histidine-isoleucine (PHI) in a number of other species (23, 24); such fibers are more abundant in larger airways. Sympathetic nerves are seen in most experimental animals, but there is significant species variation, and there is relatively little sympathetic innervation of bronchial smooth muscle in the human airway (10,25).

4. *Sensory Afferents*

The gastrointestinal and respiratory tracts and vessels which supply the skin, joint, and other organs are richly innervated by sensory afferents. In vertebrates, the peripheral projections of these neurons are called *primary afferent nociceptors* (reviewed in Ref. 26). The cell bodies of primary afferent nociceptors are located in dorsal root ganglia (DRG), and have both a peripheral axon and a central process that makes synaptic contact with sensory transmission neurons in the central nervous system (Fig. 1). Intense chemical, thermal, or mechanical stimulation is normally required to activate the afferent nociceptor, but it can be sensitized by a number of peptide and nonpeptide chemicals generated at a site of injury and released by neighboring cells. Bradykinin, interleukin-1, prostaglandins E_2 and I_2, 8*R*,15*S*-diHETE, and leukotriene B_4 are but a few of the endogenous chemicals known to activate or sensitize primary afferent nociceptors. A characteristic feature of nociceptors is their sensitivity to capsaicin, the nonpeptide pain-producing ingredient of chili peppers. Sensory afferents are discharged by capsaicin applied acutely, depleted of transmitters by capsaicin applied chronically, and apparently destroyed by capsaicin administered during the neonatal period (27). Thus events that are triggered by acute application of capsaicin, or attenuated in animals that have received chronic or neonatal capsaicin treatment, are considered to be mediated by nociceptive primary afferents.

The number of peptides detected in primary afferents exceeds 15 and continues to increase. As in sympathetic and parasympathetic motor efferents, several peptides may coexist in a single sensory system. The tachykinins (SP, NKA, neuropeptide K) and CGRP are the most commonly described sensory neuropeptides (28–31). To date, as many as four neuropeptides have been detected in various combinations in a single sensory nerve (32). SP, the most extensively studied sensory neuropeptide, is present in 20% of DRG neurons in the rat (33). Of these SP-containing neurons, up to 80% also contain CGRP and about 25% contain SOM, but a significant fraction of DRG neurons contain neither SP nor SOM.

In view of the coexistence of several neuropeptides, it is not surprising that the release of a particular peptide, or the complement of neuropeptides contained in a sensory nerve, has so far defied correlation with the activity of a particular physiologically defined class of primary afferents. The pattern of peptide content differs widely between species, is developmentally regulated, and can be altered by tissue injury or nerve damage. Most pertinent to this discussion of mast cell–nerve interactions in peripheral tissues, the peptide content of sensory nerves is determined to no small degree by the tissue they innervate. In the rat, for example, up to 85% of DRG neurons innervating the stomach antrum contain CGRP and up to 60% contain SP, whereas fewer than 15% of DRG neurons innervating the skin contain either peptide (34). When a sensory nerve from the skin is forced to innervate the stomach antrum, levels of SP and CGRP increase significantly in the reinnervating DRG neurons (35).

C. Neurogenic Inflammation and the Axon Reflex

A number of studies have shown that electrical stimulation of nerves (e.g., the superior laryngeal, vagus, or saphenous nerves, which carry both autonomic efferents and afferents) causes vasodilation and increases vascular permeability in the tissues they innervate. A variable but significant portion of the response is not blocked by cholinergic or adrenergic blockade, nor by hexamethonium, which blocks orthodromic synaptic transmission through ganglia. Thus a cholinergic- and adrenergic-independent system that does not require ganglionic transmission is implicated in some tissue responses (36,37). This local tissue response, known as *neurogenic inflammation,* is thought to be mediated by sensory fibers (38). When stimulated appropriately electrically or by chemical or mechanical irritants, the peripheral terminals of sensory neurons not only transmit signals centrally but may also release into the tissues they innervate substances that cause local vascular permeability, edema, and infiltration with inflammatory cells (39,40). Because it can occur without transmission to or back from the central nervous system, this phenomenon is thought to involve a "short circuit" through peripheral axons—an *axon reflex* (41) (Fig. 2). SP is the best characterized of the many mediators released by sensory nerves in response to electrical, chemical, or irritant stimulation. Injection of SP alone reproduces the vascular changes of neurogenic inflammation (42), but other neurotransmitters, either localized to the same nerve or released from neighboring cells, contribute to the phenomenon. For instance, prostaglandins released from sympathetic postganglionic neurons contribute to neurogenic inflammation in an experimental model of arthritis (43,44).

Experimental evidence suggests that mast cells participate in neurogenic inflammation, either as sensory-like cells which, when stimulated by antigenic,

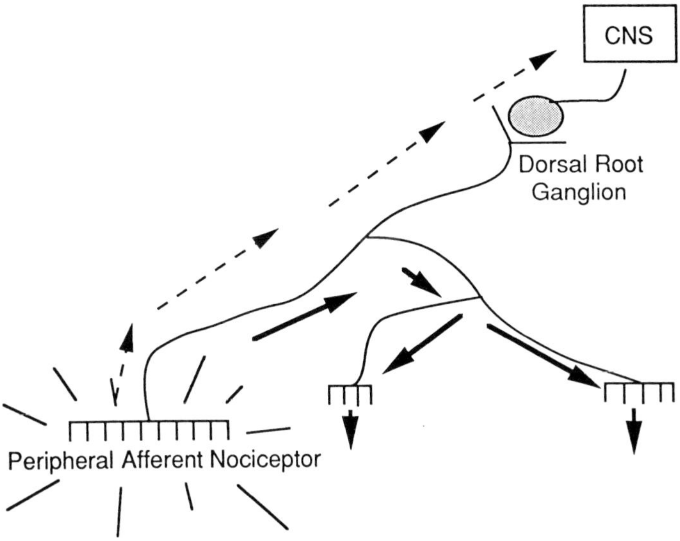

Figure 2 Axon reflex. Stimulation of the peripheral afferent nociceptor initiates not only transmission to the central nervous system (CNS) via the dorsal root ganglion, but also release of neurotransmitters from local branches of the sensory afferent fiber.

mechanical, or physical stimuli, release mediators that in turn irritate sensory nerve endings and trigger the axon reflex (45), or as neuroeffectors, signaled by nerves to release mediators with potent effects on tissue targets (46,47). Thus, in some models, plasma extravasation and vasodilation evoked by peripheral nerve stimulation can be attenuated by histamine receptor antagonists or by prior depletion of mast cell mediators with compound 48/80 (36,48).

To summarize, the peptidergic nervous system, once sought as the nonadrenergic, noncholinergic nervous system, is perhaps better described as *also-adrenergic, also-cholinergic*. Immunohistochemical studies confirm that several neuropeptides with different, perhaps even opposing, biological effects may coexist with a classical transmitter in the same neuron. The relative expression of neuropeptides also changes depending on species, tissue, developmental stage, or tissue injury and disease. In the tissues in which they have best been described, peptidergic nerves generally do not form synaptic contacts, but release their transmitters *en passant*. These features are admittedly at odds with former constructs of neurological precision. Its apparent ambiguity and enigmatic organization notwithstanding, the peptidergic system is considered important in the homeostatic control of bronchial and vascular smooth muscle tone, epithelial fluid and electrolyte transport, and in protection against infestation or mechanical and chemical intrusion. With respect to the latter function, peptidergic sensory nerves appear to play an important role in neurogenic inflammation. We examine next the evidence that mast cells also participate in this and other neurally mediated phenomena.

III. ANATOMICAL ASSOCIATIONS BETWEEN MAST CELLS AND NERVES

Incidental reports and a number of systemic morphological studies indicate that mast cells are often distributed in the vicinity of peptidergic nerves. In some sites or in certain experimentally induced or pathological conditions, mast cells and nerves apparently form more specific and frequent anatomical associations. Data for this from studies of rat and human tissues are reviewed here and summarized in Tables 1 and 2. We begin with the rat, in which both the heterogeneity of mast cell populations and the extent of their interactions with peripheral nerves have been described most systematically.

Table 1 Percentage of Rat Mast Cells Associated with Nerves

Anatomical site	Infected or inflamed	Percent mast cells near nerves	Refs.
Intestine			
Mucosa	–	20–30	19, 56
	+	30–40	19, 56
Muscularis	–	0	19
	+	3	19
Mesentery	–	[a]	58
Respiratory tract			
Larynx	–	[a]	65
Trachea	–	[a]	64
Bronchi	–	[a]	64
Alveoli, parenchyma	–	0	19
Pleura	+	0	19
Diaphragm	–	[a]	58
Heart	+	15	19
Thymus	–	15	99
Synovium	–/+	[a]	67
Skin	–	0	19

[a]Nerve–mast cell associations described, but not quantitated.

Table 2 Percentage of Human Mast Cells Associated with Nerves

Anatomical site	Infected or inflamed	Percent mast cells near nerves	Refs.
Intestine			
Mucosa	–	50–75	77, 78
	+	[a]	73, 78
Muscularis	+	[a]	73, 75
Respiratory tract[b]			
Diaphragm[b]			
Heart[b]			
Thymus[b]			
Synovium[b]			
Skin	–	~10	100–102
	+	~20	102

[a]Nerve–mast cell associations described, but not quantitated.
[b]No published description of nerve–mast cell associations in human tissues.

A. Rat

1. *Mast Cell Heterogeneity*

There are at least two distinct mast cell populations in the rat, or as some investigators prefer to describe it, there are two phenotypic extremes of mast cells between which there is a gradation of mast cell phenotypes (49,50). "Typical" mast cells predominate characteristically in the peritoneal cavity, intestinal serosa, skin, and connective tissue; hence their designation as *peritoneal, serosal,* or *connective tissue mast cells* (CTMCs). The cells retain their metachromatic properties even after formalin fixation, are degranulated by compound 48/80 as well as by antigen and anti-IgE, and their numbers in tissues do not decrease with systemic corticosteroids. Generally, mast cells with these attributes contain rat chymase I.

Another, "atypical" population of mast cells is distributed in the epithelium, subepithelium, and lamina propria of the gastrointestinal and respiratory tracts; hence their designation as *mucosal mast cells* (MMCs). Their metachromasia is altered by formalin fixation; mast cells in tissues fixed in formalin require prolonged staining with metachromatic stains at acid pH to elicit the metachromasia (51). Although stimulated by antigen and anti-IgE, these cells are not degranulated by compound 48/80, and their numbers decrease with systemic corticosteroids (52). Most notably, this population of mast cells undergoes a massive expansion during nematode infection. Such cells contain a rat chymase II, which shares with rat chymase I its primary specificity for aromatic amino acids but is antigenically and biochemically distinct (53,54).

2. *Gastrointestinal Tract*

a. Mucosal Mast Cells. Specific associations with nerves have been described most systematically for intestinal mucosal mast cells. In an ultrastructural study of rat ileum, investigators first noted that nerve processes surrounded mucosal crypts in the lamina propria and that "neuron-like" cells were distributed below the epithelium outside the intramural nervous plexuses (55). Close contacts between mucosal mast cells and these nerve processes and cells were seen frequently, and electron microscopy suggested that at least some of the axons contained large, dense-core vesicles typical of peptide-storage organelles (16).

These observations were later confirmed and extended by an investigation using quantitative immunohistochemical as well as ultrastructural techniques (56). Both uninfected and nematode-infected rats, in which there is a striking intestinal MMC hyperplasia, were studied. By light microscopy, more than 80% of MMCs in infected rats were within 2 μm of a nerve, a percentage calculated to be five times greater than would be expected by chance alone, and unlikely to be due solely to higher density or skewed distribution of nerves. Electron microscopy confirmed the association between MMCs and nerves, often demonstrating direct contact between plasma membranes of MMCs and axons. Many of the nerves contained SP and/or CGRP. Nerves with 95- to 140-nm electron-dense vesicles or with pleomorphic vesicles up to 300 nm in diameter were less commonly seen in contact with mast cells. The nature of these nerves was not defined with specific nerve depletion or histochemical studies, and may be adrenergic or cholinergic efferents. Thus distinct and specific anatomical associations between rat chymase II–containing mucosal mast cells, largely with SP/CGRP–containing unmyelinated sensory afferents, were demonstrated.

A subsequent ultrastructural study confirmed the finding that a significant fraction of jejunal MMCs are near neural processes (19). Immunoreactivity to neuropeptides was not determined in this study, but electron microscopy indicated that the majority of processes with which MMCs were associated were unmyelinated and contained neurosecretory granules, consistent with terminals of sensory afferents (16).

b. Muscularis and Serosal Mast Cells. Mast cell numbers also increase in the muscularis propria of nematode-infected rats, but of 92 such mast cells observed in one study, only about 14% were within 500 nm of a nerve (19). Others have found a higher percentage (60%) of mast cells within one-half a cell diameter of the nerve (57). In rat mesentery, associations were described between SP-immunoreactive nerve fibers and serotonin-immunoreactive cells, presumed to be mast cells (58). The two cell types were most often seen near each other in the vicinity of blood vessels, where both types of cells were more numerous. Based on their distribution in the mesentery and abundance of serotonin, these cells were likely to have been CTMCs. While it appears that they can be associated with peptidergic nerves and blood vessels, the specificity of these associations are not yet as clearly defined as that between rat chymase II-containing mast cells and unmyelinated sensory afferents in the mucosa.

3. *Respiratory Tract*

At least two types of mast cells are found in the rat airway and lung (59,60), as summarized in Table 1. Mast cells in the epithelium (also called globule leukocytes) and lamina propria contain rat chymase II. They are in this respect similar to the mucosal mast cells described in the gastrointestinal tract and are the predominant phenotype in the epithelium and lamina propria of the trachea and major conducting airways of the lung at all ages. CTMCs (rat chymase I-positive) predominate in the submucosa, posterior membrane, and adventitia of the trachea and conducting airways. They are also the predominant mast cell type in rat lung parenchyma and pleura. A unique population of peribronchial mast cells that contain both rat chymases I and II has been described (61) whose functional characteristics are not yet known.

Immunohistochemical and retrograde tracing of airway nerves indicates that the airway epithelium and mucosal arterioles—the sites in which mucosal mast cells are most heavily distributed—are densely innervated by tachykinin- and CGRP-containing fibers (40,62,63). Not surprisingly, then, investigators noted close associations between such fibers and mast cells (detected by serotonin immunoreactivity) in the mucosa of the larynx, trachea, and bronchus (64,65), a finding that was not as pronounced or frequent in more abluminal airway (64).

Because the alveoli are so sparsely innervated, mast cell–nerve associations are less likely in this site. Indeed, investigators found that of 45 mast cells from the lungs of uninfected and 69 from nematode-infected rats, none were found to be near nerves (19).

4. *Joint*

Mast cells are distributed in the normal rat synovial membrane (66). Their location, sensitivity to compound 48/80 (44), and metachomatic properties (67) suggest that they are CTMCs. They have not yet been characterized by protease content, however, and there is evidence to suggest that the synovial mast cell may be functionally different from the skin CTMCs (68). Simultaneous staining for mast cells and nerves shows their parallel distribution in all parts of normal synovium, particularly in the highly vascularized zone immediately beneath the synovial lining (67). Synovial nerves are largely CGRP-immunoreactive; SP-immunoreactive nerves are less frequent. The nearly total disappearance of both mast cells and nerves in adjuvant arthritis suggests that they are a functional unit (67), other evidence for which is reviewed later in the chapter.

5. *Skin*

Although mast cells are numerous in the skin, specific anatomical associations between them and cutaneous nerves appear to be less frequent than in the gastrointestinal and respiratory mucosa and synovium. In the skin of the ear pinna, not one of 114 mast cells detected by electron microscopy was seen within 500 nm of a nerve (19). Similarly, other investigators observed that in the dorsum of the paw, whereas more than 75% of CGRP-, SP-, or NKA-containing nerves are in the superficial dermis, fewer than 15% of mast cells are in that vicinity (69).

6. *Other Tissues*

Anatomical associations, coincidental or otherwise, are likely to occur throughout the body, particularly in perivascular sites. Thus there are reports of mast cells associated with nerves in the rat heart, thymus, diaphragm, and other sites (Table 1). Clearly, the frequency of nerve–mast cell associations varies markedly with anatomical site and local disease. These anatomical variations suggest that the likelihood and nature of functional interactions between nerves and mast cells will differ between tissues.

B. Human

1. *Mast Cell Heterogeneity*

The heterogeneity of human mast cells is well recognized, although it is often not as clear cut as in the rat. All human mast cells contain tryptase; thus a major distinction in the protease content of different types of mast cells is the presence or absence of *chymase* (70). Mast cells containing only tryptase (MC_T) are the predominant mast cell type in the lung alveoli and intestinal mucosa, and with some risk of oversimplification, may be the functional counterpart of rat intestinal MMCs. Mast cells that contain both tryptase and chymase (MC_{TC}) also contain carboxypeptidase (71). These cells predominate in skin, conjunctiva, and intestinal serosa and are considered the counterpart of rat connective tissue mast cells.

Studies of mast cell–nerve associations in human tissues are understandably limited by the availability and variability of clinical or autopsy specimens. Obtaining truly "normal" control tissues is a particular difficulty. With these limitations in mind, studies of nerve–mast cell anatomical associations in human tissues are reviewed (Table 2).

2. *Gastrointestinal Tract*

The population of intestinal mucosal mast cells proliferates in a number of inflammatory bowel diseases, including Crohn's disease, ulcerative colitis, scleroderma, polymyositis, dermatomyositis, mixed connective tissue disease, and appendicitis/fibrosis (72–74). There appears to be a particular association between mast cells and nerves in Crohn's disease, in which mast cells are focused near proliferating nerves, particularly in the deep muscularis, and are associated with widespread axonal necrosis (72,75,76). A similar predilection of mast cells for nerves was described in patients with systemic connective tissue diseases involving the gut (73); infiltrates of mast cells and eosinophils penetrating into the lamina propria were focused in clusters associated with the superficial, submucosal, and deep neural plexuses. Occasional apposition of mast cell and neural cellular membranes was observed.

To determine the frequency with which nerve–mast cell associations occur normally, investigators studied macroscopically normal pieces of bowel taken from patients with cancer or inflammatory bowel disease. One study found that approximately 50 to 75% of mast cells were within 500 nm of, and often directly apposed to, nerves (77). However, another study found very few mast cells within 500 nm of a nerve and virtually no evidence of direct contact between the cells (78). Only when using a more relaxed definition of association [i.e., close enough to appear on the same photographic print (actual distance undisclosed)] did investigators find that 30 to 60% of mucosal mast cells were associated with enteric nerves. Even in apparently uninvolved tissues, the frequency of nerve–mast cell associations in tissues from patients with inflamed bowel (ulcerative colitis, Crohn's disease) was higher than in sections from patients without inflammatory disease (malignancy, familial polyposis). In both studies, nerves near mast cells generally had the ultrastructural appearance of sensory afferents and cholinergic efferents.

3. *Respiratory Tract*

In the human airway, separate studies of mast cell types in the airway and of the tracheobronchial peptidergic innervation suggest that airway mast cells are likely to be in the vicinity of peptide-containing nerves. Mucosal mast cells (MC_Ts) predominate in the lamina propria, below the respiratory epithelium, in or near tracheobronchial smooth muscle, and in the alveolar wall (70,79–81). Connective tissue mast cells (MC_{TC}s, detected either by immuno- or enzyme histochemistry) generally account for 25 to 35% of human airway mast cells, but on closer inspection, their portion of the mast cell population varies with the tissue compartment of the airway: in the smooth muscle, for instance, only 14% of mast cells are MC_{TC}s, whereas more than 70% of mast cells near glands are MC_{TC}s (79). In the lung parenchyma, mast cells with characteristics consistent with MC_Ts predominate (80,82,83), accounting for more than 80% (some estimates are as high as 93%) of lung mast cells (70,82).

Parallel studies map the innervation of human airways and parenchyma. *Parasympathetic nerve fibers,* which contain VIP and VIP-related peptides as well as NPY and GAL, are distributed in tracheobronchial smooth muscle, around blood vessels and glands, and in nerve cell bodies of airway ganglia of human airways (23,25,84). VIP-immunoreactive nerves are more abundant in proximal airways but can be traced peripherally at least as far as bronchioles 200 μm in diameter. Single VIP-containing fibers can also be found along the basal surface of respiratory epithelial cells and the basement membrane (84). *Adrenergic nerves* are typically described in vascular smooth muscle and glands in the human. Although functional studies and special stains for catecholamines suggest that there is little adrenergic innervation of human bronchial (i.e., nonvascular) smooth muscle (10,85), more recent studies indicate that NPY- and tyrosine hydroxylase-immunoreactive sympathetic nerves are the most abundant neuropeptide-containing fibers in the bronchial muscle of newborns and young children and are as common as VIP-immunoreactive nerves, even in older children and adults (86). There is little sympathetic innervation beyond terminal bronchioli or of the epithelium. As for *sensory afferents,* SP-containing nerves are described in the lamina propria of the human airway, along blood vessels, and in ganglia, glands, and smooth muscle (25,31). Single short fibers in the basement membrane branch into the epithelium (87). CGRP-immunoreactive nerve fibers generally follow the distribution of SP-immunoreactive nerves, except for relative sparing of bronchial smooth muscle and glands (31). The paucity of SP-immunoreactive nerves in human airways (25,86,88) appears very different from the rich sensory innervation of rat airways. Some of the difference may be due to the age or smoking habits of the patients from whom tissues were resected (86,89,90). The number and morphology of SP-immunoreactive nerves are also very different in the airways of subjects with asthma (87). In this condition, fibers are longer, more numerous, more branched, and tend to run in large fiber bundles instead of as single fibers. As in the airways of nonasthmatics, they tend to innervate vessels and glands in the lamina propria, but fibers are not seen in areas of abnormal epithelium.

These studies thus suggest that mast cells and peptidergic nerves are likely to occur in the same microenvironments in human airway, particularly around blood vessels. The predominance of different types of mast cells and different types of nerve fibers varies between different tissue compartments, such as seromucous glands, smooth muscle, and epithelium. Finally, the likelihood of interactions between mast cells and nerves may be higher in some conditions, such as asthma, in which mast cell number and activation and the density of sensory afferents apparently increase.

4. *Joint*

Mast cells comprise 3% of the cells in normal human synovium (91) and their numbers increase in a number of arthritides (92,93). Protease content of these cells has not yet been reported, but human synovial mast cells may be more similar in a number of other morphological and functional characteristics to mucosal mast cells of the gastrointestinal and respiratory tracts (94). Systematic correlation between the distribution of mast cells and peptidergic nerves in the human synovium has not yet been reported.

5. *Skin*

Mast cells are abundant in human skin (95,96), which is richly innervated by peptidergic nerves. SP is localized to the nerve endings in dermal papillae and epidermis of healthy human skin (97). VIP- and NT-immunoreactive nerves supply eccrine and apocrine sweat glands, arterial segments of the deep and superficial vascular plexus, and arteriovenous anastomoses (98,99). Nerve–mast cell contacts have been described in glomus tumors of the skin (100) (Fig. 3) and in human subcutis (101), but it is unclear whether such association is a general feature. One quantitative study of nerve–mast cell contacts in diseased human skin found that both the number of mast cells and of contacts between nerves and mast cells is increased in psoriasis, particularly in the papillary dermis (102). A similar increase in nerve-mast cell contacts was not observed in lichen planus, in which there is also dermal mast cell hyperplasia, suggesting that there is a neurogenic component particular to the pathogenesis of psoriasis (103).

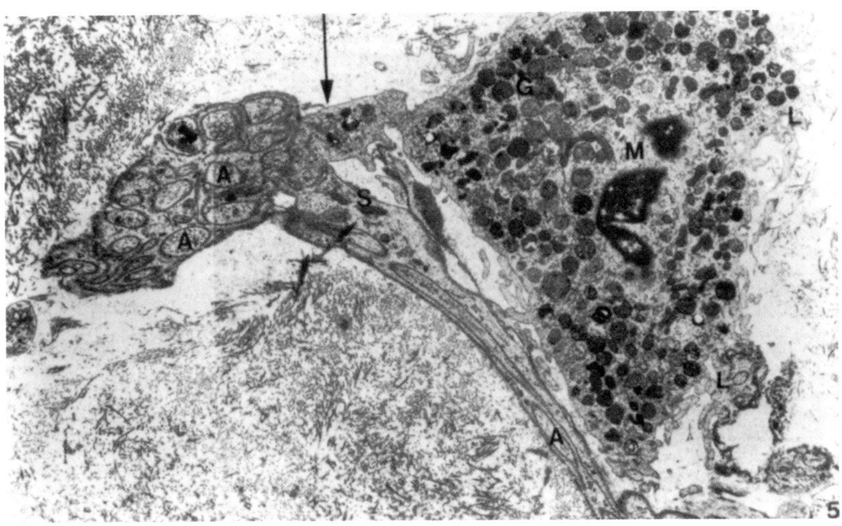

Figure 3 Mast cell in contact with a nerve fiber in a glomus tumor of human skin. A large cytoplasmic process (arrow) of a mast cell (M) containing immature granules (G) is inserted into the nerve fiber bundle. L, lamellopodia; A, axons; S, Schwann cell; × 6100. (From Ref. 100.)

Increased numbers of mast cells have also been described in scleroderma (104), but their anatomical association with nerves in this condition has not yet been specifically studied.

In summary, the frequency of nerve–mast cell associations, and the type of nerve or mast cell that may be anatomically juxtaposed, differs from one tissue site to the next. The nature and frequency of such associations within a particular tissue may also change with disease or injury. In the intestinal mucosa, extensive association between mucosal mast cells and enteric nerves, particularly cholinergic nerves or peptidergic sensory afferents, has been documented by a number of light and ultrastructural studies in rat and human tissues. While it appears that mucosal mast cells in rat and human airway and synovium are also distributed in areas innervated by peptidergic nerves, in other sites, such as skin or lung parenchyma, the anatomical associations are not as tight, and functional interactions, if any, may occur by different mechanisms. A number of diseases, including parasitic infection, inflammatory bowel disease, asthma, and psoriasis, are associated with more frequent nerve–mast cell associations.

IV. COMMUNICATION BETWEEN MAST CELLS AND NERVES

A. Signals from Nerve to Mast Cell

Structural associations between nerves and mast cells are meaningful to the extent that they allow functional interactions. In fact, an enlarging body of experimental data suggests that dialogue between nerves and mast cells is important in mediating many physiologic events. Some of these observations are reviewed here to illustrate a few central themes.

We first examine the evidence that nerves influence the activation, growth, and differentiation of mast cells. A particularly compelling demonstration of this is that parasite-infested rats, which have developed a hyperplasia of intestinal mucosal mast cells, can be conditioned to release the mast cell–specific chymase, rat chymase II, on an audiovisual cue alone (105). The neural circuits and reflexes underlying this phenomenon are complex, but simpler systems have similarly demonstrated that neural discharge can activate mast cells. For example, stimulation of autonomic fibers in or ex vivo causes, or potentiates, histamine release and morphological evidence of mast cell degranulation in a number of tissues (106–109). The major neural pathway or neurotransmitter implicated in the phenomenon depends on the innervated tissue, but the contribution of cholinergic (106,110) and primary sensory afferents (107,109) has been demonstrated in a number of sites and species. Depending on experimental conditions, sympathetic stimulation during and after antigenic stimulation in allergic animals can inhibit mast cell degranulation (111). Thus the activation state of mast cells may be modulated by all pathways of the autonomic nervous system (108).

In more reduced experimental systems, individual neurotransmitters have been shown to degranulate mast cells, both in vitro and in situ. The effects of nonpeptide neurotransmitters are described elsewhere (112,113). Our focus here is on the effects of several neuropeptides on isolated rat or human mast cells, summarized in Tables 3 and 4, respectively. As is true of their responses to classical transmitters, mast cell populations exhibit a range of responses to neuropeptides, depending on their anatomical site and species of origin (114). The neuropeptide most consistently shown to have effects in several isolated mast cell systems, both connective tissue and mucosal, is SP, whose effective concentrations are in the range 10 to 30 μ*M*, depending on the species and tissues from which the mast cells derive. Rat peritoneal mast cells are quite sensitive and are degranulated by SP and a number of other neuropeptides. SP-induced histamine release from these cells is rapid, noncytotoxic, independent of cell-bound IgE, and modulated by extracellular cations (115). In contrast, rat intestinal mucosal mast cells—and their presumed counterparts, human lung mast cells—show little histamine release even in response to SP (49,116,117). This is unexpected, given that it is the mucosal mast cell that is most often associated with peptidergic neurons in situ. Technical differences in mast cell isolation were reasonably excluded as a cause for the diminished responses in mucosal mast cells, and it is postulated that the relatively low level of SP-induced histamine release is due to selective mediator release or to the derivation of these cells from inflamed tissue whose abundance of mediators may decrease the sensitivity of the cell to a single agent (49). The low potency of SP effects on mucosal mast cells would appear to diminish its importance as a signal to mast cells. Although it may be only weakly effective alone, however, preliminary data suggest that SP may potentiate mast cell responses to other stimuli (118,119).

Table 3 Neuropeptide-Induced Secretion from Isolated Rat Mast Cells

		Mast cell secretion		
Neuropeptide	Dose (m*M*)	Peritoneal	Mucosal	Refs.
Bradykinin	10–100	+	?	49, 140
CGRP	1–100	+	?	120
NPY	1–100	+	?	141, 142
Neurotensin	0.01–100	+	?	49, 114, 140
Peptide YY	1–10	+	?	141
SOM	0.5–100	++	±	49, 143
SP	1–100	++	+	49, 115, 140
VIP	5–10	++	±	49

Table 4 Neuropeptide-Induced Secretion from Isolated Human Mast Cells

		Mast cell secretion		
Neuropeptide	Dose (m*M*)	Skin	Mucosal	Refs.
Bombesin	10–100	?	0	144
Bradykinin	30	±	?	121
CGRP	30	±	?	121, 145
Endorphins	10–100	?	0	144
Enkephalins	10–100	?	0	144
Neurokinin A	30	±	?	121
Neurokinin B	30	±	?	121
Neurotensin	30	±	?	121
Somatostatin	10–30	++	?	121, 146
Substance P	1–100	++	±	121, 144, 146
VIP	1–30	±	?	117, 121, 145

Early observations suggested that SP-induced mast cell secretion is mediated by a low-specificity, low-affinity, receptor-independent interaction at a membrane site requiring a positively charged aminoterminus and a lipophilic anchor (120–122). It has since been shown that SP (and a number of other positively charged secretagogues, such as mastoparan, neomycin, and compound 48/80) trigger exocytosis by binding a heterotrimeric G protein, G_{i3}, present in the plasma and Golgi membranes of rat peritoneal mast cells (123). Whether other neuropeptides or other types of mast cells can interact similarly is not yet described.

Perhaps because of an additive or potentiating effect with other signals or cells in the mast cell's natural milieu, mast cell degranulation—and tissue effects of secreted mast cell products—can be detected after application of much lower doses of SP in vivo or in tisu. In femto- to micromolar doses, SP causes a mast cell–dependent local neutrophilic infiltration in mouse skin (124,125), mast cell secretion of proteases (which limit the activity of co-injected CGRP in rat and human skin) (126,127), mast cell–dependent bronchoconstriction in rat lung (128,129), and expression of endothelial-leukocyte adhesion molecule in microvascular endothelium induced by mast cell secretion of tumor necrosis factor (130).

A number of histologic and in vitro observations indicate that nerves exert a trophic influence on mast cells. For instance, surgical or chemical denervation of the diaphragmatic muscle, nasal mucosa, and synovium decreases the number or changes the distribution of mast cells in those tissues (131–133). Coculture of sympathetic neurons and rat basophilic leukemic cells, studied because of their similarities to mucosal mast cells, confirmed a specific trophic interaction between the cells which requires bidirectional signaling. Neuron-derived nerve growth factor causes mast cell secretion of yet unidentified chemical signals that direct the cells' growth toward each other (134–136).

Clearly, then, exogenous neuropeptides can cause mast cell secretion of products that mediate tissue effects. In vivo, the neuroeffector role of the mast cell was suggested by Kiernan, who found that depletion of mast cells with compound 48/80, or treatment with H_1-receptor antagonists, reduces the severity of the axon reflex in the skin (47). Mast cells also appear to be important in gastrointestinal neurogenic inflammation and homeostasis (137), where the configuration of nerves, mast cells, and blood vessels is considerably different. For instance, capsaicin-induced mucosal hyperemia is increased in rats that have developed an intestinal mastocytosis in response to parasitic inflection, and is decreased by H_1-receptor antagonists, by agents that stabilize mast cells or prevent the mucosal mast cell hyperplasia, and by prior treatment with capsaicin (138).

Other observations, however, refute the importance of the mast cell in the early stages of neurogenic inflammation. In rat skin, degranulated mast cells were not seen near the postcapillary venules made permeable by electrical stimulation, and mast cell degranulation was evident only with prolonged nerve stimulation, after vascular events mediating inflammation had already occurred (69). In related experiments, neurogenic inflammation was not attenuated in mast cell–deficient mice compared to controls (139).

These disparities may suggest that the degree to which mast cells contribute to neurally mediated physiologic phenomena, or the mechanisms by which they do so, differ with the activation state or distribution of mast cells and nerves in different anatomical sites and species. That the "chain of command" can be quite complex was recently demonstrated in an experimental model of arthritis (43,44). In the synovium, activation of sensory afferents initiates a cascade of events in which mast cell release discharges local sympathetic terminals. Prostaglandins released from these sympathetic efferents finally mediate the increase in vascular permeability and plasma extravasation.

B. Signals from Mast Cell to Nerve

These observations add to an already substantial body of evidence that mast cell mediators modulate nerve function. In peripheral tissues, mast cells are a major source of histamine and, in some species (but not humans), serotonin (147). The stimulatory effects of both these mediators on primary afferent nociceptors and synaptic transmission have been well described (148–152).

Primary activation of mast cells by antigenic, physical, or chemical stimulation can effect physiological changes mediated at least in part by nerves. At the electrophysiologic level, it has been shown that antigen- or IgE-induced histamine release from isolated lung mast cells or ganglionic mast cells enhances the excitability and transmission of nerves in sympathetic ganglia (148,153), parasympathetic ganglia (154), and in vagal sensory nerves (155). In vivo, antigen-induced mast cell histamine release is enhanced by vagal stimulation, which can potentiate tissue responses to airway antigen challenge (110). In rat models of intestinal and respiratory allergy, antigen-induced epithelial permeability is significantly attenuated by tetrodotoxin (156,157) or after capsaicin depletion of sensory nerves (158), indicating the importance of neural amplification of mast cell–initiated events in these responses.

V. NEUROPEPTIDE HYDROLYSIS BY MAST CELL PROTEASES

A. Control of Peptide Hydrolysis: General Considerations

In addition to its role as an effector or a sensor for peripheral nerves, the mast cell may serve as a terminator of neuropeptide activity. As is true of other potent biological signals, the efficiency of a peptide neurotransmitter requires both its appropriate release and timely termination. There is little evidence for significant reuptake of neuropeptides. Enzymatic degradation thus appears to be a principal mechanism to inactivate neuropeptides.

Two lines of evidence suggest that mast cell proteases fulfill a critical function to terminate or modify neuropeptide signals. Mast cell degranulation in situ, or the addition of mast cell lysates/supernatants to experimental systems, may diminish or alter the bioactivity of a number of neuropeptides. There is also in vitro biochemical confirmation that purified mast cell proteases selectively hydrolyze neuropeptides at physiologically meaningful rates. Such observations for a number of neuropeptide substrates will be reviewed here, and the reader is referred to Table 5 for a compilation of neuropeptides specifically tested as potential substrates for purified proteases. While this biochemical evidence provides some insights, it is important to keep in mind other variables that change or limit potential biochemical reactions in situ (159,160). Usual determinants include anatomical proximity, local release or concentration of enzyme and substrate, and the specificity of the enzyme. Another consideration is structural modification of neuropeptides which protect against proteolytic attack. For example, cyclization via disulfide bonds (e.g., CGRP and SOM) or C-terminal amidation (e.g., VIP and SP) limits proteolytic attack against enfolded peptide bonds or by carboxypeptidases, respectively. Finally, other proteases, proteoglycans, or inhibitors which share the physiological milieu can significantly alter biochemical interactions described in vitro. As we shall see, neuropeptide hydrolysis can be increased by the concerted action of two or more proteases or by stabilization of a protease by heparin, which is co-released with proteases during mast cell degranulation.

Table 5 In Vitro Hydrolysis of Neuropeptides by Mast Cell Proteases

Neuropeptide	Protease[a]	Hydrolysis	K_m (μM)	k_{cat} (s^{-1})	k_{cat}/K_m ($10^6\ s^{-1}\ M^{-1}$)	Refs.
Calcitonin gene–related peptide	T, human	+	69	18.6	0.27	162
	T, human	+	6.8			163
Cholecystokinin-8	CPA, human	–				172
Dynorphin A	CPA, human	–				172
[Leu[5]]enkephalin	T, human	–				172
	Ch, human	–				172
	CPA, human	+	36	44	1.2	172, 182
[Met[5]]enkephalin	CPA, human	–				182
Neurokinin A	T, human	–				162
Neurokinin B	T, human	–				162
Neuromedin B	CPA, human	–				182
Neuromedin N	CPA, human	+				182
Neurotensin	T, rat	+				185
	T, human	–	182			
	Ch, human	+				182
	Ch, rat I	+				170
	CPA, human	+	16	51	3.2	172, 182, 183

(continued)

Table 5 Continued

Neuropeptide	Protease[a]	Hydrolysis	K_m (μM)	k_{cat} (s^{-1})	k_{cat}/K_m ($10^6\ s^{-1}\ M^{-1}$)	Refs.
Peptide histidine methionine	T, human	+	230	4.4	0.019	162
Somatostatin	Ch, rat I	+				170
	CPA, human	–				172
Substance P	T, human	–				162
	T, dog	–				169
	Ch, dog	+	560	22	0.039	169
	Ch (rat I, pure)	–				170
	Ch (rat I, granule)	+				170
	CPA, human	+	182			
[Arg[8]]vasopressin	T, human	+				184
	T, rat	+				185
Vasoactive intestinal peptide	T, human	+	52	7.8	0.15	162
	T, dog	+	200	44	0.22	169
	Ch, dog	+	3300	179	0.054	169
	Ch (rat I, granule)	+				170
	CPA, human	–				172

[a]T, tryptase; Ch, Chymase; CPA, carboxypeptidase.

B. Neuropeptide Hydrolysis: Specific Examples

1. *Calcitonin Gene–Related Peptide*

This 37-amino acid peptide is localized to several discrete areas of the central and peripheral nervous system. Of particular interest to this discussion is its abundance in primary sensory afferents innervating arterioles in skin, gastrointestinal and respiratory tracts, and other sites. CGRP is a potent vasodilator 1000 times more potent than ACh or SP, 10 to 100 times more potent than isoprenaline, and equipotent with prostaglandin E_2. Injection of femto- to picomolar doses into the skin causes local vasodilation and reddening that persists for as long as 10 to 12 h (161). Notably, this prolonged effect is converted to a transient response when CGRP is injected into the skin with SP, with which it is often co-localized or released (127). Injection with compound 48/80 also abbreviates CGRP's effects, and protease inhibitors block the SP-induced attenuation of CGRP's effects, indicating that CGRP's activity is terminated by proteases released from SP-stimulated mast cells. In vitro data subsequently confirmed that CGRP is hydrolyzed very effectively by purified human mast cell tryptase (162), with loss of its vasodilator activities (163). Furthermore, the comparable inhibition profiles of CGRP hydrolysis by human mast cell lysates and purified tryptase suggest that tryptase is the major CGRP-degrading activity in human mast cells (163). Whether the fragments generated by tryptase retain or assume actions ascribed to partial sequences of CGRP, including eosinophil chemotaxis (164) and antagonism of vasodilator activity (165), remains to be seen. Clearly, CGRP's vasodilator effects are terminated by tryptase. By this mechanism, tryptase may limit the extent of local inflammation or may contribute to the singular depletion of CGRP-immunoreactive nerves and abnormal vasomotor regulation associated with Raynaud's phenomenon and scleroderma, in which there is often a dermal mast cell hyperplasia (104,166).

2. *Tachykinins: Substance P, Neurokinin A, Neurokinin B*

The tachykinins, of which SP is the best characterized, are localized in sensory nerves, which have been implicated in neurogenic inflammation. SP is a potent direct vasodilator (although not of the same order as CGRP), and to the extent that it can cause mast cell histamine release, it may also cause vasodilation indirectly (167). SP contracts all muscle layers of the gastrointestinal tract and induces bronchoconstriction in the airway (168).

SP is hydrolyzed by dog chymase (169), rat chymase I stabilized by heparin (170), and to a lesser extent by human heart chymase (171). It has not been tested as a substrate for other chymases. Despite a potential tryptic site at Lys^3, SP is not hydrolyzed by either dog or human tryptase (162,169). Its amidated C terminus apparently protects it from hydrolysis by mast cell carboxypeptidase (172).

suggest that neuropeptides work in concert with other transmitters to degranulate mast cells. Whatever the signals may be, functional communication between nerves and mast cells occurs in vivo.

Mast cells may serve as sensors and effectors in this communication and are thought to mediate some of the changes in neurogenic inflammation. We have examined another potential function of mast cells: their proteases may serve to terminate or modify the peptide messages transmitted by nerves to target tissues. In most instances this terminates the biological signal, but sometimes a new message is created. The normal modulation of peptide activity preserves homeostasis of local vascular, smooth muscle, glandular, and inflammatory functions, but derangements in the activity or preponderance of mast cell proteases may disrupt the normal balance of neuropeptides and contribute to the edema, protein extravasation, smooth muscle changes, and inflammation characteristic of diverse diseases such as asthma or arthritis.

ACKNOWLEDGMENTS

The author wishes to thank Xiao Ying Hua for valuable review and discussion of this work and gratefully acknowledges support from the Veterans Administration Research Service, University of California Tobacco-Related Disease Research Program, American Lung Association, Hawaii Heart Association, Leahi Trust, and the Robert Perry Fund of the Hawaii Community Foundation.

REFERENCES

1. J. Neuman, Über das Vorkommen der sogenannten "Mastzellen" bei patholgischen Veränderungen des Gehirns, *Arch. Pathol. Anat. Physiol. Virchows 122*:378 (1890).
2. S. Metal'nikov and V. Chorine, Rôle des réflexes conditionnels dans l'immunité, *Ann. Inst. Pasteur Paris 40*:893 (1926).
3. Y. Olsson, Mast cells in the nervous system, *Int. Rev. Cytol. 24*:27 (1968).
4. R. Ader, D. Felten, and N. Cohen, Interactions between the brain and the immune system, *Annu. Rev. Pharmacol. Toxicol. 30*:561 (1990).
5. D. Johnson and W. Krenger, Interactions of mast cells with the nervous system: recent advances, *Neurochem. Res. 17*:939 (1992).
6. T. C. Theoharides, Mast cells: the immune gate to the brain, *Life Sci. 46*:607 (1990).
7. P. J. Barnes, J. N. Baraniuk, and M. G. Belvisi, Neuropeptides in the respiratory tract, *Am. Rev. Respir. Dis. 144*:1391 (1991).
8. P. Baluk, J. Nadel, and J. M. McDonald, Calcitonin gene–related peptide is a marker of secretory cells of the rat tracheal epithelium, *FASEB J. 5*:A1268 (1992).
9. D. E. Johnson and M. K. Georgieff, Pulmonary neuroendocrine cells: their secre-

tory products and their potential roles in health and chronic lung disease in infancy, *Am. Rev. Respir. Dis. 140*:1807 (1989).
10. J. Richardson and J. Beland, Nonadrenergic inhibitory nervous system in human airways, *J. Appl. Physiol. 41*:764 (1976).
11. L. Diamond and J. B. Richardson, Inhibitory innervation to airway smooth muscle, *Exp. Lung Res. 3*:379 (1982).
12. J. M. Lundberg, L. Lundblad, C.-R. Martling, A. Saria, P. Stjarne, and A. Anggard, Coexistence of multiple peptides and classic transmitters in the airway neurons: functional and pathophysiologic aspects, *Am. Rev. Respir. Dis. 136*:16 (1987).
13. B. Lindh and T. Hokfelt, Structural and functional aspects of acetylcholine peptide coexistence in the autonomic nervous system, *Prog. Brain Res. 84*:175 (1990).
14. T. Hokfelt, Neuropeptides in perspective: the last ten years, *Neuron 7*:867 (1991).
15. F. Sundler, E. Ekblad, and R. Hakanson, Projections of enteric peptide-containing neurons in the rat, *Arch. Histol. Cytol. S2 Suppl*:181 (1989).
16. A. J. Smolen, Morphology of synapses in the autonomic nervous system, *J. Electron Microsc. Tech. 10*:187 (1988).
17. J. C. Bornstein and J. B. Furness, Correlated electrophysiological and histochemical studies of submucous neurons and their contribution to understanding enteric neural circuits, *J. Auton. Nerv. Syst. 25*:1 (1989).
18. J. B. Furness, J. C. Bornstein, T. K. Smith, R. Murphy, and S. Pompolo, Correlated functional and structural analysis of enteric neural circuits, *Arch. Histol. Cytol. 52S*:161 (1989).
19. N. Arizono, S. Matsuda, T. Hattori, Y. Kojima, T. Maeda, and S. J. Galli, Anatomical variation in mast cell nerve associations in the rat small intestine, heart, lung, and skin, *Lab. Invest. 62*:626 (1990).
20. K. M. Allen, J. Wharton, J. M. Polak, and S. G. Haworth, A study of nerves containing peptides in the pulmonary vasculature of healthy infants and children and of those with pulmonary hypertension, *Br. Heart J. 62*:353 (1989).
21. J. B. Palmer, F. M. C. Cuss, P. K. Mulderry, M. A. Ghatei, D. R. Springall, A. Cadieux, S. R. Bloom, J. M. Polak, and P. J. Barnes, CGRP is localized to human airway nerves and potently constricts human airway smooth muscle, *Br. J. Pharmacol. 91*:95 (1987).
22. J. Wharton, S. Gulbenkian, A. Merighi, D. M. Kuhn, R. Jahn, K. M. Taylor, and J. M. Polak, Immunohistochemical and ultrastructural localisation of peptide-containing nerves and myocardial cells in the human atrial appendage, *Cell Tissue Res. 254*:155 (1988).
23. R. D. Dey, D. W. J. Shannon, and S. I. Said, Localization of VIP-immunoreactive nerves in airways and pulmonary vessels of dogs, cats, and human subjects, *Cell Tissue Res. 2220*:(1981).
24. J. M. Lundberg, J. Fahrenkrug, T. Hokfelt, C.-R. Martling, O. Larsson, K. Tatemoto, and A. Anggard, Co-existence of peptide HI (PHI) and VIP in nerves regulating blood flow and bronchial smooth muscle tone in various mammals including man, *Peptides 5*:(1984).
25. A. Luts, R. Uddman, P. Alm, J. Basterra, and F. Sundler, Peptide-containing

nerve fibers in human airways: distribution and coexistence pattern, *Int. Arch. Allergy Immunol. 101*:52 (1993).
26. J. D. Levine, H. L. Fields, and A. I. Basbaum, Peptides and the primary afferent nociceptor, *J. Neurosci. 13*:2273 (1993).
27. S. H. Buck and T. F. Burks, The neuropharmacology of capsaicin: review of some recent observations, *Pharmacol. Rev. 38*:179 (1986).
28. S. Y. Hua, E. Theodorsson-Norheim, E. Brodin, J. M. Lundberg, and T. Hokfelt, Multiple tachykinins (neurokinin A, neuropeptide K and substance P) in capsaicin-sensitive sensory neurons in the guinea-pig, *Regul. Pept. 13*:1 (1985).
29. J. M. Lundberg, A. Franco-Cereceda, X. Hua, T. Hokfelt, and J. A. Fischer, Coexistence of substance P and calcitonin gene–related peptide-like immunoreactivities in sensory nerves in relation to cardiovascular and bronchoconstrictor effects of capsaicin, *Eur. J. Pharmacol. 108*:315 (1985).
30. R. D. Dey, J. B. Altemus, I. Zervos, and J. Hoffpauir, Origin and colocalization of CGRP- and SP-reactive nerves in cat airway epithelium, *J. Appl. Physiol. 68*:770 (1990).
31. T. Komatsu, M. Yamamoto, K. Shimokata, and H. Nagura, Distribution of substance P-immunoreactive and calcitonin gene–related peptide-immunoreactive nerves in normal human lungs, *Int. Arch. Allergy Appl. Immunol. 1991*:23 (1991).
32. A. A. Cameron, J. D. Leah, and P. J. Snow, The coexistence of neuropeptides in feline sensory neurons, *Neuroscience 27*:969 (1988).
33. Z. Weisenfeld-Hallin, T. Hokfelt, J. M. Lundberg, W. G. Forssmann, M. Reinecke, F. A. Tschopp, and J. A. Fisher, Immunoreactive calcitonin gene--related peptide and substance P coexist in sensory neurons to the spinal cord and interact in spinal behavioral responses of the rat, *Neurosci. Lett. 52*:199 (1984).
34. T. Green and G. J. Dockray, Characterization of the peptidergic afferent innervation of the stomach in the rat, mouse and guinea pig, *Neuroscience 25*:181 (1988).
35. K. Horgan and D. Van der Kooy, Visceral targets specify calcitonin gene–related peptide and substance P enrichment in trigeminal afferent projections, *J. Neurosci. 12*:1135 (1992).
36. F. Lembeck and P. Holzer, Substance P as a neurogenic mediator of antidromic vasodilation and neurogenic plasma extravasation, *Naunyn Schmeideberg's Arch. Pharmacol. 310*:175 (1979).
37. L. A. Laitinen, A. Laitinen, R. O. Salonen, and J. G. Widdicombe, Vascular actions of airway neuropeptides, *Am. Rev. Respir. Dis. 136*:(1987).
38. N. Jansco, A. Jansco-Gabor, and J. Szolcsanyi, Direct evidence for neurogenic inflammation and its prevention by denervation and by pretreatment with capsaicin, *Br. J. Pharmacol. 31*:138 (1967).
39. D. M. McDonald, Neurogenic inflammation in the rat trachea. I. Changes in venules, leucocytes and epithelial cells, *J. Neurocytol. 17*:583 (1988).
40. D. M. McDonald, R. A. Mitchell, G. Gabella, and A. Haskell, Neurogenic inflammation in the rat trachea. II. Identity and distributon of nerves mediating the increase in vascular permeability, *J. Neurocytol. 17*:605 (1988).
41. A. N. Bruce, Über die Beziehung der sensiblen Nervenendigungen zum Entzündungsvorgang, *Arch. Exp. Pathol. Pharmakol. 63*:424 (1910).

42. B. Pernow, Role of tachykinins in neurogenic inflammation, *J. Immunol. 135*:812 (1985).
43. A. I. Basbaum and J. D. Levine, The contribution of the nervous system to inflammation and inflammatory disease, *Can. J. Physiol. Pharmacol. 69*:647 (1990).
44. T. J. Coderre, A. I. Basbaum, and J. D. Levine, Neutral control of vascular permeability: interactions between primary afferents, mast cells, and sympathetic efferents, *J. Neurophysiol. 62*:48 (1989).
45. A. Saria, C.-R. Martling, Z. Yan, E. Theodorsson-Norheim, R. Gamse, and J. M. Lundberg, Release of multiple tachykinins from capsaicin-sensitive sensory nerves in the lung by bradykinin, histamine, dimethylphenyl piperazinium and vagal nerve stimulation, *Am. Rev. Respir. Dis. 137*:1330 (1988).
46. J. A. Kiernan, A pharmacological and histological investigation of the involvement of mast cells in cutaneous axon reflex vasodilation, *Q. J. Exp. Physiol. 60*:123 (1975).
47. J. A. Kiernan, The involvement of mast cells in vasodilation due to axon reflexes in injured skin, *Q. J. Exp. Physiol. 57*:311 (1972).
48. P. T. Arvier, L. A. Chahl, and R. J. Ladd, Modification by capsaicin and compound 48/80 of dye leakage induced by irritants in the rat, *Br. J. Pharmacol. 59*:61 (1977).
49. F. Shanahan, J. A. Denburg, J. Fox, J. Bienenstock, and D. Befus, Mast cell heterogeneity: effects of neuroenteric peptides on histamine release, *J. Immunol. 135*:1331 (1985).
50. K. R. Tainsh and F. L. Pearce, Mast cell heterogeneity: evidence that mast cells isolated from various connective tissue locations in the rat display markedly graded phenotypes, *Int. Arch. Allergy Immunol. 98*:26 (1992).
51. L. Enerback, Mast cells in the rat gastrointestinal mucosa. I. Effect of fixation, *Acta Pathol. Microbiol. Scand. 66*:303 (1966).
52. B. J. Heap and J. A. Kiernan, Histological, histochemical and pharmacological observations on mast cells in the stomach of the rat, *J. Anat. 115*:315 (1973).
53. R. G. Woodbury and H. Neurath, Purification of an atypical mast cell protease and its levels in developing rats, *Biochemistry 17*:4298 (1978).
54. R. G. Woodbury, G. M. Gruzenski, and D. Lagunoff, Immunofluorescent localization of a serine protease in rat small intestine, *Proc. Natl. Acad. Sci. USA 75*:2785 (1978).
55. B. Newson, A. Dahlstrom, L. Enerback, and H. Ahlman, Suggestive evidence for a direct innervation of mucosal mast cells, *Neuroscience 10*:565 (1983).
56. R. H. Stead, M. Tomioka, G. Quinonez, G. T. Simon, S. Y. Felten, and J. Bienenstock, Intestinal mucosal mast cells in normal and nematode-infected rat intestines are in intimate contact with peptidergic nerves, *Proc. Natl. Acad. Sci. USA 84*:2975 (1987).
57. A. M. Rothschild, E. L. Gomes, and M. A. Rossi, Reversible rat mesenteric mast cell swelling caused by vagal stimulation or sham-feeding, *Agents Actions 34*:295 (1991).
58. G. Skofitsch, J. M. Savitt, and D. M. Jacobowitz, Suggestive evidence for a functional unit between mast cells and substance P fibers in the rat diaphragm and mesentery, *Histochemistry 82*:5 (1985).

59. E. K. Tam, L. D. Calonico, J. A. Nadel, and D. M. McDonald, Globule leukocytes and mast cells in the rat trachea: their number, distribution, and response to compound 48/80 and dexamethasone, *Anat. Embryol. 178*:107 (1988).
60. L. K. Wilkes, C. McMenamin, and P. G. Holt, Postnatal maturation of mast cell subpopulations in the rat respiratory tract, *Immunology 75*:535 (1992).
61. J. F. Huntley, A. M. George, F. J. Newlands, J. Irvine, and H. R. P. Miller, Mapping of the rat mast cell granule proteinases RMCP I and II by enzyme-linked immunosorbent assay and paired imunofluorescence, *Acta Pathol. Microbiol. Immunol. Scand. 98*:933 (1990).
62. A. Cadieux, D. R. Springall, P. K. Mulderry, J. Rodrigo, M. A. Ghatei, G. Terenghi, S. R. Bloom, and J. M. Polak, Occurrence, distribution and ontogeny of CGRP immunoreactivity in the rat lower respiratory tract: effect of capsaicin treatment and surgical denervations, *Neuroscience 19*:605 (1986).
63. P. Baluk, J. A. Nadel, and D. M. McDonald, Substance P-immunoreactive sensory axons in the rat respiratory tract: a quantitative study of their distribution and role in neurogenic inflammation, *J. Comp. Neurol. 319*:586 (1992).
64. G. Nilsson, K. Alving, S. Ahlstedt, T. Hokfelt, and J. M. Lundberg, Peptidergic innervation of rat lymphoid tissue and lung: relation to mast cells and sensitivity to capsaicin and immunization, *Cell Tissue Res. 262*:125 (1990).
65. S. Domeij, B. Carlsoo, A. Dahlqvist, and S. Forsgren, Occurrence of mast cells in relation to the distribution of nerve fibers in the rat larynx, *Acta Otol. (Stockh.) 111*:981 (1991).
66. H. R. Schumaker, Jr., Ultrastructure of the synovial membrane, *Ann. Clin. Sci. 5*:489 (1975).
67. M. Hukkanen, M. Gronblad, R. Rees, Y. T. Kottinen, S. J. Gibson, J. Hietanen, J. M. Polak, and D. A. Brewerton, Regional distribution of mast cells and peptide-containing nerves in normal and adjuvant arthritic rat synovium, *J. Rheumatol. 18*:177 (1991).
68. D. G. Malone and P. W. Dolan, Droperidol-fentanyl inhibits mast cell histamine release in rat synovium, but not skin, after immunologic activation: evidence for mast cell heterogeneity, *J. Allergy Clin. Immunol. 85*:821 (1990).
69. J. N. Baraniuk, M. L. Kowalski, and M. A. Kaliner, Relationships between permeable vessels, nerves, and mast cells in rat cutaneous neurogenic inflammation, *J. Appl. Physiol. 68*:2305 (1990).
70. A. A. Irani, N. M. Schechter, S. S. Craig, G. DeBlois, and L. B. Schwartz, Two types of human mast cells that have distinct neutral protease compositions, *Proc. Natl. Acad. Sci. USA 83*:4464 (1986).
71. A.-M. A. Irani, S. M. Goldstein, B. U. Wintroub, T. Bradford, and L. B. Schwartz, Human mast cell carboxypeptidase: selective localization to MC_{TC} cells, *J. Immunol. 147*:247 (1991).
72. A. M. Dvorak and W. Silen, Differentiation between Crohn's disease and other inflammatory conditions by electron microscopy, *Ann. Surg. 201*:53 (1985).
73. K. DeSchryver-Kecskemeti and R. E. Clouse, Perineural and intraneural inflammatory infiltrates in the intestines of patients with systemic connective-tissue disease, *Arch. Pathol. Lab. Med. 113*:394 (1989).

74. R. H. Stead, A. J. Franks, C. H. Goldsmith, J. Bienenstock, and M. F. Dixon, Mast cells, nerves and fibrosis in the appendix: a morphological assessment, *J. Pathol. 161*:209 (1990).
75. A. M. Dvorak, R. A. Monahan, J. E. Osage, and G. R. Dickersin, Mast cell degranulation in Crohn's disease, *Lancet 1*:498 (1978).
76. A. M. Dvorak and R. A. Monahan, Crohn's disease–mast cell quantitation using one micron plastic sections for light microscopic study, *Pathol. Annu. 18*:181 (1983).
77. R. H. Stead, M. F. Dixon, N. H. Bramwell, R. H. Riddell, and J. Bienenstock, Mast cells are closely apposed to nerves in the human gastrointestinal mucosa, *Gastroenterology 97*:575 (1989).
78. A. M. Dvorak, R. S. McLeod, A. B. Onderdonk, R. A. Monahan-Earley, J. B. Cullen, D. A. Antonioli, E. Morgan, J. E. Blair, P. Estrella, R. L. Cisneros, Z. Cohen, and W. Silen, Human gut mucosal mast cells: ultrastructural observations and anatomic variation in mast cell–nerve associations in vivo, *Int. Arch. Allergy Appl. Immunol. 98*:158 (1992).
79. R. Matin, E. K. Tam, J. A. Nadel, and G. H. Caughey, Distribution of chymase-containing mast cells in human bronchi, *J. Histochem. Cytochem. 40*:781 (1992).
80. F. J. van Overveld, L. A. M. J. Houben, F. E. M. Schmitz du Moulin, P. L. B. Bruijnzeel, J. A. M. Raaijmakers, and G. K. Terpstra, Mast cell heterogeneity in human lung tissue, *Clin. Sci. 77*:297 (1989).
81. J. Baert and M. Frederix, Globule leukocytes in the respiratory epithelium of human upper airways: an ultrastructural study, *Anat. Rec. 212*:143 (1985).
82. F. Shanahan, I. MacNiven, N. Dyck, J. A. Denberg, J. Bienenstock, and A. D. Befus, Human lung mast cells: distribution and abundance of histochemically distinct subpopulations, *Int. Arch. Allergy Appl. Immunol. 83*:329 (1987).
83. G. L. Brinkman, The mast cell in normal human bronchus and lung, *J. Ultrastruct. Res. 23*:115 (1968).
84. S. Ollerenshaw, D. Jarvis, A. Woolcock, C. Sullivan, and T. Scheibner, Absence of immunoreactive vasoactive intestinal polypeptide in tissue from the lungs of patients with asthma, *N. Engl. J. Med. 320*:1244 (1989).
85. A. Laitinen, M. Paranen, A. Hervonen, and L. A. Laitinen, Electron microscopic study of the innervation of human lower respiratory tract: evidence of adrenergic nerves, *Eur. J. Respir. Dis. 67*:209 (1985).
86. A. A. Hislop, J. Wharton, K. M. Allen, J. M. Polak, and S. G. Haworth, Immunohistochemical localization of peptide-containing nerves in human airways: age-related changes, *Am. J. Respir. Cell Mol. Biol. 3*:191 (1990).
87. S. L. Ollerenshaw, D. Jarvis, C. E. Sullivan, and A. J. Woolcock, Substance P immunoreactive nerves in airways from asthmatics and nonasthmatics, *Eur. Respir. J. 4*:673 (1991).
88. L. A. Laitinen, A. Laitinen, P. A. Panula, M. Partanen, K. Tervo, and T. Tervo, Immunohistochemical demonstration of substance P in the lower respiratory tract of the rabbit and not of man, *Thorax 38*:531 (1983).
89. J. M. Lundberg, T. Hokfelt, C.-R. Martling, A. Saria, and C. Cuello, Substance P-immunoreactive sensory nerves in the lower respiratory tract of various mammals including man, *Cell Tissue Res. 235*:(1984).

90. A. Saria and J. M. Lundberg, Activation of sensory substance P neurons in the respiratory tract by cigarette smoke, mechanical, and chemical irritants, *Front. Horm. Res. 12*:123 (1984).
91. C. W. Castor, The microscopic structure of normal human synovial tissue, *Arthritis Rheum. 3*:140 (1960).
92. S. I. Wasserman, The mast cell and synovial inflammation, or what's a nice cell like you doing in a joint like this? *Arthritis Rheum. 27*:841 (1984).
93. D. G. Malone, A.-M. Irani, L. B. Schwartz, K. E. Barrett, and D. D. Metcalfe, Mast cell numbers and histamine levels in synovial fluids from patients with diverse arthritides, *Arthritis Rheum. 29*:956 (1986).
94. J. A. Kopicky-Burd, A. Kagey-Sobotka, S. P. Peters, A. M. Dvorak, D. W. Lennox, L. M. Lichtenstein, and F. M. Wigley, Characterization of human synovial mast cells, *J. Rheumatol. 15*:1326 (1988).
95. G. R. Mikhail and A. Miller-Milinska, Mast cell population in human skin, *J. Inv. Dermatol. 43*:249 (1964).
96. R. A. J. Eady, T. Cowen, T. F. Marshall, V. Plummer, and M. W. Greaves, Mast cell population density, blood vessel density and histamine content of human skin, *Br. J. Dermatol. 100*:623 (1979).
97. A. Naukkarinen, B. J. Nickoloff, and E. M. Farber, Quantification of cutaneous sensory nerves and their substance P content in psoriasis, *J. Invest. Dermatol. 92*:126 (1989).
98. S. S. Karanth, D. R. Springall, S. Lucas, D. Levy, P. Ashby, M. M. Levene, and J. M. Polak, Changes in nerves and neuropeptides in skin from 100 leprosy patients investigated by immunocytochemistry, *J. Pathol. 157*:15 (1989).
99. E. Weihe, W. Hartschuh, and M. Reinecke, Peptidergic (neurotensin, VIP) innervation of the skin, *Acta Anat. 111*:161 (1981).
100. L. Wiesner-Menzel, B. Schulz, F. Vakilzadeh, and B. M. Czarnetzki, Electron microscopical evidence for a direct contact between nerve fibres and mast cells, *Acta Dermato-Venereol. (Stockh.) 61*:465 (1981).
101. H. Heine and J. F. Forster, Relationships between mast cells and preterminal nerve fibers, *Z. Mikrosk. Anat. Forsch. Leipz. 89*:934 (1975).
102. A. Naukkarinen, I. T. Harvima, M. L. Aalto, R. J. Harvima, and M. Horsmanheimo, Quantitative analysis of contact sites between mast cells and sensory nerves in cutaneous psoriasis and lichen planus based on a histochemical double staining technique, *Arch. Dermatol. Res. 283*:433 (1991).
103. I. T. Harvima, H. Viinamaki, A. Naukkarinen, K. Paukkonen, H. Neittaanmaki, R. J. Harvima, and M. Horsmanheimo, Association of cutaneous mast cells and sensory nerves with psychic stress in psoriasis, *Psychother. Psychosom. 60*:168 (1993).
104. H. N. Claman, On scleroderma: mast cells, endothelial cells, and fibroblasts, *JAMA 262*:1206 (1989).
105. G. MacQueen, J. Marshall, M. Perdue, S. Siegel, and J. Bienenstock, Pavlovian conditioning of rat mucosal mast cells to secrete rat mast cell protease II, *Science 243*:83 (1989).
106. T. Bani-Sacchi, M. Barattini, S. Bianchi, P. Blandina, S. Brunelleschi, R.

Fantozzi, P. F. Mannaioni, and E. Masini, The release of histamine by parasympathetic stimulation in guinea-pig auricle and rat ileum, *J. Physiol. 371*:29 (1986).

107. J. A. Kiernan, Degranulation of mast cells in the trachea and bronchi of the rat following stimulation of the vagus nerve, *Int. Arch. Allergy Appl. Immunol. 91*:398 (1990).
108. R. H. Stead, M. H. Perdue, M. G. Blennerhasset, Y. Kakuta, P. Sestini, and J. Bienenstock, The innervation of mast cells, in *The Neuroendocrine–Immune Network* (S. Freir, ed.), CRC Press, Boca Raton, Fla., 1990.
109. V. Dimitriadou, M. G. Buzzi, M. A. Moskowitz, and T. C. Theoharides, Trigeminal sensory fiber stimulation induces morphological changes reflecting secretion in rat dura mater mast cells, *Neuroscience 44*:97 (1991).
110. A. R. Leff, N. P. Stimler, N. M. Munoz, T. Shioya, J. Tallet, and C. Dame, Augmentation of respiratory mast cell secretion of histamine caused by vagus nerve stimulation during antigen challenge, *J. Immunol. 136*:1066 (1986).
111. E. R. Garrity, N. P. Stimler, N. M. Munoz, J. Tallet, A. C. David, and A. R. Leff, Sympathetic modulation of biochemical and physiological response to immune degranulation in canine broncial airways in vivo, *J. Clin. Invest. 75*:2038 (1985).
112. V. Dimitriadou, M. Lambracht-Hall, J. Reichler, and T. C. Theoharides, Histochemical and ultrastructural characteristics of rat brain perivascular mast cells stimulated with compound 48/80 and carbachol, *Neuroscience 39*:209 (1990).
113. R. Fantozzi, E. Masini, P. Blandina, P. F. Mannaioni, and T. Bani-Sacchi, Release of histaine from rat mast cells by acetylcholine, *Nature 273*:(1978).
114. D. E. Cochrane, Peptide regulation of mast cell function, in *Progress in Medicinal Chemistry* (G. P. Ellis and G. B. West, eds.), Elsevier, New York, 1990, p. 143.
115. C. M. S. Fewtrell, J. C. Foreman, C. C. Jordan, P. Oehme, H. Renner, and J. M. Stewart, The effects of substance P on histamine and 5-hydroxytryptamine release in the rat, *J. Physiol. 330*:393 (1982).
116. M. K. Church, R. C. Benyon, M. A. Lowman, P. A. Hutson, and S. T. Holgate, Allergy or inflammation? From neuropeptide stimulation of human skin mast cells to studies on the mechanism of the late asthmatic response, *Agents Actions 26*:22 (1989).
117. M. K. Church, M. A. Lowman, C. Robinson, S. T. Holgate, and R. C. Benyon, Interaction of neuropeptides with human mast cells, (1989).
118. E. K. Tam and X.-Y. Hua, Substance P augments tracheal mast cell release induced by electrical field stimulation, *Am. Rev. Respir. Dis. 147*:A711 (1993).
119. H. O. Handwerker, What peripheral mechanisms contribute to nociceptive transmission and hyperalgesia? in *Towards a New Pharmacotherapy of Pain* (A. I. Basbaum and J. M. Besson, eds.), Wiley, New York, 1991, p. 5.
120. W. Piotrowski and J. C. Foreman, On the actions of substance P, somatostatin, and vasoactive intestinal polypeptide on rat peritoneal mast cells and in human skin, *Naunyn Schmeideberg's Arch. Pharmacol. 331*:364 (1985).
121. M. A. Lowman, R. C. Benyon, and M. K. Church, Characterization of neuropeptide-induced histamine release from human dispersed skin mast cells, *Br. J. Pharmacol. 95*:121 (1988).

122. H. Repke and M. Bienert, Structural requirements for mast cell triggering by substance P-like peptides, *Agents Actions 23*:207 (1988).
123. M. Aridor, G. Rajmilevich, M. A. Beaven, and R. Sagi-Eisenberg, Activation of exocytosis by the heterotrimeric G protein G_{i3}, *Science 262*:1569 (1993).
124. H. Yano, B. K. Wershil, N. Arizono, and S. J. Galli, Substance P–induced augmentation of cutaneous vascular permeability and granulocyte infiltration in mice is mast cell dependent, *J. Clin. Invest. 84*:1276 (1989).
125. H. Matsuda, K. Kawakita, Y. Kiso, T. Nakano, and Y. Kitamura, Substance P induces granulocyte infiltration through degranulation of mast cells, *J. Immunol. 142*:921 (1989).
126. S. D. Brain and T. J. Williams, Interactions between the tachykinins and calcitonin gene–related peptide lead to the modulation of oedema formation and blood flow in rat skin, *Br. J. Pharmacol. 96*:(1989).
127. S. D. Brain and T. J. Williams, Substance P regulates the vasodilator activity of CGRP, *Nature 86*:855 (1988).
128. G. F. Joos, R. A. Pauwels, and M. E. van der Straeten, The mechanism of tachykinin-induced bronchoconstriction in the rat, *Am. Rev. Respir. Dis. 137*:1038 (1988).
129. G. Joos and R. Pauwels, The in vivo effect of tachykinins on airway mast cells of the rat, *Am. Rev. Respir. Dis. 148*:922 (1993).
130. W. L. Matis, R. M. Lavker, and G. F. Murphy, Substance P induces the expression of an endothelial–leukocyte adhesion molecule by microvascular endothelium, *J. Invest. Dermatol. 94*:492 (1990).
131. G. Sanchez-Mejorada and F. Alonso-deFlorida, Changes in mast-cell distribution in skeletal muscle after denervation, *Muscle Nerve 15*:(1992).
132. L. Rucci, E. Masini, R. Arbi-Riccardi, E. Gianella, C. Fioretti, P. F. Mannaioni, M. B. Borghi-Cirri, and O. Fini-Storchi, Vidian nerve resection, histamine turnover and mucosal mast cell function in patients with chronic hypertophic non-allergic rhinitis, *Agents Actions 28*:224 (1989).
133. J. D. Levine, T. J. Coderre, K. Covinsky, and A. I. Basbaum, Neural influences on synovial mast cell density in rat, *J. Neurosci. Res. 26*:301 (1990).
134. J. S. Marshall, R. H. Stead, C. McSharry, L. Nielsen, and J. Bienenstock, The role of mast cell degranulation products in mast cell hyperplasia. I. Mechanism of action of nerve growth factor, *J. Immunol. 144*:1886 (1990).
135. M. G. Blennerhassett, M. Tomioka, and J. Bienenstock, Formation of contacts between mast cells and sympathetic neurons in vitro, *Cell Tissue Res. 265*:121 (1991).
136. M. G. Blennerhassett, J. Janiszewski, and J. Bienenstock, Sympathetic nerve contact alters membrane resistance of cells of the RBL-2H3 mucosal mast cell line, *Am. J. Respir. Cell Mol. Biol. 6*:504 (1992).
137. D. M. McKay and M. H. Perdue, Intestinal epithelial function: the case for immunophysiological regulation (first of two parts), *Dig. Dis. Sci. 38*:1377 (1993).
138. J. L. Wallace, G. W. McKnight, and A. D. Befus, Capsaicin-induced hyperemia in the stomach: possible contribution of mast cells, *Am. J. Physiol. (Gastrointest. Liver Physiol.) 263*:G209 (1992).

139. M. L. Kowalski, M. Sliwinska-Kowalska, and M. Kaliner, Neurogenic inflammation, vascular permeability, and mast cels. II. Additional evidence indicating that mast cells are not involved in neurogenic inflammation, *J. Immunol. 145*:1214 (1990).
140. F. Levi-Schaffer and M. Shalit, Differential release of histamine and prostaglandin D_2 in rat peritoneal mast cells activated with peptides, *Int. Arch. Allergy Appl. Immunol. 90*:352 (1989).
141. L. Grundemar and R. Hakanson, Neuropeptide Y, peptide YY and C-terminal fragments release histamine from rat peritoneal mast cells, *Br. J. Pharmacol. 104*:776 (1991).
142. C. Arzubiaga, J. Morrow, J. Robers, and I. Biaggioni, Neuropeptide Y, a putative cotransmitter in noradrenergic neurons, induces mast cell degranulation but not prostaglandin D_2 release, *J. Allergy Clin. Immunol. 87*:88 (1991).
143. T. A. Kassessinoff and F. L. Pearce, Histamine secretion from mast cells stimulated with somatostatin, *Agents Actions 23*:123 (1988).
144. A. Miadonna, E. Leggieri, A. Tedeschi, M. Lorini, M. Froldi, and C. Zanussi, Study of the effect of some neuropeptides and endogenous opioid peptides on in vitro histamine release from human lung mast cells and peripheral blood basophils, *Agents Actions 25*:11 (1988).
145. J. C. Foreman, Substance P and calcitonin gene–related peptide: effects on mast cells and in human skin, *Int. Arch. Allergy Appl. Immunol. 82*:(1987).
146. J. P. Caulfield, S. El-Lati, G. Thomas, and M. K. Church, Dissociated human foreskin mast cells degranulate in response to anti-IgE and substance P, *Lab. Invest. 63*:502 (1990).
147. M. A. Kai and P. Keen, Localization of 5-hydroxytryptamine (5HT) to neurons and endoneruial mast cells in rat sensory ganglia, *J. Neurocytol. 14*:63 (1985).
148. E. P. Christian, B. J. Undem, and D. Weinreich, Endogenous histamine excites neurons in the guinea pig superior cervical ganglion in vitro, *J. Physiol. 409*:297 (1989).
149. D. P. Prell and J. P. Green, Histamine as a neuroregulator, *Annu. Rev. Neurosci. 9*:209 (1986).
150. D. Weinreich and B. J. Undem, Immunological regulation of synaptic transmission in isolated guinea pig autonomic ganglia, *J. Clin. Invest. 79*:1529 (1987).
151. X.-Y. Hua and T. L. Yaksh, Pharmacology of the effects of bradykinin, serotonin, and histamine on the release of calcitonin gene-related peptide from c-fiber terminals in the rat trachea, *Neuroscience* (1993).
152. E. Lang, A. Novak, P. W. Reeh, and H. O. Handwerker, Chemosensitivity of fine afferents from rat skin in vitro, *J. Neurophysiol. 63*:887 (1990).
153. B. J. Undem, W. C. Hubbard, E. P. Christian, and D. Weinrich, Mast cells in the guinea pig superior cervical ganglion: a functional and histological assessment, *J. Auton. Nerv. Syst. 30*:75 (1990).
154. A. C. Myers, B. J. Undem, and D. Weinrich, Influence of antigen on membrane properties of guinea pig bronchial ganglion neurons, *J. Appl. Physiol. 71*:970 (1991).
155. R. J. Greene, D. Fowler, D. MacGlashan, and D. Weinreich, IgE-challenged

human lung mast cells excite vagal sensory neurons in vitro, *J. Appl. Physiol. 64*:2249 (1988).
156. S. E. Crowe and M. H. Perdue, Functional abnormalities in the intestine associated with mucosal mast cell activation, *Reg. Immunol. 4*:113 (1992).
157. S. E. Crowe, K. Soda, M. Stanisz, and M. H. Perdue, Intestinal permeability in allergic rats: nerve involvement in antigen-induced changes, *Am. J. Physiol. (Gastrointest. Liver Physiol.) 264*:G617 (1993).
158. P. Sestini, M. Dolovich, C. Vancheri, R. H. Stead, J. S. Marshall, M. H. Perdue, J. Gauldie, and J. Bienenstock, Evidence of mast cell–nerve interactions: antigen-induced lung epithelial permeability in rats is dependent on capsaicin-sensitive nerves, *Am. Rev. Respir. Dis. 139*:401 (1989).
159. J. F. McKelvy and S. Blumberg, Inactivation and metabolism of neuropeptides, *Annu. Rev. Neurosci. 9*:415 (1986).
160. N. W. Bunnett, 2. Release and breakdown. Postsecretory metabolism of peptides, *Am. Rev. Respir. Dis. 136*:27 (1987).
161. S. D. Brain, T. J. Williams, J. R. Tippins, H. R. Morris, and I. MacIntyre, Calcitonin gene–related peptide is a potent vasodilator, *Nature 13*:(1985).
162. E. K. Tam and G. H. Caughey, Degradation of airway neuropeptides by human lung tryptase, *Am. J. Respir. Cell Mol. Biol. 3*:27 (1990).
163. A. F. Walls, S. D. Brain, A. Desai, P. J. Jose, E. Hawkings, M. K. Church, and T. J. Williams, Human mast cell tryptase attenuates the vasodilator activity of calcitonin gene–related peptide, *Biochem. Pharmacol. 43*:1243 (1992).
164. H. C. Manley and L. W. Haynes, Eosinophil chemotactic response to rat CGRP-1 is increased after exposure to trypsin or guinea pig lung particulate fraction, *Neuropeptides 13*:29 (1989).
165. S. R. Hughes and S. D. Brain, A calcitonin gene–related peptide (CGRP) antagonist ($CGRP_{8\text{-}37}$) inhibits microvascular responses induced by CGRP and capsaicin in skin, *Br. J. Pharmacol. 104*:(1991).
166. C. B. Bunker, G. Terenghi, D. R. Springall, J. M. Polak, and P. M. Dowd, Deficiency of calcitonin gene–related peptide in Raynaud's phenomenon, *Lancet 336*:1530 (1990).
167. O. Hagermark, T. Hokfelt, and B. Pernow, Flare and itch induced by substance P in human skin, *J. Invest. Dermatol. 71*:233 (1978).
168. J. Solway and A. R. Leff, Sensory neuropeptides and airway function, *J. Appl. Physiol. 71*:2077 (1991).
169. G. H. Caughey, F. Leidig, N. F. Viro, and J. A. Nadel, Substance P and vasoactive intestinal polypeptide degradation by mast cell tryptase and chymase, *J. Pharmacol. Exp. Ther. 244*:133 (1988).
170. H. Le Trong, H. Neurath, and R. G. Woodbury, Substrate specificity of the chymotrypsin-like protease in secretory granules isolated from rat mast cells, *Proc. Natl. Acad. Sci. USA 84*:364 (1987).
171. U. Urata, A. Kinoshita, K. Misono, F. M. Bumpus, and A. Husain, Identification of a highly specific chymase as the major angiotensin II–forming enzyme in the human heart, *J. Biol. Chem. 265*:22348 (1991).
172. N. W. Bunnett, S. M. Goldstein, and P. Nakazato, Isolation of a neuropeptide-

degrading carboxypeptidase from the human stomach, *Gastroenterology 102*:76 (1992).
173. R. A. Skidgel and E. G. Erdos, Novel activity of human angiotensin I converting enzyme: release of the NH_2- and COOH-terminal tripeptides from the luteinizing hormone-releasing hormone, *Proc. Natl. Acad. Sci. USA 82*:1025 (1985).
174. R. Matsas, A. J. Kenny, and A. J. Turner, The metabolism of neuropeptides: the hydrolysis of peptides, including enkephalins, tachykinins, and their analogues, by endopeptidase 1, *Biochem. J.* (1984).
175. K. Sekizawa, J. Tamaoki, J. Nadel, and B. Borson, Enkephalinase inhibitor potentiates substance P- and electrically induced contraction in ferret trachea, *J. Appl. Physiol. 63*:1401 (1987).
176. S. I. Said, Neuropeptides as modulators of injury and inflammation, *Life Sci. 47*:19 (1990).
177. A. Stanisz, R. Scicchitano, R. Stead, H. Matssuda, M. Tomioka, J. Denburg, and J. Bienenstock, Neuropeptides and immunity, *Am. Rev. Respir. Dis. 136*:S48 (1987).
178. G. Franconi, P. D. Graf, S. C. Lazarus, J. A. Nadel, and G. H. Caughey, Mast cell tryptase and chymase reverse airway smooth muscle relaxation induced by vasoactive intestinal peptide, *J. Pharmacol. Exp. Ther. 248*:947 (1989).
179. E. K. Tam, G. M. Franconi, J. A. Nadel, and G. H. Caughey, Protease inhibitors potentiate smooth muscle relaxation induced by vasoactive intestinal peptide in isolated human bronchi, *Am. J. Respir. Cell Mol. Biol. 2*:449 (1990).
180. C. M. Lilly, M. A. Martins, and J. M. Drazen, Peptidase modulation of vasoactive intestinal peptide pulmonary relaxation in tracheal superfused guinea pig lungs, *J. Clin. Invest. 91*:235 (1993).
181. P. H. Howarth, R. Djukanovic, J. Wilson, D. Springall, J. Polak, and S. T. Holgate, The influence of asthma on the presence and distribution of neuropeptide containing nerves within the endobronchial mucosa, *Am. Rev. Respir. Dis.* A622 (1990).
182. S. M. Goldstein, J. Leong, and N. W. Bunnett, Human mast cell proteases hydrolyze neurotensin, kinetensin and Leu^5-enkephalin, *Peptides 12*:995 (1991).
183. D. E. Cochrane, R. E. Carraway, W. Boucher, and R. S. Feldberg, Rapid degradation of neurotensin by stimulated rat mast cells, *Peptides 12*:1187 (1991).
184. J. A. Cromlish, N. G. Seidah, M. Marcinkiewicz, J. Hamelin, D. A. Johnson, and M. Chretien, Human pituitary tryptase: molecular forms, NH_2-terminal sequence, *J. Biol. Chem. 262*:1363 (1987).
185. V. J. Braganza and W. H. Simmons, Tryptase from rat skin: purification and properties, *Biochemistry 30*:4997 (1991).

6

Mast Cell Carboxypeptidase: Structure and Regulation of Gene Expression

Sanford M. Goldstein
University of California, San Francisco, San Francisco, California

I. INTRODUCTION

In this chapter we discuss recent studies that further characterize mast cell carboxypeptidase A (CPA), the third secretory granule protease identified and purified from human mast cells. In little more than a decade, our understanding of this protease has progressed from the initial detection of CPA activity in rodent tissues to the cloning of the human cDNA and gene encoding this enzyme, the mapping of the human gene to a specific chromosome, an elucidation of the genealogy of the CPA family of enzymes, and an initial understanding of factors that regulate the expression of the CPA gene.

II. PHYSICOCHEMICAL CHARACTERIZATION

The initial detection and characterization of this zinc metalloexopeptidase in rat tissues, in murine bone marrow and serosal mast cells, and in human mast cells have been reviewed (1). Briefly, CPA-like activities were first identified in rat tissues and in rat peritoneal mast cells using spectrophotometric assays for the cleavage of carboxy-terminal hydrophobic residues (2–4). A more sensitive high-performance liquid chromatography (HPLC) assay allowed the detection of the enzyme activity in human lung and skin mast cells (5). The protease was purified to homogeneity using affinity chromatography, yielding a single polypeptide with an M_r value of about 34,500 on SDS-PAGE (6). An apparently identical enzyme was purified from human gastric tissue (7). The amino-terminal amino acid sequence of the enzyme purified from skin confirmed the homology of mast cell CPA with other CPAs. However, the mast cell CPA sequence demonstrated 65% positional identity with human pancreatic carboxypeptidase B and only 19% with human pancreatic carboxypeptidase A, despite the fact that hydrolytic activity of mast cell CPA was that of a carboxypeptidase A (hydrolyzing hydrophobic carboxy-terminal residues) and not a carboxypeptidase B (which cleaves basic and not hydrophobic carboxy-terminal amino acids) (6). The discrepancy between the CPA-like substrate preferences of the mast cell enzyme and its closer homology to the carboxypeptidase B enzymes was finally resolved after the enzyme was cloned and evolutionary tree analysis was performed as described in Section IX.

More recent studies demonstrated that mast cell CPA is a glycoprotein (8), like human mast cell tryptase and chymase (9,10). Evidence for glycosylation included a reduction in calculated M_r from 39,800 to 36,300 on SDS-PAGE after treatment with glycosidase F. The purified enzyme also binds to Con A-Sepharose, and after immobilization on nitrocellulose may be stained with biotin-labeled lentil lectin before, but not after, treatment with glycosidase (8). The fact that human mast cell CPA, but not pancreatic CPA, is glycosylated was attributed to the presence of two consensus sequences for N-linked glycosylation sites in mast cell CPA and not in the bovine (and presumably human) pancreatic enzymes (8).

III. FUNCTIONAL CHARACTERIZATION

A. Use of Synthetic Substrates

As reviewed previously (1), rat mast cell CPA cleaves carboxy-terminal hydrophobic amino acids such as phenylalanine from peptide and ester linkages. Potato tuber CPA inhibitor blocks this activity (2). The human mast cell CPA has similar peptidase activity, which was maximal between pH 8.5 and 9.5 (5), somewhat higher than for the rat enzyme (2). Recent studies using synthetic tripeptide substrates demonstrated that human mast cell CPA and pancreatic CPA, represented by bovine CPA, have similar but distinct substrate specificities (8). Neither hydrolyzes the small hydrophobic amino acid, alanine. Hydrolytic activity of mast cell CPA is maximal with phenylalanine (k_{cat}/K_m 8.9 × 10^6 s^{-1} M^{-1}) followed by tyrosine (k_{cat}/K_m 7.0 × 10^6 s^{-1} M^{-1}). The bulkiest hydrophobic carboxy-terminal amino acid, tryptophan, was not hydrolyzed. However, the synthetic tripeptide substrate used in these studies, carbobenzoxy-glycylglycyl-L-tryptophan, does not permit measurement with maximal sensitivity. In addition, if one extrapolates from various substrates with carboxy-terminal leucines, more complex physiologic peptide substrates are hydrolyzed by mast cell CPA with lower K_m by up to one order of magnitude and twofold-higher k_{cat} than the tripeptide substrate carbobenzoxyglycylglycyl-L-leucine (Table 1) (11). Thus it is possible that the use of synthetic tripeptide substrates does not predict the activity of the enzyme against more complex peptides, and that naturally occurring substrates with carboxy-terminal tryptophan residues are hydrolyzed in vivo.

B. Substrates of Potential Physiological Importance: Neuropeptides

The physiological function of mast cell CPA is unknown, as are its physiological or pathophysiological substrates. However, some substrates of potential interest have been studied. Rat mast cell chymase I and rat mast cell CPA act sequentially in degrading apolipoprotein B of low-density lipoprotein (12). Other studies have focused on the activity of mast cell CPA against neuropeptides and other vasoactive peptides. It was originally demonstrated that angiotensin was hydrolyzed to des Leu^{10}–angiotensin I by mast cell CPA (5). It was then shown that the carboxy-terminal leucine residue of the peptides kinetensin, [Leu^5]-enkephalin, neuromedin N, neurotensin, and xenopsin were hydrolyzed by the mast cell CPA purified from human gastric tissues (7). Penultimate phenylalanine residues were subsequently hydrolyzed from kinetensin and [Leu^5]enkephalin after a prolonged incubation of the enzyme and substrates. Studies using mast cell CPA purified from skin replicated these results with kinetensin, [Leu^5]enkephalin, and neurotensin, but hydrolysis of the penultimate phenylalanine residues was not detected after a shorter incubation (11). The kinetics of mast cell CPA hydrolysis of these substrates (Table 1) were comparable to those of tryptase against neuropeptide substrates such as calcitonin gene–related pep-

Table 1 Kinetic Constants for Substrate Hydrolysis by Mast Cell Carboxypeptidase A[a]

Substrate[b]	k_{cat} (s^{-1})	K_m (μM)	k_{cat}/K_m (10^{-6} s^{-1} M^{-1})
Cbz–Gly–Gly–Ala	ND	ND	ND
Cbz–Gly–Gly–Leu	21 ± 0.7	160 ± 14.5	0.13
Cbz–Gly–Gly–Phe	23.5 ± 1.6	26.5 ± 2.3	0.89
Cbz–Gly–Gly–Tyr	19 ± 0.4	27 ± 0.1	0.70
Cbz–Gly–Gly–Trp	ND	ND	ND
Angiotensin I	37.3 ± 3.7	60 ± 19	0.62
Leu^5–enkephalin	44 ± 3.5	36 ± 1	1.2
Neurotensin	51 ± 12	16 ± 1	3.2
Kinetensin	53 ± 9.5	15 ± 3	3.5

[a]ND, no hydrolysis detected.
[b]Cbz, carbobenzoxy.

tide, peptide histidine methionine, and vasoactive intestinal peptide (13). To study the concerted or sequential activities of the mast cell proteases tryptase, chymase, and CPA against kinetensin, [Leu^5]enkephalin, and neurotensin, dispersed skin mast cells were activated with calcium ionophore. The mast cell supernatants that contained mast cell secretory granule proteases were then incubated with these neuropeptides in the presence or absence of specific inhibitors of each of the mast cell proteases. Hydrolytic products were separated by HPLC and identified by amino acid composition or by liquid secondary-ion mass spectrometry. [Leu^5]Enkephalin was cleaved only by mast cell CPA. Kinetensin was cleaved by tryptase, chymase, and CPA, yielding five different fragments, one of which was produced by the concerted or sequential activity of tryptase and chymase and one by the activity of tryptase and CPA. Neurotensin was cleaved only by chymase and CPA individually, yielding the two expected hydrolytic products.

A number of substrates are not hydrolyzed by mast cell CPA. These include [Met^5]enkephalin, substance P, gastrin-releasing peptide 10, neuromedin B, bradykinin, dynorphin A, cholecytokinin 8, somatostatin 14, and vasoactive intestinal polypeptide. The carboxy-terminal amino acid residues of these peptides are either methionine, arginine, phenylalanine linked to a penultimate proline or aspartic acid, glutamine, cysteine, or isoleucine (7).

IV. LOCALIZATION OF MAST CELL CPA IN SITU

Earler studies suggested that mast cell CPA is not distributed uniformly among all mast cells (5,6). Mast cells dispersed from human skin demonstrated different CPA activity levels depending on the age of the donor and the anatomical site from which the specimen was obtained. Thus there was 173 U of activity (1 U = 1 nmol of des-Leu^{10}–angiotensin I product generated per minute) per 10^6 adult facial skin mast cells compared to 564 U per 10^6 mast cells from adult foreskin, but only 17 U per 10^6 neonatal foreskin mast cells, and 406 U per per 10^6 mast cells from the foreskin of an 11-month-old child (1,6). Based on specific activity, it has been estimated that adult foreskin mast cells may contain as much as 16 pg of mast cell CPA per cell (6). The amount of mast cell CPA in lung mast cells is low compared to skin, with an average of 15 U per 10^6 mast cells from lung (5). This pattern of expression was similar to that of mast cell chymase (14) and suggested that mast cell CPA might also be a phenotypic marker of the chymase-positive, tryptase-positive (MC_{TC}) mast cell (15,16).

A. Localization of Immunoreactive Mast Cell CPA: A Marker of the MC_{TC} Phenotype

Two monoclonal antibodies (Mab) were produced against mast cell CPA purified from skin (17). One of these anti-CPA Mab was utilized with an antitryptase and an antichymase Mab in a histochemical double-labeling technique to characterize mast cells in skin, lung, and small intestine. In this technique an antitryptase Mab conjugated to alkaline phosphatase is used with a biotinylated anti-CPA Mab. Alternatively, a biotinylated antichymase Mab is used with an anti-CPA Mab conjugated to alkaline phosphatase. This permits immunohistochemical staining of either of two enzymes in the same section. From 97 to 100% of mast cells in skin contained immunoreactive CPA. In all cases in skin and other tissues, mast cells either co-labeled with both chymase and CPA antibodies, or stained only with antitryptase Mab. Immunostaining for CPA, therefore, was found only in MC_{TC} mast cells; thus mast cell CPA is a marker of this mast cell phenotype.

Another study used the same anti-CPA and antitryptase Mab with a rabbit polyclonal antichymase antibody, to phenotype mast cells in lung alveoli and bronchi, bowel mucosa and submucosa, axillary nodes, breast skin, and breast panechyma (18). The investigators used a double-immunolabeling technique by staining tissues with either mouse antitryptase or anti-CPA Mab and with rabbit polyclonal antichymase antibody. The first antibodies were followed by second fluorescein-conjugated antimouse and rhodamine-conjugated antirabbit antibodies. Thus, using different excitation wavelengths, two different proteases could be detected in single mast cells in situ. CPA and chymase always co-localized in cells that also stained with avidin. However, some cells were detected that contained these proteases but no tryptase. These were called MCc cells and were rare in skin and lung alveoli but constituted 12% of cells in bronchi and 17% of mast cells in bowel submucosa. These results stand in contrast to those of previous studies (15,16), which detected tryptase in virtually all human mast cells.

B. Detection of Mast Cell CPA mRNA in Cultured Cells and Cell Lines

A recent study demonstrated mast cell CPA mRNA transcripts in dispersed lung cells enriched for mast cells and in the myelomonocytic tumor cell lines KG-1 and HL-60 (19). The T-cell lymphoma line Jurkat expressed low levels of CPA mRNA, but the T-lymphocytic tumor cell lines CEM and H9, an Epstein-Barr virus-transformed B lymphoblastoid cell line, and the monocytic tumor cell lines THP-1 and U937 did not. No transcripts were found in eosinophils, unfractionated peripheral blood mononuclear cells, monocytes, neutrophils, $CD56^+CD6^-$ polyclonal NK cells, $CD56^-CD6^+$ polyclonal T cells activated by IL-2, polyclonal T cells activated with concanavalin A, primary cultures of dermal fibroblasts, three cloned $CD8^+CD4^-$ cytotoxic T-lymphocyte cell lines, the Raji-Burkitt lymphoma cell line, or in HeLa cells. Interestingly, a megakaryocytic cell line CMK expressed mast cell CPA mRNA before and after induction of phenotypic differentiation with phorbol myristate acetate. It has not been established whether the cell lines that express mRNA transcripts also produce active enzyme, but studies of mouse bone marrow mast cells suggest that separate transcriptional and translational controls may be important in the final expression of active enzyme (20).

V. CHARACTERIZATION OF CPA-CONTAINING PROTEASE COMPLEXES RELEASED FROM HUMAN MAST CELLS

A. Diffusion and Cell Association of CPA-Containing Complexes After Human Skin Mast Cell Activation

Mast cell proteases are released from rat and mouse mast cells as part of a protease–proteoglycan complex (21,22) that remains cell associated to some degree (23), whereas histamine diffuses freely into the medium after release. Thus the ratio of the net percent release of a protease to that of histamine is always less than 1; that is, more histamine than protease is found in the release medium because of diffusion and cell association, respectively. When mast cells dispersed from human skin were activated by calcium ionophore, the net percent release of tryptase, chymase, and CPA were determined in the same preparations and were plotted against the net percent release of histamine. A simple regression was then used to determine the release ratio of each protease to that of histamine (i.e., the slope of the regression line) (24) (Fig. 1). Interestingly, the release ratios of CPA and chymase to histamine were virtually identical, and lower than that of tryptase to histamine. Tryptase was then shown to diffuse away from the mast cell as part of a complex stable in low salt and, therefore, presumably bound to proteoglycan (25,26). CPA and chymase remained cell associated to a greater degree than did tryptase. These experiments suggested that CPA and chymase may be released from activated human skin mast cells in a complex or complexes separate from those containing tryptase.

B. Separation of CPA and Chymase Complexes from Tryptase-Containing Complexes

The initial characterization of mast cell CPA demonstrated that enzyme released from human lung and skin mast cells eluted in high-molecular-weight fractions when gel filtration was performed in low-ionic-strength buffer (5). At high concentrations of salt, the protease eluted at a molecular weight close to 30,000 Da (5). Thus human mast cell CPA appeared to be ionically bound in some type of complex. These observations and the release ratios of the three secretory granule proteases relative to histamine, described above, led to further characterization of the complexes released from human skin mast cells. Dispersed skin mast cells were activated with ionophore, then sedimented. When the resulting mast cell supernatants were subjected to get filtration at low ionic strength, CPA and chymase co-filtered in fractions of 400 to 560 kDa molecular size (Fig. 2).

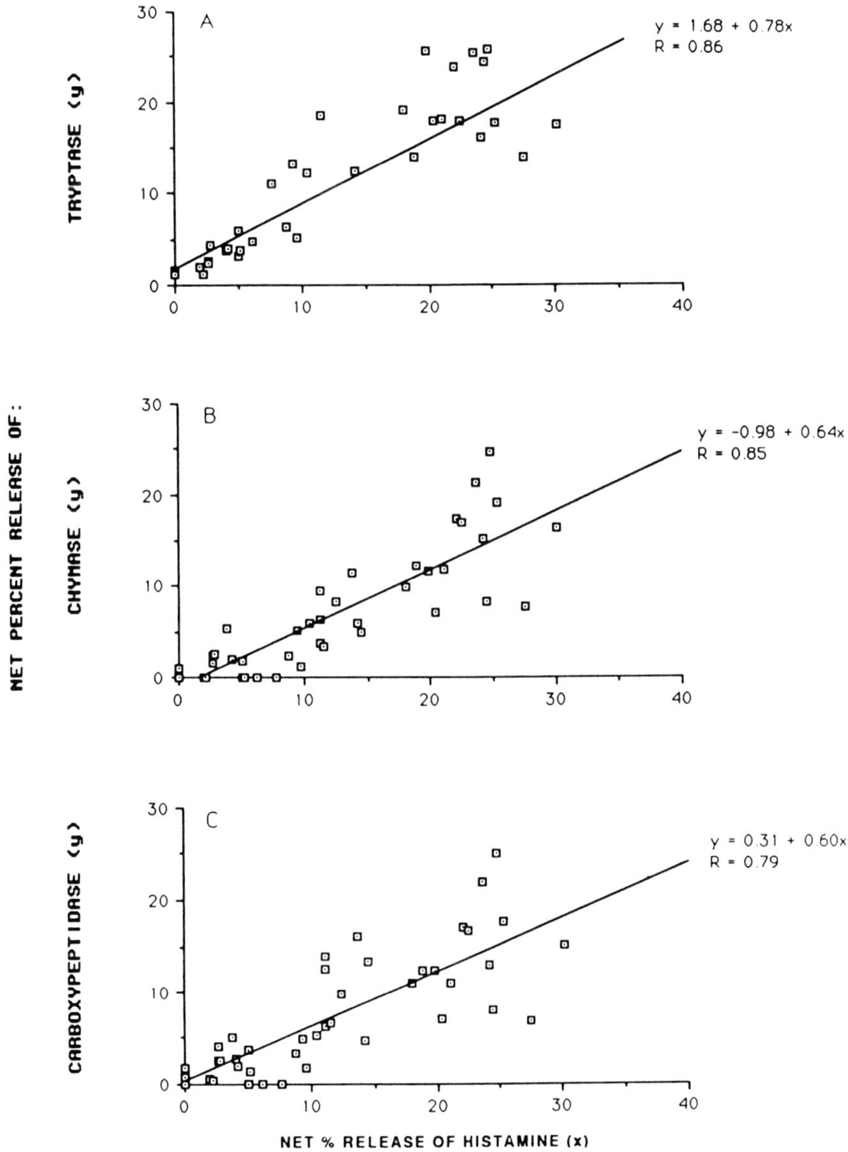

Figure 1 Net percent of release values for tryptase (a), chymase (b), and carboxypeptidase (c) plotted against those for histamine from eight experiments with human skin mast cells stimulated by A23187 or anti-IgE. (From Ref. 24.)

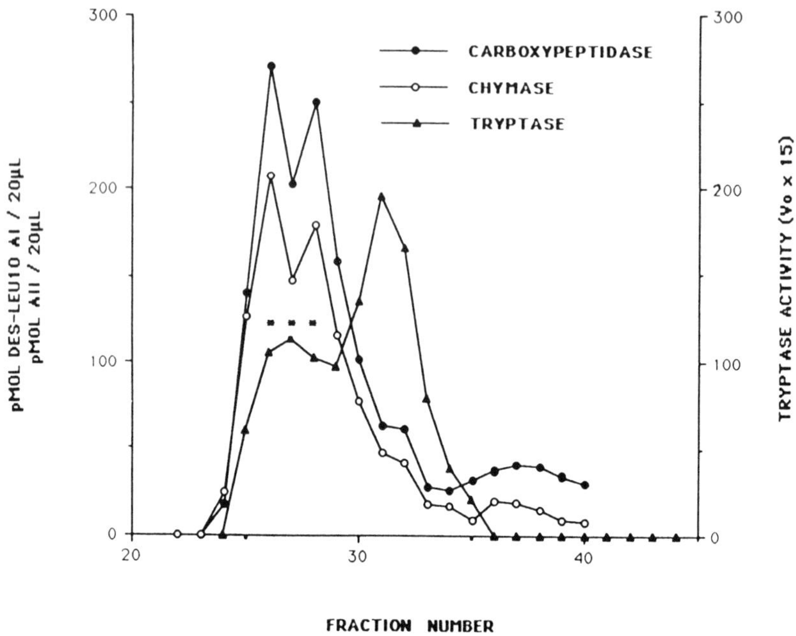

Figure 2 Fast protein liquid chromatography, in 0.15 *M* NaCl-containing buffer, of tryptase, chymase, and carboxypeptidase activities in released mast cell supernatants. Void volume was 10.7 mL, and total volume was 24.15 mL. Total recovery of starting activity was 66% for tryptase, 68% for chymase, and 57% for carboxypeptidase. (From Ref. 24.)

A separate peak of tryptase activity was seen at 200 to 250 kDa, which is larger than the molecular weight of tryptase after purification away from heparin complexes. Released mast cell supernatants were applied to a column of immobilized antitryptase mAb (Fig. 3). CPA and chymase activities were found in the effluent and wash fractions, but tryptase was found only in the eluate, with only small amounts of the other proteases. This demonstrated that tryptase was retained by the antitryptase column but that the other proteases were not also immobilized. Therefore, they were not contained in complexes with tryptase. Conversely, when supernatants were applied to an affinity column for CPA (potato tuber CPA-Sepharose), tryptase was found in the effluent and washes and eluted at an apparent molecular mass of 200 kDa on fast protein liquid chromatography (Fig. 4). CPA and chymase were found primarily in the eluate. That immobilization of CPA on an affinity column also immobilized chymase but not tryptase demonstrated that CPA is released as part of a high-

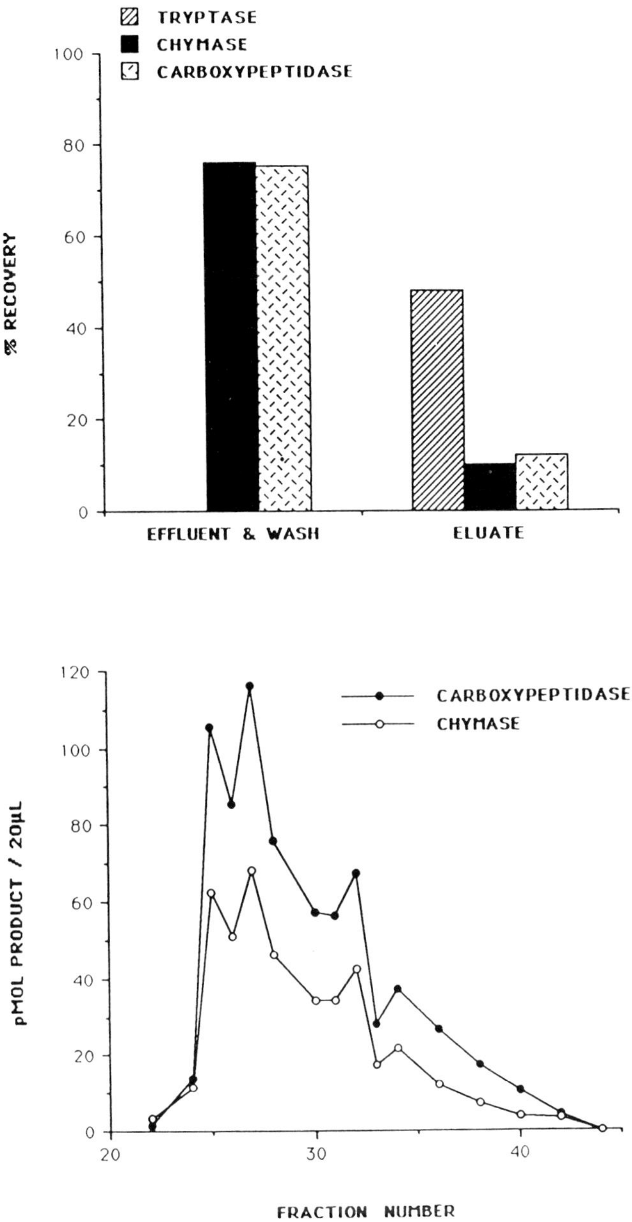

Figure 3 (*Top*) Affinity chromatography of released mast cell supernatants with immobilized antitryptase B2 antibody. The percentage of starting activity of tryptase, chymase, and carboxypeptidase appearing in effluent and wash or eluate fractions is shown. (*Bottom*) FPLC profile of effluent fractions containing chymase and carboxypeptidase activities. The activities eluted in a bimodal peak at 560 and 400 kDa. (From Ref. 24.)

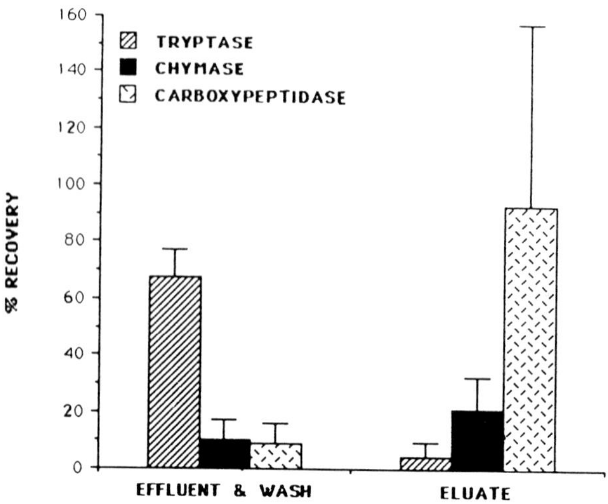

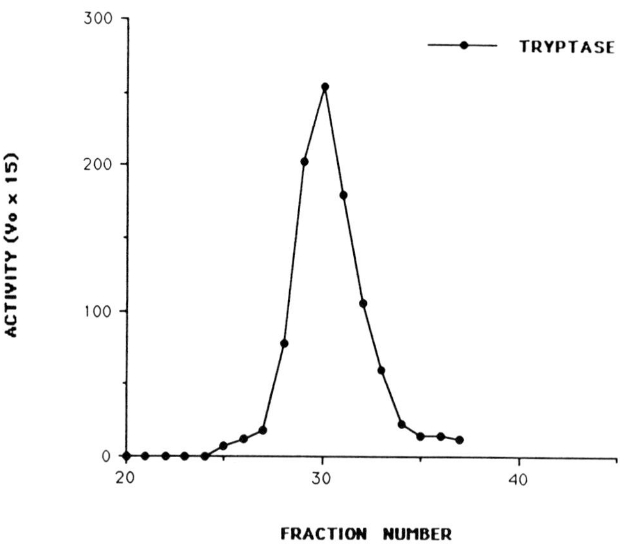

Figure 4 (*Top*) Affinity chromatography of released mast cell supernatants with potato tuber carboxypeptidase inhibitor-Sepharose. The percentage of starting activity of tryptase, chymase, and carboxypeptidase recovered in effluent and wash or eluate fractions is shown. (*Bottom*) Fast protein liquid chromatography profile of PCl-Sepharose column effluent fractions containing tryptase activity. The peak of eluted activity has a molecular mass of about 200 kDa. (From Ref. 24.)

molecular-weight complex with chymase, separate from tryptase. It is not yet known whether chymase and CPA are complexed individually to the same proteoglycan component of the complex, or to each other, or both. However, the data from rat mast cell granules suggest that positively charged residues of CPA bind to negatively charged groups of heparin of the granular matrix, leaving the active site of the enzyme exposed to substrates (27).

VI. CLONING OF MOUSE AND HUMAN MAST CELL CPA CDNA

Murine mast cell CPA was purified from a Kirsten sarcoma virus (KiSV)—immortalized mast cell line and from mouse peritoneal mast cells (28). The resulting amino-terminal amino acid sequence was used to prepare an oligonucleotide probe, which was used to screen a KiSV-immortalized mast cell cDNA library. A 1470-base pair cDNA was isolated that encoded a 15-amino acid signal peptide, 94-amino acid activation peptide, and 308-amino acid mature enzyme with a predicted M_r value of 35,780. Several amino acids thought to be critical for zinc binding, substrate binding, and intrachain disulfide bond formation in the pancreatic CPAs were conserved. Interestingly, sequence identities of the mouse mast cell CPA compared to rat pancreatic CPAs A1 and A2 were lower, at 43% and 41%, than to pancreatic CPA B, at 53%. The more recently reported amino acid sequence of the rat mast cell CPA obtained by protein sequencing (27) is very similar to that of the mouse (28).

Two cDNAs for human mast cell CPA have been isolated and sequenced. First, the mouse cDNA was used to screen a human lung cDNA library, and a full-length cDNA for human mast cell CPA was obtained (29). This cDNA encodes a 15-amino acid signal peptide, a 94-amino acid activation peptide, and a 308-amino acid mature enzyme with 86% positional identity with mouse mast cell CPA. A cDNA encoding the mature human mast cell CPA was also isolated from a human skin cell library using an oligonucleotide based on the amino-terminal sequence of the human protease (8). This cDNA was identical in sequence to that obtained from the lung library, supporting the evidence from Southern blot analysis for a single mast cell CPA gene in humans (see below).

Sequencing of the human cDNAs revealed several substitutions in the S1′ active site, a binding pocket that is critical in determining substrate specificity of the enzyme. Noteworthy was the substitution at position 255, which is thought to be critical for determining substrate preferences either for hydrophobic (CPA) or basic (CPB) amino acid carboxy-terminal residues. In pancreatic CPAs this is isoleucine, but in the mast cell CPAs this is leucine, suggesting that mast cell and pancreatic CPAs may not be directly related in an evolutionary sense (8,29). In comparing the amino acid sequence of the mast cell and pancreatic enzymes, it was also noted that there were S1′ subsite replacements of two bulkier, one isomeric, and one smaller residue in mast cell CPA (8). Replacements of smaller residues in rat pancreatic CPA are associated with substrate preferences skewed toward bulkier carboxy-terminal residues such as tryptophan (30), and replacements of bulkier residues skew specificities toward smaller carboxy-terminal residues such as alanine. It was proposed that compared to its pancreatic homologs, mast cell CPA would probably demonstrate larger backbone differences than seen among pancreatic CPAs, in order to accommodate these replacements and still demonstrate substrate specificity similar to that of pancreatic CPA. However, the enzyme has not yet been crystallized and its three-dimensional structure is unknown.

VII. CLONING AND CHROMOSOMAL LOCALIZATION OF THE HUMAN GENE FOR MAST CELL CPA

Southern blot analysis suggested the presence of only one mast cell CPA gene (29), a finding consistent with the observation that the sequence of cDNAs from human skin and lung mast cells is identical. A human genomic library was screened with several human cDNAs encoding different fragments of mast cell CPA, yielding four genomic clones whose inserts spanned the 32-kb mast cell CPA gene (19). The gene contained 11 exons. The exon–intron organization of the mast cell gene conserved and closely resembled that of the pancreatic CPA B gene. However, there was no significant homology between the 5′ flanking region of the mast cell gene compared to those of the genes encoding rat pancreatic carboxypeptidases A1, A2, and B. The mast cell CPA gene 5′ flanking DNA did contain the sequence CAGTCTCCAAGTGC, a motif related to the pancreatic enhancer-like motif CAGTCTCC-AGTGC, found in the human tryptase gene (31), as well as CAGCTGCTC, a motif found in the human cathepsin G gene. Finally, a GATA-binding transcription factor recognition site was found 51 bp upstream from the transcription initiation site, immediately 5′ to the TATAAA-like sequence.

Polymerase chain reaction (PCR) amplification of a 129-bp sequence from the mast cell CPA gene was used to screen 18 somatic cell hybrids containing human chromosomes. This localized the human mast cell CPA gene to chromosome 3 (19).

VIII. REGULATION OF MAST CELL CPA GENE EXPRESSION

A. GATA Motif for the Activity of the Mast Cell CPA Gene Promoter

The factors that regulate the very specific and perhaps coordinated expression of mast cell proteases are not understood. Both the human mast cell CPA and tryptase genes contain a motif that is related to the pancreatic enhancer elements of rat pancreatic exocrine genes (32). This motif is also found in the 5′ flanking DNA of the rat chymase II gene, where it acts as a binding site for mast cell–specific trans-acting factors and upregulates chymase expression (33). As noted in the discussion of the cloning of the mast cell CPA gene, the human and mouse mast cell CPA gene promotors also contain the sequence GATA, which is a DNA binding site for nuclear proteins that act as regulators of transcription (19). Several mouse and rat mast cell lines that contain mast cell CPA were shown to contain GATA-1, GATA-2, and GATA-3 transcription factor proteins that bind to GATA sites (34).

The ability of these binding proteins to regulate the expression of mouse mast cell CPA was studied in rat basophilic leukemia cells, in mouse P815 cells (a differentiated mastocytoma cell line that contains only GATA-2 and low amounts of mast cell CPA mRNA), and in P815 cells that were stably transfected to express GATA-1. First, nuclear extracts from these cells were incubated with an oligonucleotide that extended from bp –27 to –62 of the mouse mast cell CPA gene, the area of the promoter that includes the GATA motif. Gel mobility shift assays demonstrated that proteins in these nuclear extracts bound to the oligonucleotide in a sequence-specific manner (i.e., the change of one nucleotide in the GATA motif from GATA to GTTA prevented the formation of oligonucleotide DNA-transcription factor complexes detected on gel shift assays). Next, 160 bp of mouse mast cell CPA gene promoter region, containing the GATA motif, was ligated to a human growth hormone cDNA, which acts as the reporter gene for transient transfection assays. The ability of this promoter to initiate transcription of the growth hormone cDNA was then tested. Rat basophilic leukemia cells transfected with this construct produced 17.5-fold more human growth hormone than did cells transfected with an identical construct except for a single-base-pair mutation, again changing the critical GATA motif to GTTA. Thus this small piece of the mouse mast cell CPA gene 5′ flanking DNA does function as a promoter in these rat basophilic leukemia cells, and an intact GATA motif is necessary for this promoter function. P815 cells also expressed more growth hormone when transfected with the authentic DNA compared to DNA with the single-base-pair mutation. P815 transfectants that made more GATA-1 also produced two- to fourfold more growth hormone than did wild-type P815 cells. It was proposed that that GATA-1 protein regulated mast cell CPA gene expression in rat basophilic leukemia cells, but that GATA-2 regulated mast cell CPA gene expression in P815 cells. The authors speculated that the expression of different GATA-binding proteins may be important in the differentiation program of mast cells, and that different GATA proteins may be stage-specific during such a program. These important insights into the mast cell CPA gene promoter, however, are unlikely to explain all of the diversity of mast cell CPA expression in vivo. From other systems used to study the regulation of gene expression, it is likely that many other motifs and transcription factors will prove to be important regulators of mast cell CPA gene expression.

B. Effects of Cells and Growth Factors on Mast Cell CPA Gene Expression

Another approach to investigating mast cell CPA gene regulation involves use of cells in culture as they differentiate into mast cells. One such model system utilizes mouse bone marrow mast cells that express little heparin, high levels of mast cell CPA mRNA, but little enzyme protein (20). Co-culture of these cells with 3T3 fibroblasts results in the appearance of serosal-like mast cells expressing heparin and an almost 100-fold increase in mast cell CPA enzyme activity. Treatment of bone marrow mast cells with recombinant c-*kit* ligand (stem cell factor/mast cell growth factor/steel factor) (35–39) produced increased amounts of heparin proteoglycans but only minimal amounts of mouse mast cell CPA (40). Culture of bone marrow mast cells with recombinant IL-3 had little effect on levels of mast cell CPA mRNA transcripts (40). These results suggest that the program of mast cell differentiation is likely to be quite complex and to be comprised of multiple independently regulated phenotypic features.

IX. GENEALOGY OF MAST CELL CPA

As noted above, mast cell CPA varies significantly in several respects from the pancreatic enzyme which it was thought to most closely resemble on the basis of enzymatic activity (i.e., pancreatic CPA). First, amino-terminal amino acid sequencing revealed less positional identity between the mast cell enzyme and pancreatic CPA than between mast cell CPA and pancreatic CPB (6). Analysis of the full amino acid sequences confirmed and extended this observation (28, 29). Second, the critical amino acid in the substrate-binding site of mast cell CPA isolated from all species is leucine, and not isoleucine, as in CPA (8,27,28). Finally, the location of intron 3 in human mast cell CPA is homologous to that of rat CPB, whereas codons encoding seven additional amino acids precede intron 3 in the rat pancreatic CPA2 gene, and no equivalent third intron is found in the rat pancreatic CPA1 gene. Yet the substrate specificity of mast cell CPA is clearly related to CPA and not to CPB. These findings led to evolutionary tree analysis for the mast cell enzyme (8,19) after alignment of the amino acid sequences of the 10 known CPAs from *S. griseus,* crayfish, cow, rat, mouse, and human. Amino acid parsimony trees (minimal nucleotide replacement trees) were built using computer programs, and the shortest trees (i.e., those requiring the fewest amino acid replacements and gene duplications to explain the modern sequence diversity and known distribution) were analyzed (Fig. 5). The three shortest trees documented that mast cell CPA probably arose from a gene duplication among the pancreatic CPB lineage, not along the CPA lineage.

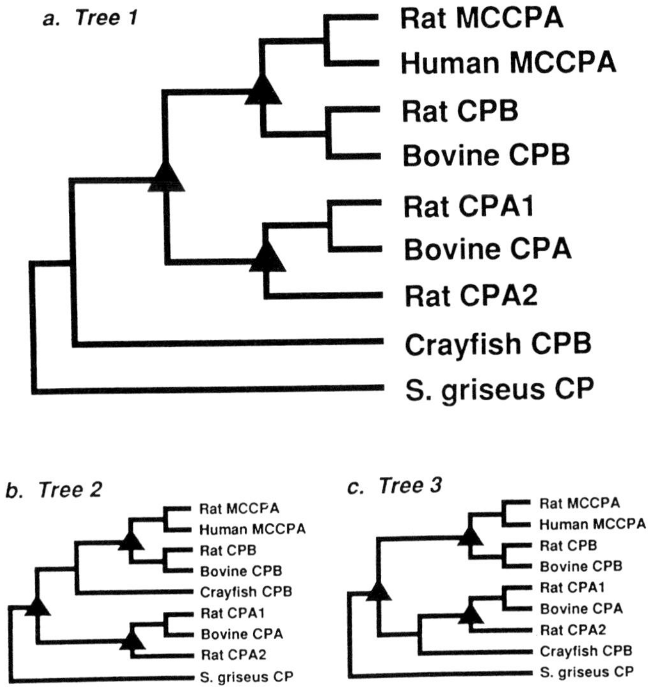

Figure 5 Genealogy of the carboxypeptidase gene family. (a) The evolutionary tree depicted here is the most parsimonious evolutionary tree determined by amino and parsimony analysis (requiring 1054 amino acid replacements). (b) and (c) The next two shortest trees for the nine carboxypeptidase sequences (each requiring 1062 amino acid replacements). These trees are rooted using the bacterial carboxypeptidase as an outgroup. Triangles indicate gene duplications, and the other divergences represent speciations; the minimum number of duplications that is required to explain the modern species distribution of these proteins is shown. (From Ref. 8).

X. FUTURE STUDIES

The studies discussed in this chapter have established the primary structure of mast cell CPA, its evolutionary and functional relationship to other CPAs, and its importance as a phenotypic marker of the mast cell. Its activity is limited to specific substrates, and several peptide substrates of potential importance have been identified. However, a full understanding of its function in vivo and the reason for its conserved expression in particular subsets of mast cells is lacking. Additional studies are also required to elucidate more fully the DNA regulatory sequences that lead to its restricted expression.

REFERENCES

1. S. M. Goldstein, C. E. Kaempfer, and B. U. Wintroub, Human mast cell CPA, *Monogr. Allergy 27*:132 (1990).
2. M. Everitt and H. Neurath, Rat peritoneal mast cell carboxypeptidase: localization, purification and enzymatic properties, *FEBS Lett. 110*:292 (1980).
3. J. Bodwell and W. Meyer, Purification and characterization of carboxypeptidase from rat skeletal muscle, *Biochemistry 20*:2767 (1981).
4. R. Haas, P. Heinrich, and D. Sasse, Proteolytic enzymes of rat liver mitochondria: evidence for a mast cell origin, *FEBS Lett. 103*:168 (1979).
5. S. M. Goldstein, C. E. Kaempfer, D. Proud, L. B. Schwartz, A.-M. Irani, and B. U. Wintroub, Detection and partial characterization of a human mast cell carboxypeptidase, *J. Immunol. 139*:2724 (1987).
6. S. M. Goldstein, C. E. Kaempfer, J. Kealey, and B. U. Wintroub, Human mast cell carboxypeptidase: purification and characterization, *J. Clin. Invest. 83*:1630 (1989).
7. N. W. Bunnett, S. M. Goldstein, and P. Nakazato, Isolation of a neuropeptide-degrading carboxypeptidase from the human stomach, *Gastroenterology 102*:76 (1992).
8. M. Natsuaki, C.-B. Stewart, P. Vanderslice, L. B. Schwartz, M. Natsuaki, B. U. Wintroub, W. J. Rutter, and S. M. Goldstein, Human skin mast cell carboxypeptidase: functional characterization, cDNA cloning, and genealogy, *J. Invest. Dermatol. 99*:138 (1992).
9. J. Cromlish, J. Seidah, M. Marcinkiewicz, J. Hamelin, D. Johnson, and M. Chretien, Human pituitary tryptase: molecular forms, NH_2-terminal sequence, immunocytochemical localization, and specificity with prohormone and fluorogenic substrates, *J. Biol. Chem. 262*:1363 (1987).
10. N. Schechter, Human chymase, *Monogr. Allergy 27*:114 (1990).
11. S. M. Goldstein, J. Leong, and N. W. Bunnett, Human mast cell proteases hydrolyze neurotensin, kinetensin and Leu^5-enkephalin, *Peptides 12*:995 (1991).
12. J. O. Kokkonen, M. Vartiainen, and P. T. Kovanen, Low density lipoprotein degradation by secretory granules of rat mast cells: sequential degradation of apolipoprotein B by granule chymase and carboxypeptidase A, *J. Biol. Chem. 261*:16067 (1986).
13. E. K. Tam and G. H. Caughey, Degradation of airway neuropeptides by human lung tryptase, *Am. J. Respir. Cell Mol. Biol. 3*:27 (1990).
14. L. B. Schwartz, A. Irani, K. Roller, M. Castells, and N. Schechter, Quantitation of histamine, tryptase and chymase in dispersed human T and TC mast cells, *J. Immunol. 138*:2611 (1987).
15. A. Irani, N. Schechter, S. Craig, G. DeBlois, and L. B. Schwartz, Two types of human mast cells that have distinct neutral protease compositions, *Proc. Natl. Acad. Sci. USA 83*:4464 (1986).
16. A.-M. Irani, T. Bradford, C. Kepley, N. Schechter, and L. B. Schwartz, Detection of MCT and MCTC types of human mast cells by immunohistochemistry using new monoclonal anti-tryptase and anti-chymase antibodies, *J. Histochem. Cytochem. 37*:1509 (1989).

17. A.-M. Irani, S. M. Goldstein, B. U. Wintroub, T. Bradford, and L. B. Schwartz, Human mast cell carboxypeptidase: selective localization to MCTC cells, *J. Immunol. 147*:247 (1991).
18. N. Weidner and K. F. Austen, Heterogeneity of mast cells at multiple body site: fluorescent determination of avidin binding and immunofluorescent determination of chymase, tryptase, and carboxypeptidase content, *Pathol. Res. Pract. 189*:156 (1993).
19. D. S. Reynolds, D. S. Gurley, and K. F. Austen, Cloning and characterization of the novel gene for mast cell carboxypeptidase A, *J. Clin. Invest. 89*:273 (1992).
20. W. Serafin, E. Dayton, R. Gravallese, K. Austen, and R. Stevens, Carboxypeptidase A in mouse mast cells: identification, characterization and use as a differentiation marker, *J. Immunol. 139*:3771 (1987).
21. L. B. Schwartz, B. Riedel, J. Schratz, and K. F. Austen, Localization of carboxypeptidase A to the macromolecular heparin proteoglycan–protein complex in secretory granules of rat serosal mast cells, *J. Immunol. 128*:1128 (1982).
22. W. Serafin, H. Katz, K. F. Austen, and R. Stevens, Complexes of heparin proteoglycans, chondroitin sulfate E proteoglycans, and [^{3}H]diisopropyl fluorophosphate–binding proteins are exocytosed from activated mouse bone marrow–derived mast cells, *J. Biol. Chem. 261*:15017 (1986).
23. L. B. Schwartz, J. Reidel, J. Caulfield, S. Wasserman, and K. F. Austen, Cell association of complexes of chymase, heparin proteoglycan, and protein after degranulation by rat mast cells, *J. Immunol. 126*:2071 (1981).
24. S. M. Goldstein, J. Leong, L. B. Schwartz, and D. Cooke, Protease composition of exocytosed human skin mast cell protease-proteoglycan complexes: tryptase resides in a complex distinct from chymase and carboxypeptidase, *J. Immunol. 148*:2475 (1992).
25. L. B. Schwartz and T. Bradford, Regulation of tryptase from human lung mast cells by heparin: stabilization of the active tetramer, *J. Biol. Chem. 261*:7372 (1986).
26. S. Alter, D. Metcalfe, T. Bradford, and L. B. Schwartz, Regulation of human mast cell tryptase: effects of enzyme concentration, ionic strength and the structure and negative charge density of polysaccharides, *Biochem. J. 248*:821 (1987).
27. K. R. Cole, S. Kumar, H. L. Trong, R. G. Woodbury, K. A. Walsh, and H. Neurath, Rat mast cell CPA: amino acid sequence and evidence of enzyme activity within mast cell granules, *Biochemistry 30*:648 (1991).
28. D. S. Reynolds, R. L. Stevens, D. S. Gurley, W. S. Lane, K. F. Austen, and W. E. Serafin, Isolation and molecular cloning of mast cell carboxypeptidase A: a novel member of the carboxypeptidase gene family, *J. Biol. Chem. 264*:20094 (1989).
29. D. S. Reynolds, D. S. Gurley, R. L. Stevens, D. J. Sugarbaker, K. F. Austen, and W. E. Serafin, Cloning of cDNA encoding human mast cell carboxypeptidase A, and comparison with mouse mast cell carboxypeptidase A and rat pancreatic carboxypeptidase, *Proc. Natl. Acad. Sci. USA 86*:9480 (1989).
30. S. Gardell, C. S. Craik, E. Clauser, E. J. Goldsmith, C.-B. Stewart, M. Graf, and W. J. Rutter, A novel rat Carboxypeptidase A, Carboxypeptidase 2: characterization, molecular cloning, and evolutionary implications on substrate specificity in the CPA gene family, *J. Biol. Chem. 263*:17828 (1988).

31. P. Vanderslice, S. M. Ballinger, E. K. Tam, S. M. Goldstein, C. Craik, and G. H. Caughey, Human mast cell tryptase: multiple cDNAs and genes reveal a multigene serine protease family, *Proc. Natl. Acad. Sci. USA 87*:3811 (1990).
32. A. Boulet, C. Erwin, and W. Rutter, Cell-specific enhancers in the rat exocrine pancreas, *Proc. Natl. Acad. Sci. USA 83*:3599 (1986).
33. J. Sarid, P. N. Benfey, and P. Leder, The mast cell–specific expression of a protease gene, RMCP II, is regulated by an enhancer element that binds specifically to mast cell trans-acting factors, *J. Biol. Chem. 264*:1022 (1989).
34. L. I. Zon, M. F. Gurish, R. L. Stevens, C. Mather, D. S. Reynolds, K. F. Austen, and S. H. Orkin, GATA-binding transcription factors in mast cells regulate the promoter of the mast cell carboxypeptidase A gene, *J. Biol. Chem. 266*:22948 (1991).
35. J. Flanagan and P. Leder, The *kit* ligand: a cell surface molecule altered in steel mutant fibroblasts, *Cell 63*:185 (1990).
36. N. Copeland, D. Gilbert, B. Cho, P. J. Donovan, N. A. Jenkins, D. Cosman, D. Anderson, S. D. Lyman, and D. E. Williams, Mast cell growth factor maps near the steel locus on mouse chromosome 10 and is deleted in a number of steel alleles, *Cell 63*:175 (1990).
37. J. Flanagan, D. Chan, and P. Leder, Transmembrane form of the *kit* ligand growth factor is determined by alternative splicing and is missing in the Sl^d mutant, *Cell 64*:1025 (1991).
38. K. Nocka, J. Buck, E. Levi, and P. Besmer, Candidate ligand for the c-*kit* transmembrane kinase receptor: KL, a fibroblast derived growth factor stimulates mast cells and erythroid progenitors, *EMBO J. 9*:3287 (1990).
39. E. Huang, K. Nocka, D. Beier, T. Y. Chu, J. Buck, H. W. Lahm, D. Wellner, P. Leder, and P. Besmer, The hematopoeitic growth factor KL is encoded by the *Sl* locus and is the ligand of the c-*kit* receptor, the gene product of the *W* locus, *Cell 63*:225 (1990).
40. M. F. Gurish, N. Ghildyal, H. P. McNeil, K. F. Austen, S. Gillis, and R. L. Stevens, Differential expression of secretory granule proteases in mouse mast cells exposed to interleukin 3 and c-*kit* ligand, *J. Exp. Med. 175*:1003 (1992).

7

Tissue and Developmental Variation of Protease Expression in Human Mast Cells

Anne-Marie A. Irani
Medical College of Virginia, Richmond, Virginia

I. INTRODUCTION

Although mast cell involvement as effector cells in immediate-type hypersensitivity reactions is well recognized, their physiological functions are less clear. Complicating this issue is the recognition of mast cell heterogeneity in the rodent, and more recently, in the human system as well. In this regard, the neutral proteases of mast cells have become markers used to define various types of mast cells and to study the development and differentiation of mast cells. In this chapter we review data on protease expression seen in human mast cells obtained from various tissues in vivo as well as from in vitro–developed human mast cells.

II. DISTRIBUTION OF NEUTRAL PROTEASES IN HUMAN MAST CELLS IN VIVO

A. Tryptase

Tryptase was purified to homogeneity from dispersed human lung mast cell preparations by Schwartz et al. in 1981 (1). In Carnoy's-fixed sections of human lung, skin, and small intestine, sequential double labeling with alcian blue followed by immunohistochemistry with a murine monoclonal antitryptase antibody showed a one-to-one correlation between cells stained metachromatically with alcian blue and cells reacting with the antitryptase antibody (2). These results indicated that tryptase was present in all mast cells identified by classic histochemistry where metachromasia occurs with acidic dyes. Furthermore, because only mast cells stained with antitryptase antibody, tryptase was shown to be a selective marker for human mast cells (3). Basophils contain very small amounts of tryptase and are not stained by the standard immunohistochemical technique for tryptase. Recently, expression of tryptase messenger RNA was detected in normal peripheral blood monocytes and in two monocytic cell lines, U937 and Mono Mac 6, but tryptase immunoreactivity was found in only 10% of the Mono Mac 6 cells (4). Tryptase immunoreactivity in formalin-fixed tissue is diminished compared to Carnoy's fluid-fixed tissue, leading to an apparent decrease in total numbers of mast cells in lung and small intestine, but not in skin (5). The presence of tryptase in all mast cells of the human small intestine was confirmed by enzyme histochemistry using the synthetic substrates D-Val–Leu–Arg–MNA and Z–Ala–Ala–Lys–MNA (6). Using a sandwich ELISA technique for measurement of immunoreactive tryptase in cell extracts, dispersed human lung mast cells were found to contain 11 pg per cell, while mast cells dispersed from adult foreskin were found to contain 35 pg per cell (Table 1), indicating quantitative differences in the distribution of tryptase between lung and skin mast cells (7).

B. Chymase

Chymase was purified from human skin by Schechter et al. in 1983 (8) and was localized to human skin mast cells by immunofluorescence using a polyclonal rabbit antichymase antibody (9). Chymase immunoreactivity is lost in formalin-fixed tissues. Immunohistochemical staining of adjacent, Carnoy's-fixed tissue

Table 1 Characteristics of Human MC_{TC} and MC_T Cells

Characteristic	MC_{TC}	MC_T
Protease (pg/cell)	Tryptase (35) Chymase (4.5) Carboxypeptidase (10) Cathepsin G	Tryptase (10)
Histamine (pg/cell)	1.9	1.5
Proteoglycan	Heparin Chondroitin sulfates A and E	Heparin Chondroitin sulfates A and E
Granule morphology	Grating/lattice Complete scroll-poor	Complete scroll-rich

sections with antitryptase antibody and antichymase antibody, respectively, demonstrated that chymase-positive cells were found in the skin and in the submucosa of the small intestine, but not in the alveolar walls of the lung and in the small intestinal mucosa, while tryptase-positive cells were abundant in all tissues examined (10). Double-labeling experiments were conducted by incubating the various tissue sections simultaneously with the rabbit polyclonal antichymase antibody and the murine monoclonal antitryptase antibody. These experiments clearly established the presence of tryptase in all chymase-positive cells found in the skin and small intestinal submucosa. In addition, tryptase-positive, chymase-negative mast cells were visualized in lung and small intestinal mucosa. Thus it became evident that tryptase was present in all mast cells, while chymase was present in a subset of mast cells. These results were confirmed by electron microscopy using immunogold labeling with both antitryptase and antichymase antibodies to demonstrate the simultaneous presence of tryptase in all chymase-containing cells and the absence of chymase from a subset of mast cells (11). In this regard it is important to point out that chymase-negative mast cells appeared mature by morphologic criteria and did not display any signs of activation or regeneration. Thus the presence or absence of chymase is not correlated with the stage of maturation or activation of mast cells. In addition, in mast cells that contained both neutral proteases, the enzymes were found in the *same* cytoplasmic granules. Independent confirmation of these results was provided by Osman et al. using the synthetic substrates Suc–Ala–Ala–Phe–MNA and *N*-acetyl-L-methionine-α-naphtyl ester to demonstrate the presence of chymase in mast cells by enzyme histochemistry (6). These authors demonstrated that chymase activity was absent from the human small intestinal mucosal mast cells but was found in most mast cells located in the submucosa. In addition, tryptase enzymatic activity was found in all chymase-positive cells. In contrast, chymase-positive, tryptase-negative mast cells have recently been reported by Weidner et al. using immunofluorescence staining with monoclonal antibodies (12). In our laboratory we have been unable to detect such cells. Whether tryptase-negative mast cells truly exist or not will need to be verified using more sensitive immunohistochemical stains.

Levels of immunoreactive chymase measured by radioimmunoassay in extracts of dispersed mast cells obtained from human adult foreskin averaged 4.5 pg per mast cell and were below the limit of sensitivity of the assay (0.3 μg per 10^6 mast cells) in extracts of dispersed human lung mast cells (7). Measurement of chymase-like enzymatic activity in cell extracts obtained from adult skin and lung demonstrated 1.01 ± 0.54 U per 10^6 mast cells, and 0.04 ± 0.02 U per 10^6 mast cells, respectively. The small amounts of enzymatic activity found in extracts of dispersed human lung mast cells is consistent with the small numbers of chymase-positive cells found in these preparations by immunohistochemistry. Furthermore, the deficiency of chymase enzymatic activity in cell preparations containing a majority of tryptase-positive, chymase-negative mast cells argues against the presence of an antigenically distinct chymotrypsin-like enzyme in these cells.

C. Chymase and Tryptase as Markers of Distinct Types of Human Mast Cells

To distinguish between tryptase-positive, chymase-positive mast cells, referred to as MC_{TC} cells, and tryptase-positive, chymase-negative mast cells, referred to as MC_T cells, in the same tissue section, a *sequential* double-labeling method was developed, whereby Carnoy's-fixed tissue sections were incubated first with the rabbit polyclonal antichymase antibody, and chymase-positive cells were visualized in brown using indirect immunoperoxidase. The same tissue sections were then incubated with the monoclonal murine antitryptase antibody, which bound only to previously unstained chymase-negative mast cells. The latter cells were then visualized in blue using an indirect immunoalkaline phosphatase technique (10). This method was later modified to include the use of a biotinylated murine monoclonal antichymase antibody and an alkaline phosphatase-conjugated murine antitryptase antibody (13). Using the sequential double-labeling method, the distribution of MC_T and MC_{TC} cells was examined in normal adult human tissues. As shown in Table 2, MC_T cells predominate in the alveolar walls of the lung and in the small intestinal mucosa, while MC_{TC} cells are the exclusive type of mast cell seen in the skin and in the synovium and predominate in the small intestinal submucosa and in the conjunctival substantia propria. In the nasal subepithelium and in the connective tissue of tonsils, a mixture of the two types of human mast cells is seen. Epithelial mast cells are scant but when present

Table 2 Distribution of Human MC_{TC} and MC_T Cells in Normal Tissues

Tissue	MC_{TC} (%)	MC_T (%)
Skin	>99	<1
Small intestine		
Mucosa	19	81
Submucosa	77	23
Lung		
Alveoli	7	93
Bronchial epithelium	0	100
Bronchial subepithelium	25	75
Nasal mucosa	34	66
Conjunctiva	95	5

are predominantly of the MC_T type. Thus it appears that the distribution of the MC_T and MC_{TC} types of human mast cells varies among different organs such that location alone cannot be used to distinguish between these two cell types. Finally, the normal distribution of MC_T and MC_{TC} cells is altered in disease states, as reviewed later.

D. Cathepsin G

The presence of a cathepsin G–like protease was demonstrated in MC_{TC} but not in MC_T cells by immunohistochemistry and by enzymatic activity (14). These experiments showed that all chymase-containing mast cells also contained cathepsin G. The staining of mast cells with the anticathepsin G antibody was blocked by preincubating the antibody with purified cathepsin G but not with purified chymase, indicating lack of cross-reactivity between these two neutral proteases. The anticathepsin G also stains neutrophils and monocytes and thus is not specific for mast cells.

E. Carboxypeptidase

Human mast cell carboxypeptidase (HMC-CP) was detected in lysates of unstimulated human lung and skin mast cells (15). The enzyme was purified to homogeneity by Goldstein et al. (16). Using a murine monoclonal anti-HMC-CP and the murine monoclonal antichymase antibody in a *simultaneous* double-labeling technique, a one-to-one correlation was demonstrated between chymase-positive mast cells and mast cells containing HMC-CP (17). Thus HMC-CP, together with chymase and a cathepsin G–like protease, serve as markers for distinguishing MC_{TC} from MC_T cells. Furthermore, HMC-CP, like chymase, is found exclusively in MC_{TC} cells and can be used as a specific marker for this type of mast cell. A murine monoclonal antibovine carboxypeptidase A antibody (Sigma Chemical Co., St. Louis) failed to stain MC_{TC} cells in adult skin, indicating lack of cross-reactivity between the bovine pancreatic enzyme and the human mast cell carboxypeptidase enzyme (A.-M. A. Irani, unpublished observations).

F. Elastase

Like cathepsin G, elastase is a neutral protease found in neutrophil leukocytes and in monocytes and macrophages (18). Immunohistochemical studies using a polyclonal rabbit antielastase antibody demonstrated the absence of elastase from all mast cells in human skin, lung, and small intestine (19). Previous reports of the detection of elastase in dispersed human lung mast cells by immunoassay and enzymatic assay were probably due to contamination of the mast cell preparations with neutrophil leukocytes and macrophages (20). Dispersed human skin mast cell preparations that do not contain appreciable numbers of neutrophils and macrophages are devoid of elastase enzymatic activity (21).

III. ULTRASTRUCTURAL CHARACTERISTICS OF HUMAN MC_T AND MC_{TC} CELLS

The intracellular distribution of tryptase and chymase was analyzed at the ultrastructural level by performing double immunogold staining as described above. All granules of mature MC_{TC} cells seen in the skin and the small intestinal submucosa contained both tryptase and chymase, while granules of MC_T cells found in the lung and the small intestinal mucosa contained tryptase alone (11). Adjacent tissue sections were then processed for optimal analysis of ultrastructural detail. These studies demonstrated several distinguishing features of the two types of mast cells. As shown in Fig. 1, secretory granules of MC_{TC} cells appeared uniform in shape and homogeneously electron dense, and displayed a mean cross-sectional area of 0.130 ± 0.088 μm^2, while secretory granules of MC_T cells appeared much more variable in shape and electron den-

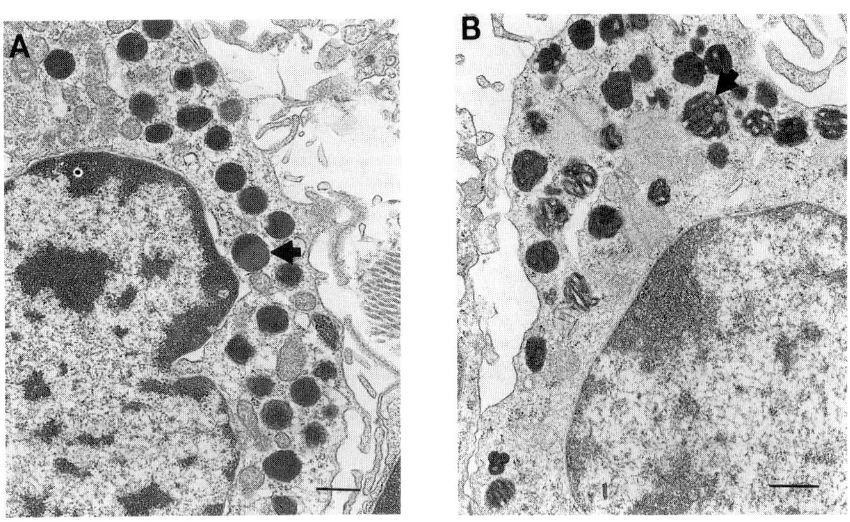

Figure 1 Ultrastructural features of MC_{TC} and MC_T cells. (A) MC_{TC} cell from human skin showing electron-dense granules with regions of grating/lattice structure (arrow). (B) MC_T cell from human lung showing granules of variable size and shape with numerous complete scrolls (arrow). X, 17,500; Bars, 0.5 μm. (Courtesy of Shirley Craig, Virginia Commonwealth University.)

sity and were smaller in size with a mean cross-sectional area of 0.053 ± 0.028 μm². At higher magnification, differences in the granular substructure between the two types of mast cells were appreciated. Thus a grating and lattice configuration was seen exclusively in granules of MC_{TC} cells, together with incomplete concentric lamellae, referred to as nondiscrete or incomplete scrolls, and electron-dense particulate material. The characteristic feature of secretory granules of MC_T cells consisted of complete concentric lamellae, also referred to as discrete or complete scrolls, with varying degrees of electron density. Nondiscrete scrolls were also observed as well as particulate or beaded material. In less than 1% of MC_{TC} granules, discrete scrolls were observed in peripheral regions of the granule, where focal areas of chymase deficiency were demonstrated by immunogold labeling (22). These studies show that certain morphologic features are associated with the different protease compositions found in secretory granules of MC_T and MC_{TC} types of human mast cells.

Immature mast cells identified by morphologic criteria, including a high nuclear/cytoplasmic ratio, small cell size, fewer and smaller granules, and lack of characteristic features of mast cell activation or regeneration, comprise essentially all mast cells of the newborn foreskin tissue and between 5 and 15% of mast cells in adult lung, skin, and small intestinal mucosa and submucosa. Using the double immunogold labeling technique, the compositional difference between MC_T and MC_{TC} cells was reproduced in immature mast cells, and the distribution of these cells was found to parallel that of their adult counterparts (23). Furthermore, the secretory granules of immature MC_T cells contained the same characteristic discrete scrolls found in mature MC_T cells, while the granules of immature MC_{TC} cells lacked the grating and lattice substructure seen in mature MC_{TC} cells, and displayed amorphous electron-dense cores. Transitional forms of MC_{TC} granules were observed, containing electron-dense cores as well as gratings and lattices.

In summary, electron microscopy studies indicate that the presence of gratings and lattices in secretory granules of mast cells identify these cell as belonging to the MC_{TC} phenotype, while the presence of discrete scrolls suggests chymase deficiency but cannot be used exclusively to identify MC_T types of mast cells. Furthermore, the identification of compositional and morphological differences in immature mast cells suggests that these two types of mast cells follow distinct developmental pathways and do not simply reflect various stages of maturation of a single cell type.

IV. DISTRIBUTION OF MC_T AND MC_{TC} CELLS IN DISEASE

The availability of the immunohistochemical technique for the identification of MC_T and MC_{TC} types of human mast cells has facilitated evaluation of mast cell

25. A. A. Irani, H. A. Sampson, and L. B. Schwartz, Mast cells in atopic dermatitis, *Allergy 44, Suppl. 9*:31 (1989).
26. A. A. Irani, B. L. Gruber, L. D. Kaufman, M. B. Kahaleh, and L. B. Schwartz, Mast cell changes in scleroderma: presence of MC_T cells in the skin and evidence of mast cell activation, *Arthritis Rheum. 35*:933 (1992).
27. A. A. Irani, N. Golzar, G. DeBlois, B. Gruber, and L. B. Schwartz, Distribution of mast cell subsets in rheumatoid arthritis and osteoarthritis synovia, *Arthritis Rheum. 30*:66 (1987).
28. A.-M. A. Irani, S. I. Butrus, K. F. Tabbara, and L. B. Schwartz, Human conjunctival mast cells: distribution of MC_T and MC_{TC} in vernal conjunctivitis and giant papillary conjunctivitis, *J. Allergy Clin. Immunol. 86*:34 (1990).
29. A. K. Bhan, L. S. Fijikawa, and C. S. Foster, T cell subsets and Langerhans cells in normal and diseased conjunctiva, *Am. Ophthalmol. 94*:205 (1982).
30. S. I. Butrus, D. Y. M. Leung, S. Gellis, J. Baum, K. Kenyon, and M. B. Abelson, Vernal conjunctivitis in the hyperimmunoglobulinemia E syndrome, *Ophthalmology 91*:1213 (1984).
31. A. M. Bentley, M. R. Jacobson, V. Cumberworth, J. R. Barkans, R. Moqbel, L. B. Schwartz, A.-M. A. Irani, A. B. Kay, and S. R. Durham, Immunohistology of the nasal mucosa in seasonal allergic rhinitis: increases in activated eosinophils and epithelial mast cells, *J. Allergy Clin. Immunol. 89*:877 (1992).
32. S. Kawabori, J. A. Denburg, L. B. Schwartz, A. A. Irani, D. Wong, G. Jordana, S. Evans, and J. Dolovich, Histochemical and immunohistochemical characteristics of mast cells in nasal polyps, *Am. J. Respir. Cell Mol. Biol. 6*:37 (1992).
33. B. L. Bradley, M. Azzawi, M. Jacobson, B. Sooufi, J. V. Collins, A. A. Irani, L. B. Schwartz, S. R. Durham, P. K. Jeffery, and A. B. Kay, Eosinophils, T-lymphocytes, mast cells, neutrophils, and macrophages in bronchial biopsies from atopic asthmatics: comparison with atopic non-asthma and normal controls and relationship to bronchial hyperresponsiveness, *J. Allergy Clin. Immunol. 88*:661 (1991).
34. R. Djukanovic, J. W. Wilson, K. M. Britten, S. J. Wilson, A. F. Walls, W. R. Roche, P. H. Howarth, and S. T. Holgate, Quantitation of mast cells and eosinophils in the bronchial mucosa of symptomatic atopic asthmatics and healthy control subjects using immunohistochemistry, *Ann. Rev. Respir. Dis. 142*:863 (1990).
35. A. M. Bentley, Q. Meng, D. S. Robinson, Q. Hamid, A. B. Kay, and S. R. Durham, Increases in activated T lymphocytes, eosinophils, and cytokine mRNA expression for interleukin-5 and granulocyte/macrophage colony-stimulating factor in bronchial biopsies after allergen inhalation challenge in atopic asthmatics, *Am. J. Respir. Cell Mol. Biol. 8*(1):35 (1993).
36. R. Djukanovic, C. K. Lai, J. W. Wilson, K. M. Britten, S. J. Wilson, W. R. Roche, P. H. Howarth, and S. T. Holgate, Bronchial mucosal manifestations of atopy: a comparison of markers of inflammation between atopic asthmatics, atopic nonasthmatics and health controls, *Eur. Respir. J. 5*(5):538 (1992).
37. A. Pesci, A. Foresi, G. Bertorelli, A. Chetta, and D. Oliveri, Histochemical characteristics and degranulation of mast cells in epithelium and lamina propria of bronchial biopsies from asthmatic and normal subjects, *Am. Rev. Respir. Dis. 147*:684 (1993).

38. L. A. Laitinen, A. Laitinen, and T. Haahtela, Airway mucosal inflammation even in patients with newly diagnosed asthma, *Am. Rev. Respir. Dis. 147*(3):697 (1993).
39. S. L. Ollerenshaw and A. J. Woolcock, Characteristics of the inflammation in biopsies from large airways of subjects with asthma and subjects with chronic airflow limitation, *Am. Rev. Respir. Dis. 145*(4 Pt. 1):922 (1992).
40. A. Di Stefano, M. Saetta, P. Maestrelli, G. Milani, F. Pivirotto, C. E. Mapp, and L. M. Fabbri, Mast cells in the airway mucosa and rapid development of occupational asthma induced by toluene diisocyanate, *Am. Rev. Respir. Dis. 147*:1005 (1993).
41. A.-M. Irani, S. S. Craig, G. DeBlois, C. O. Elson, N. M. Schechter, and L. B. Schwartz, Deficiency of the tryptase-positive, chymase-negative mast cell type in gastrointestinal mucosa of patients with defective T lymphocyte function, *J. Immunol. 138*:4381 (1987).
42. J. N. Ihle, J. R. Keller, S. Oroszlan, L. E. Henderson, T. D. Copeland, F. W. Fitch, M. B. Prystowsky, E. Goldwasser, J. W. Schrader, E. W. Palaszynski, M. Dy, and B. Lebel, Biologic properties of homogeneous interleukin 3. I. Demonstration of WEHI-3 growth factor activity, mast cell growth factor activity, p cell--stimulating factor activity, colony-stimulating factor activity, and histamine-producing cell-stimulating factor activity, *J. Immunol. 131*:282 (1983).
43. T. Ishizaka, D. H. Conrad, T. F. Huff, D. D. Metcalfe, R. L. Stevens, and R. A. Lewis, Unique features of human basophilic granulocytes developed in in vitro culture, *Int. Arch. Allergy Appl. Immunol. 77*:137 (1985).
44. P. Valent, G. Schmidt, J. Besemer, P. Mayer, G. Zenke, E. Liehl, W. Hinterberger, K. Lechner, D. Maurer, and P. Bettelheim, Interleukin-3 is a differentiation factor for human basophils, *Blood 73*:1763 (1989).
45. P. Mayer, P. Valent, G. Schmidt, E. Liehl, and P. Bettelheim, Two in vivo effects of recombinant human interleukin-3: demonstration of basophil differentiation factor, histamine-producing activity, and priming of GM-CSF–responsive progenitors in nonhuman primates, *Blood 74*:613 (1989).
46. P. Valent, J. Besemer, C. Sillaber, J. H. Butterfield, R. Eher, O. Majdic, K. Kishi, W. Klepetko, F. Eckersberger, K. Lechner, and P. Bettelheim, Failure to detect IL-3–binding sites on human mast cells, *J. Immunol. 145*:3432 (1990).
47. K. Tsuji, T. Nakahata, M. Takagi, T. Kobayashi, A. Ishiguro, T. Kikuchi, K. Naganuma, K. Koike, A. Miyajima, K. Arai, and T. Akabane, Effects of interleukin-3 and interleukin-4 on the development of "connective tissue-type" mast cells: interleukin-3 supports their survival and interleukin-4 triggers and supports their proliferation synergistically with interleukin-3, *Blood 75*:421 (1990).
48. K. M. Zsebo, J. Wypych, I. K. McNiece, H. S. Lu, K. A. Smith, S. B. Karkare, R. K. Sachdev, V. N. Yuschenkoff, N. C. Birkett, L. R. Williams, et al., Identification, purification, and biological characterization of hematopoietic stem cell factor from buffalo rat liver–conditioned medium, *Cell 63*:195 (1990).
49. F. H. Martin, S. V. Suggs, K. E. Langley, H. S. Lu, J. Ting, K. H. Okino, C. F. Morris, I. K. McNiece, F. W. Jacobsen, E. A. Mendiaz, et al., Primary structure and functional expression of rat and human stem cell factor DNAs, *Cell 63*:203 (1990).

50. J. G. Flanagan and P. Leder, The *kit* ligand: a cell surface molecule altered in steel mutant fibroblast, *Cell 63*:185 (1990).
51. E. Huang, E. Nocka, D. R. Beir, T.-Y. Chu, J. Buck, H.-W. Lahm, D. Wellner, P. Leder, and P. Besmer, The hematopoietic growth factor *k*I is encoded by the *sI* locus and is the ligand of the c-*kit* receptor, the gene product of the *W* locus, *Cell 63*:225 (1990).
52. D. M. Anderson, S. D. Lyman, A. Baird, J. M. Wignall, J. Eisenman, C. Rauch, C. J. March, H. S. Boswell, S. D. Gimpel, D. Cosman, and D. E. Williams, Molecular cloning of mast cell growth factor, a hematopoietin that is active in both membrane bound and soluble forms, *Cell 63*:235 (1990).
53. T. Furitsu, H. Saito, A. Dvorak, L. B. Schwartz, A. A. Irani, J. F. Burdick, K. Ishizaka, and T. Ishizaka, Development of human mast cells in vitro, *Proc. Natl. Acad. Sci. USA 86*:10039 (1989).
54. T. Ishizaka, H. Mitsui, M. Yanagida, T. Miura, and A. M. Dvorak, Development of human mast cells from their progenitors, *Curr. Opin. Immunol. 5*:937 (1993).
55. H. Mitsui, T. Furitsu, A. M. Dvorak, A. A. Irani, L. B. Schwartz, N. Inagaki, M. Takei, K. Ishizaka, K. M. Zsebo, S. Gillis, and T. Ishizaka, Development of human mast cells from umbilical cord blood cells by recombinant human and murine c-*kit* ligand, *Proc. Natl. Acad. Sci. USA 90*:735 (1993).
56. A. A. Irani, S. S. Craig, G. Nilsson, T. Ishizaka, and L. B. Schwartz, Characterization of human mast cells developed in vitro from fetal liver cells cocultured with murine 3T3 fibroblasts, *Immunology 77*:136 (1992).
57. A. A. Irani, G. Nilsson, U. Miettinen, S. C. Craig, L. K. Ashman, T. Ishizaka, K. M. Zsebo, and L. B. Schwartz, Recombinant human stem cell factor stimulates differentiation of mast cells from dispersed human fetal liver cells, *Blood 80*:3009 (1992).
58. A. S. Kirschenbaum, J. P. Goff, S. C. Dreskin, A. A. Irani, L. B. Schwartz, and D. D. Metcalfe, IL-3–dependent growth of basophil-like cells and mast-like cells from human bone marrow, *J. Immunol. 142*:2424 (1989).
59. A. S. Kirschenbaum, J. P. Goff, S. W. Kessler, J. M. Mican, K. M. Zsebo, and D. D. Metcalfe, Effect of IL-3 and stem cell factor on the appearance of human basophils and mast cells from $CD34^+$ pluripotent progenitor cells, *J. Immunol. 148*:772 (1992).
60. P. Valent, E. Spanblöchl, W. R. Sperr, C. Sillaber, K. M. Zsebo, H. Agis, H. Strobl, K. Geissler, P. Bettelheim, and K. Lechner, Induction of differentiation of human mast cells from bone marrow and peripheral blood mononuclear cells by recombinant human stem cell factor/*kit*-ligand in long-term culture, *Blood 80*:2237 (1992).
61. H. Agis, M. Willheim, W. R. Sperr, A. Wilfing, E. Krömer, E. Kabrna, E. Spanblöchl, H. Strobl, K. Geissler, A. Splittler, G. Boltz-Nitulescu, O. Majdic, K. Lechner, and P. Valent, Monocytes do not make mast cells when cultured in the presence of SCF: characterization of the circulating mast cell progenitor as a c-kit^+, $CD34^+$, Ly^-, $CD14^-$, $CD17^-$, colony-forming cell, *J. Immunol. 151*:4221 (1993).
62. B. M. Czarnetzki, G. Krüger, and W. Sterry, In vitro generation of mast cell-like cells from human peripheral mononuclear phagocytes, *Int. Arch. Allergy Appl. Immunol. 71*:161 (1983).

63. B. M. Czarnetzki, C. G. Figdor, G. Kolde, T. Vroom, R. Aalberse, and J. E. DeVries, Development of human connective tissue mast cells from purified blood monocytes, *Immunology 51*:549 (1984).
64. P. Welker, J. Grabbe, J. Hakimi, A. F. Walls, H. Ostmeier, and B. M. Czarnetzki, Fibroblast-derived factors induce different mast cell characteristics in human myeloid cell lines and peripheral monocytes, *Int. Arch. Allergy Appl. Immunol. 99*:337 (1992).
65. G. Nilsson, U. Miettinen, T. Ishizaka, L. K. Ashman, A. Irani, and L. B. Schwartz, Interleukin-4 inhibits the expression of kit and tryptase during stem cell factor-dependent development of human mast cells from fetal liver cells, *Blood 84*:1519 (1994).
66. A. A. Irani, G. Nilsson, L. K. Ashman, and L. B. Schwartz, Dexamethasone inhibits the development of mast cells from dispersed human fetal liver cells cultured in the presence of recombinant human stem cell factor, *Immunology 84*: (1995).

8

Dog Mast Cell Proteinases in Models of Airway Secretion, Bronchoconstriction, Cutaneous Vascular Permeability, and Tissue Fibrosis

Christian P. Sommerhoff
Surgical Clinic City, University of Munich, Munich, Germany

I. INTRODUCTION

Mast cells are widely recognized as the central cell type in the pathophysiology of immediate hypersensitivity reactions. Considerable evidence has been presented that mast cells also play an important role in chronic inflammatory diseases such as asthma, chronic bronchitis, cystic fibrosis, and psoriasis, and in fibrotic disorders. During the last 10 years or so special interest has been focused on the granule-associated proteinases tryptase (EC 3.4.21.59) and chymase (EC 3.4.21.39), which constitute the major proteins of mast cell secretory granules (1). The abundance of these neutral serine proteinases, their rapid release by exocytosis, and their unique biochemical properties have triggered the hypothesis that they may contribute significantly to the physiology and pathology of mast cells.

However, studies pursuing various aspects of this hypothesis, particularly those relating to the physiology and cell biology of tryptase and chymase, were severely hindered by the lack of adequate experimental tools. Both proteinases (as well as human mast cells) were difficult to isolate and were not available in sufficient amounts. Furthermore, because proteinase inhibitors specific for tryptase and chymase were not known, it was difficult to study the actions of the enzymes following their release from mast cells located within tissues.

Lazarus, DeVinney, Gold, and co-workers fostered the exploration of mast cell protease biology by establishing dog mastocytoma cells as permanent lines by serial passages in athymic mice and by establishing some of these cells in continuous cell culture (2,3). These mastocytoma cells closely resemble normal dog and human mast cells in morphology, ultrastructure, and functional parameters such as the release of newly formed and preformed mediators, including tryptase and chymase (4–6). Dog mastocytoma cells serve as a convenient source for the isolation of large quantities of tryptase and chymase (7,8). The structural and functional features of the purified proteinases are highly similar to those of chymases and tryptases of other species, including humans (7–11).

As discussed below in more detail, dog mastocytoma cells and their isolated proteinases have been used successfully in many studies. The results establish that tryptase and chymase may affect airway gland serous cell secretion, shedding of tracheal epithelial cell glycocalyx, bronchial smooth muscle tone, peptidergic neurotransmission, vascular permeability, and fibroblast proliferation. These observations support the hypothesis that mast cell proteinases play significant roles in the pathophysiology of mast cell–related disorders.

II. AIRWAY GLAND SECRETION

A. Chymase Stimulates Degranulation of Serous Cells

Increased and abnormal airway secretions are a characteristic feature of inflammatory airway diseases such as asthma, chronic bronchitis, and cystic fibrosis. Proteinases derived from neutrophils and bacteria are capable of increasing airway secretions by inducing goblet cell discharge (12,13) and the release of cell surface glycoconjugates (14–16). However, in humans the largest contribution to airway secretions is thought to originate from submucosal glands (17).

To determine whether proteinases stimulate airway gland secretion, we investigated the ability of mast cell chymase and tryptase to cause degranulation of bovine airway gland serous cells (18). Incubation of cultured serous cells with chymase (10^{-10} to 10^{-8} *M*) markedly stimulated degranulation as measured by the secretion of ^{35}S-labeled chondroitin sulfate proteoglycans (Fig. 1). Chymase-induced degranulation was noncytotoxic and was blocked by active site–directed inhibitors of chymase and by inhibitors of cellular energy metabolism. The response induced by 10^{-8} *M* chymase probably reflects nearly complete degranulation, because (a) the magnitude of the response is similar to that caused by neutrophil elastase (10^{-8} *M*), which depletes $>80\%$ of the cellular pools of secretory proteoglycans (19), and (b) immunocytochemistry of cells exposed to chymase demonstrated a complete loss of the punctate staining which indicates chondroitin sulfate proteoglycans associated with secretory granules. In contrast to chymase, purified tryptase ($\leq 10^{-8}$ *M*) had no detectable effect on serous cells.

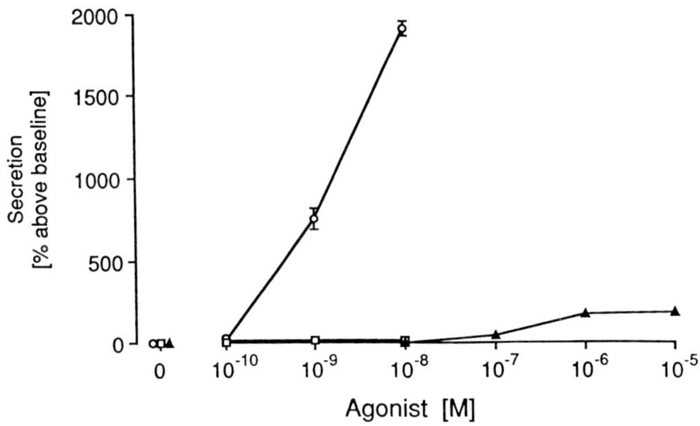

Figure 1 Effect of isolated chymase (circles), tryptase (squares), and histamine (triangles) on secretion of ^{35}S-labeled chondroitin sulfate proteoglycans from cultured bovine tracheal gland serous cells. Degranulation is stimulated by chymase and histamine whereas tryptase has no significant effect. Chymase is markedly more potent than histamine in respect to threshold concentration and magnitude of the secretory response. Values are mean ± SEM, $n \geq 5$. (Modified from Refs. 18 and 30.)

Chymase is the most potent secretagogue for serous cells identified so far. The marked secretagogue activity is shared by the serine proteinases from neutrophils, cathepsin G being somewhat less potent than elastase (19). In contrast, other agonists, including histamine, prostaglandins, and β-adrenergic agonists, require ≥100-fold higher concentrations to stimulate degranulation (20–23). Perhaps more significantly, chymase and other proteinases induce a maximal secretory response >10-fold larger than that of other agonists, causing nearly complete degranulation.

Chymase-containing mast cells represent a higher proportion of total mast cells in the bronchial epithelium than in peripheral lung tissue, and a particularly high proportion (about 73% of all mast cells) is found in the vicinity of airway submucosal glands (24,25). Furthermore, in cystic fibrosis, a disease characterized by airway hypersecretion, the proportion of chymase-containing mast cells in bronchoalveolar lavage fluid is dramatically higher than in normal subjects (87% compared to 0 to 17%) (26). It seems likely that both an increased release of chymase from mast cells present near glands as well as an increased number of chymase-containing mast cells (as in cystic fibrosis) may lead to airway hypersecretion. As serous cells are the source of mucus proteinase inhibitor (secretory leukocyte proteinase inhibitor, antileukoprotease) (27,28), which is a major determinant of protease–antiprotease balance in airway secretions (29), the sensitivity of serous cells to proteases provides a potential regulatory feedback mechanism in the defense of the airways (Fig. 2).

B. Chymase Degrades Serous Cell Proteoglycans

In addition to stimulating serous cell degranulation, chymase may cause abnormal secretions. Gel filtration chromatography of radiolabeled secretions showed that chymase not only increased the amount of chondroitin sulfate proteoglycans but also decreased the average apparent molecular weight of the secreted glycoconjugates (30). Thus like neutrophil elastase and cathepsin G (19), chymase has proteoglycanase activity. Degradation of secreted serous cell proteoglycans is likely to cause an alteration in rheological properties, increased colloid osmotic pressure, and potentially, the modification of biological functions.

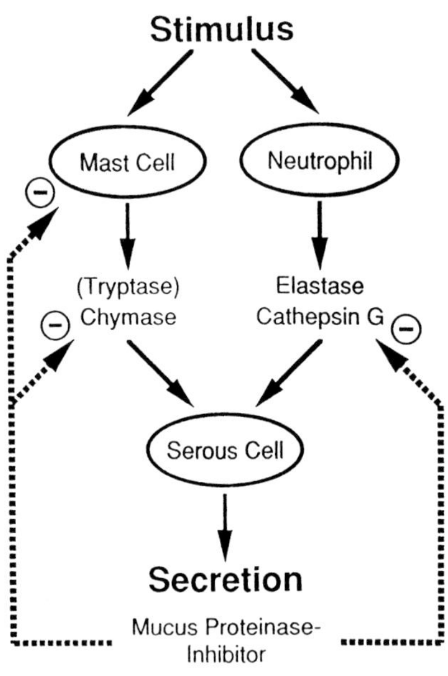

Figure 2 Potential feedback mechanism regulating protease–antiprotease balance in the airways. Various stimuli are capable of activating mast cells and neutrophils in the airways, causing the secretion of proteinases. Increased extracellular concentrations of catalytically active chymase, elastase, or cathepsin G may then stimulate secretion from serous cells, and the mucus proteinase inhibitor secreted may subsequently inhibit the activity of elastase, cathepsin G, and chymase. In addition, mucus proteinase inhibitor may block further degranulation of mast cells (38).

C. Mechanisms of Chymase-Induced Degranulation

For three decades it has been known that chymotrypsin and rat mast cell chymase cause degranulation of mast cells (31–33), but the mechanism of protease-induced mast cell activation is still unknown. To determine the mechanism of chymase-induced degranulation in serous cells, we compared the second messengers involved in secretion caused by chymase, histamine, and bradykinin (34). These particular agonists were selected because they elicit markedly different maximal secretory responses (1910 ± 47%, 177 ± 9%, and 59 ± 7% above baseline secretion for 10^{-8} *M* chymase, 10^{-6} *M* histamine, and 10^{-6} *M* bradykinin, respectively).

1. Degranulation induced by histamine, but not by bradykinin or chymase was associated with the generation of cAMP by adenylate cyclase; cellular cAMP measured by radioimmunoassay increased about 17-fold after challenge of cells with histamine. Furthermore, inhibition of phosphodiesterase by 3-isobutyl-1-methylxanthine augmented only histamine-induced separation.
2. Incorporation of [^{3}H]myoinositol into soluble inositol phosphates increased about fivefold in response to bradykinin but was not affected by histamine or chymase.
3. Depletion of protein kinase C from cells blocked the secretory response elicited by bradykinin and reduced the response caused by histamine, but did not alter chymase-induced degranulation.
4. Measurements of the intracellular calcium concentration using FURA-2 fluorescence showed that bradykinin and histamine, but not chymase, caused $[Ca^{2+}]_i$ transients. Furthermore, the intracellular calcium antagonist TMB-8 decreased only the histamine-induced secretory response.

These results indicate that serous cell degranulation induced by histamine involves cAMP-, protein kinase C-, and $[Ca^{2+}]_i$-dependent second messenger pathways. Similarly, bradykinin-induced secretion involves inositol phosphates, protein kinase C, and $[Ca^{2+}]_i$. Surprisingly, although the secretory response elicited is markedly larger, none of these classical second messengers appears to participate in chymase-induced degranulation (Fig. 3).

These data do not reveal how chymase stimulates serous cells. There is the possibility that chymase activates an as-yet-unknown second messenger pathway; studies of mast cells have suggested the existence of additional yet undefined signaling pathways (35,36). Alternatively, chymase may cause degranulation by substituting for an endogenous protease involved in stimulus–secretion coupling. This hypothesis is supported by studies of mast cells demonstrating that rat chymase I (RMCP-I) stimulates secretion (33,37) and that an endogenous

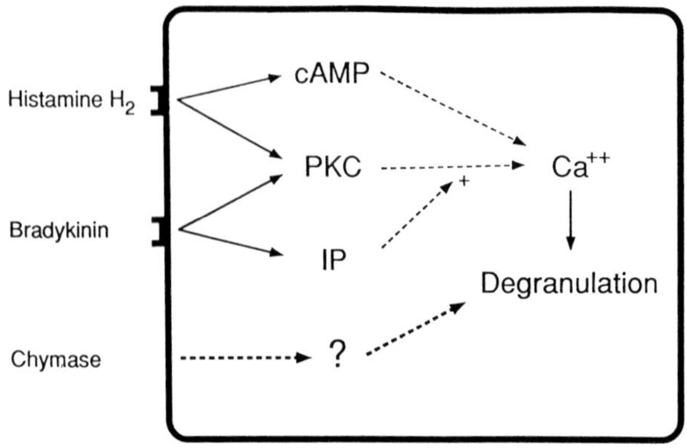

Figure 3 Stimulation–secretion coupling of cultured airway gland serous cells. Secretion induced by histamine involves the action of cAMP and of protein kinase C (PKC). Bradykinin-induced degranulation depends on the activity of protein kinase C and is accompanied by the production of soluble inositol phosphates (IP). Secretion induced by both agonists depends on intracellular calcium. In contrast, none of these classical second messengers are involved in chymase-induced degranulation of serous cells.

chymase is required for mast cell degranulation caused by other agonists (38–41). Similarly, the activity of a metalloprotease is involved in a late step of stimulus–secretion coupling in adrenal chromaffin cells (42).

To clarify whether endogenous proteases are involved in degranulation of serous cells, we studied the effects of protease inhibitors on histamine-induced secretion. Inhibitors of serine, thiol, and aspartic proteases had no significant effect. In contrast, the metalloprotease inhibitor 1,10-phenanthroline inhibited secretion with an IC_{50} of about 500 μM (Fig. 4). Secretion was also reduced by the irreversible metalloprotease inhibitor $ClCH_2$–CO (N–OH)–Phe- Ala--Ala–NH_2 and by the putative metalloprotease substrates Z–Ser–Leu, Z–Gly–Phe–NH_2, and Z–Gly–Leu–NH_2 (43). Phenanthroline also blocked secretion induced by dibutyryl cAMP and by the calcium ionophore A23187, but not the response caused by chymase. These results suggest that the activity of an endogenous metalloprotease is required in a late step of stimulus–secretion coupling of serous cells (i.e., after calcium entry or in the exocytotic mechanism). In contrast to other agonists, chymase-induced degranulation is independent of the action of this enzyme. These observations foster the interesting theory that chymase substitutes for this endogenous proteinase and directly activates degranulation, bypassing the stimulation–secretion coupling mechanisms involved in secretion caused by other agonists.

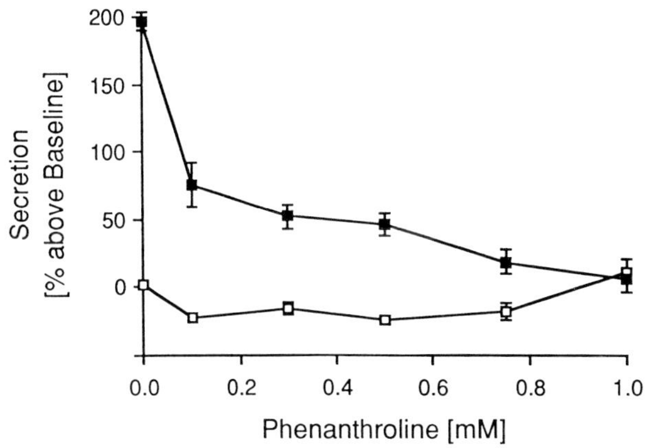

Figure 4 Effect of the metalloprotease inhibitor 1,10-phenanthroline on secretion of radiolabeled chondroitin sulfate proteoglycans from serous cells. Phenanthroline slightly reduces baseline secretion (open squares) and inhibits the secretory response caused by histamine (10^{-6} *M*; closed squares) in a concentration-dependent fashion. Values are mean ± SEM, $n \geq 5$.

D. Mast Cell Proteoglycans Modulate Chymase Activity

During degranulation, chymase is released from mast cells attached to proteoglycans (30,44). To study the interaction of chymase with proteoglycans and other mediators released, we prepared degranulation supernatant by activation of mastocytoma cells with the calcium ionophore A23187 in a two-stage reaction (18). The results obtained with this mediator-rich degranulation supernatant verified (Fig. 5) that chymase is the principal secretagogue produced by mast cells, and that the secretagogue activity of chymase is expressed in degranulation supernatant and in purified preparations. In this respect, the secretagogue effect of dog chymase on serous cells differs from that of rat mast cell chymase I (RMCP-I) on rat peritoneal mast cells, which is limited to the purified protease (37).

However, chymase present in degranulation supernatant was less potent in stimulating degranulation and causing degradation of secreted products than was purified chymase [30]. These differences were related to the proteoglycans present in degranulation supernatant and also were observed when isolated chymase was reconstituted with purified proteoglycans (Fig. 6): both heparin and low sulfation chondroitin sulfate proteoglycans partially inhibited the proteoglycanase and secretagogue activities and increased the amidolytic activity of

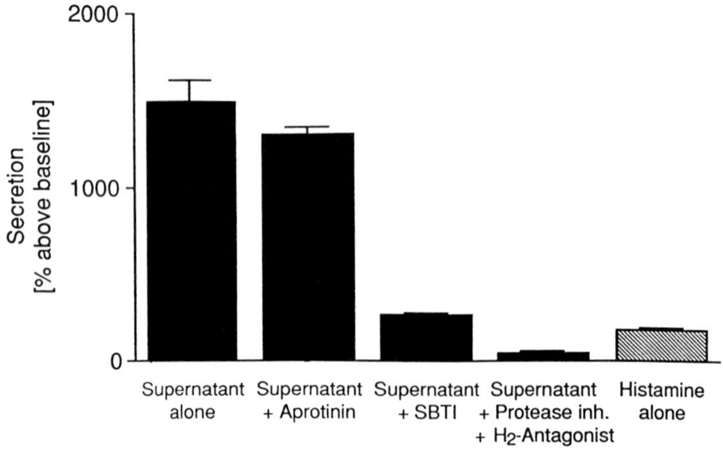

Figure 5 Effect of inhibitors on secretion from serous cells induced by mastocytoma degranulation supernatant. Degranulation supernatant causes a marked secretory response. Addition of aprotinin, an inhibitor of dog tryptase, does not significantly change the response, whereas soybean trypsin inhibitor (SBTI), an inhibitor of chymase, decreases secretion to the level induced by exogenous histamine alone in an equivalent concentration. Addition of the H_2-receptor antagonist metiamide to the proteinase inhibitors abolishes supernatant-induced secretion. Values are mean ± SEM, $n \geq 6$. (Modified from Ref. 18.)

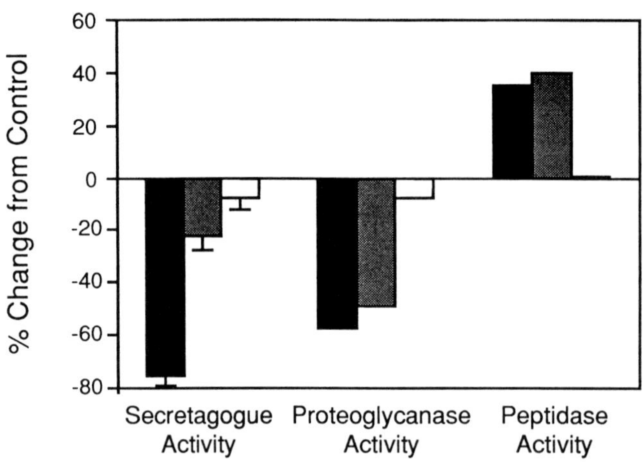

Figure 6 Comparison of the effects of proteoglycans isolated from mastocytoma cells on the secretagogue, proteoglycanase, and peptidase activities of chymase (w/w ratio 3:1). Both heparin (closed bars) and low-sulfation chondroitin sulfate proteoglycans (shaded bars) partially inhibit the proteoglycanase and secretagogue activities, whereas high-sulfation chondroitin sulfate proteoglycan (open bars) has little effect. These inhibitory actions of the glycans are not predicted by their effects on the peptidase activity of chymase. (Modified from Ref. 30.)

chymase, whereas high-sulfation chondroitin sulfate proteoglycan had little effect.

These findings suggest that mast cell proteoglycans modulate the secretagogue, proteoglycanase, and peptidase activities of chymase, and that the extent of this modulation depends on the concentration and nature of the proteoglycans present. Proteoglycan expression may vary markedly with shifts in mast cell phenotype (45,46), providing a mechanism distinct from that provided by classical proteinase inhibitors regulating the activities of chymase.

III. CELL SURFACE GLYCOCALYX

To determine whether glycoconjugates may be released by proteinases from the surface of epithelial cells and thus contribute to airway secretions, Varsano et al. studied dog tracheal epithelial cells in culture (14). These polarized epithelial cells, which do not contain secretory granules, incorporated ^{35}sulfate into macromolecular ($>10^6$ Da) glycoconjugates and released them spontaneously into the culture medium. The amount of material liberated was markedly increased by incubation of cells with proteinases [e.g., neutrophil elastase, *Pseudomonas aeruginosa* elastase, and mast cell chymase (each 10^{-7} *M*)]. The effect was related to the catalytic activity of the proteinases and was accompanied by a decrease in the glycoconjugates present on the apical cell surface.

These studies suggest that chymase may increase airway secretions not only by stimulating the degranulation of submucosal gland serous cells but also by releasing glycoconjugates from the apical surface of epithelial cells. In addition, chymase may modulate other cellular functions attributed to the surface glycocalyx (e.g., cellular adhesion, differentiation, and proliferation) (47).

IV. AIRWAY SMOOTH MUSCLE

Although airway hyperreactivity is a fundamental feature of asthma, its pathogenesis is not completely understood. To test the hypothesis that mast cell–derived mediators influence airway reactivity, Sekizawa et al. studied dog bronchial rings in vitro (48). Degranulation supernatant from mastocytoma cells caused a marked contraction (Fig. 7). The response was completely blocked by H_1-receptor antagonists, suggesting its dependence on the action of histamine. Surprisingly, the magnitude of the contractile response caused by degranulation supernatant was about sixfold larger than expected from the histamine present in supernatant. The mast cell–derived mediator responsible for the marked potentiation of the histamine-induced contraction appeared to be tryptase, because addition of aprotinin, an inhibitor of dog tryptase, reduced the supernatant-induced response, the magnitude of the remaining response being that expected from the level of histamine in the supernatant.

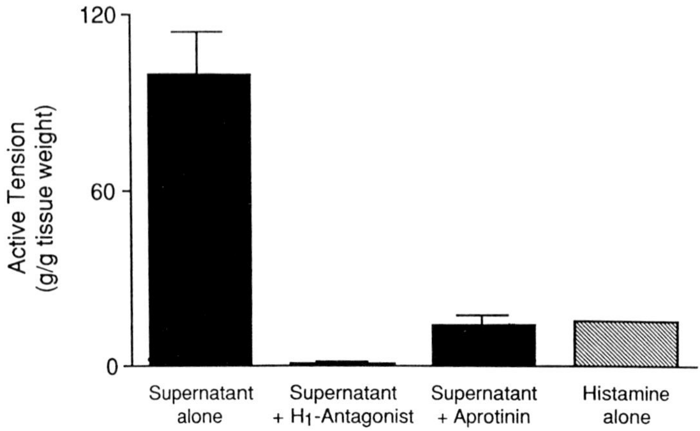

Figure 7 Effect of inhibitors on contraction of dog bronchi induced by mastocytoma degranulation supernatant. Degranulation supernatant causes a marked contraction which is abolished by addition of the H_1-receptor antagonist pyrilamine. In addition, aprotinin, an inhibitor of dog mastocytoma tryptase, greatly reduces the response; the remaining response is similar to that caused by an equivalent concentration of histamine (interpolated from histamine concentration–response curves). Data are mean ± SEM, $n = 5$. (Modified from Ref. 48.)

Subsequent experiments verified that tryptase augments the histamine-induced contractile response. Thus isolated tryptase had no effect on the resting tension of smooth muscle but decreased the ED_{50} value for histamine and increased the maximal contraction (Fig. 8). The augmenting effect of tryptase on histamine-induced contraction was concentration dependent and significant at concentrations of 15 ng or more tryptase per milliliter (e.g., about 10^{-10} *M*).

Several lines of evidence support the hypothesis that tryptase modulates voltage-dependent Ca^{2+} channels, perhaps by limited proteolysis of the extracellular protein:

1. Tryptase augmented the responses induced by KCl and serotonin, which, like histamine, contract dog airway smooth muscle by a mechanism involving voltage-gated Ca^{2+} channels.
2. Tryptase did not affect contraction induced by acetylcholine, which mobilizes intracellular calcium.
3. The Ca^{2+}-channel blockers verapamil and nifedepine reversed the tryptase-induced potentiation.
4. Tryptase exposure did not result in permanent tissue damage, as indicated by the finding that smooth muscle responsiveness returned to the control state within 3 h after tryptase removal.

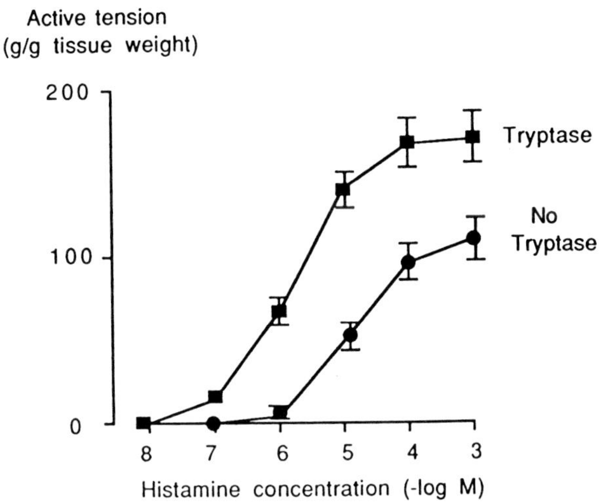

Figure 8 Effect of isolated tryptase on the histamine-induced contraction of dog bronchial rings. Addition of tryptase (90 ng/mL) greatly increases the contractility of the smooth muscle, shifting the concentration–response curve for histamine to the left and increasing the maximal contraction. Data are mean ± SEM, $n = 5$. (From Ref. 48.)

These studies reveal a novel mechanism by which tryptase may cause airway hyperresponsiveness. In addition to this effect, which probably involves direct action of tryptase on smooth muscle, mast cell proteinases may promote bronchial hyperreactivity indirectly by inactivating bronchodilatatory neuropeptides. Tryptase rapidly cleaves the bronchodilating neuropeptides vasoactive intestinal peptide (VIP) and peptide histidine-methionine without affecting bronchoconstricting tachykinins (substance P, neurokinin A, neurokinin B) (49,50), and reverses VIP-induced relaxation in ferret airways (51). The absence of immunoreactive VIP which has been observed in the airways of asthmatic patients (52) has been attributed to this action of tryptase.

The relation between airway reactivity and tryptase is corroborated by clinical studies:

1. Increased levels of tryptase have been measured in bronchoalveolar lavage fluid (BALF) of patients with asthma (53–56).
2. Tryptase concentrations in BALF are highly correlated to airway responsiveness in children with chronic asthma (56).
3. Experimental ozone exposure of subjects increases both tryptase in BALF and airway reactivity (57,58).

V. VASCULAR PERMEABILITY

A role of proteinases in the pathogenesis of edema formation has been discussed for over 30 years (59,60). To study the effect of chymase on vascular permeability in ragweed-allergic dogs, Rubinstein et al. measured wheal formation after intradermal injections (61). Injection of isolated chymase (≤5 μg) did not cause the development of wheals. However, chymase (5 μg) markedly potentiated plasma extravasation induced by subsequent injection of histamine, shifting the concentration response curve to the left. The chymase-induced potentiation was dependent on the amount injected (Fig. 9) and was blocked by soybean trypsin inhibitor, suggesting the involvement of the active site of the proteinase.

Because chymase is stored in the secretory granules of nearly all mast cells present in the skin (24,62), Rubinstein et al. also studied plasma extravasation following injections of ragweed antigen and compound 48/80 (Fig. 10). Both agonists caused wheal formation, which was abolished by local pretreatment with the H_1-receptor antagonist pyrilamine, verifying that the responses depend on histamine released from mast cells. More important, plasma extravasation was greatly reduced by soybean trypsin inhibitor, indicating that chymase (or possibly another proteinase affected by this inhibitor) contributes to the magnitude of the response.

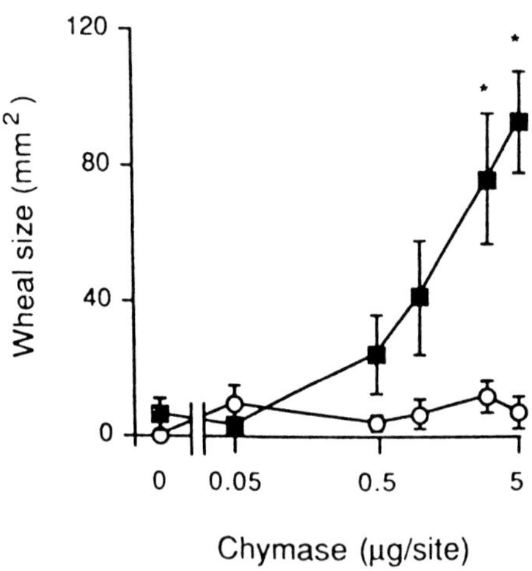

Figure 9 Effect of isolated chymase on histamine-induced vascular permeability in the skin of ragweed-allergic dogs. Injection of chymase alone (circles) does not cause a significant wheal formation. Injection of chymase together with a subthreshold amount of histamine (10^{-7} *M*, 100 μL; squares) causes a dose-dependent increase in wheal size. Values are mean ± SEM, $n = 7$. Differences in mean wheal size with $p < 0.05$ are indicated by an asterisk. (From Ref. 61.)

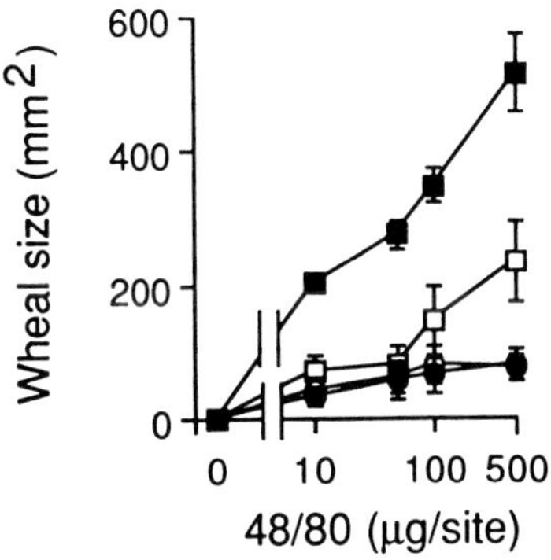

Figure 10 Effect of inhibitors on compound 48/80-induced vascular permeability in the skin of allergic dogs. Compound 48/80 (closed squares) causes a dose-dependent wheal formation. The response is reduced by pretreatment with soybean trypsin inhibitor (open squares), with the H_1-receptor antagonist pyrilamine (open circles), or with both inhibitors (closed circles). Values are mean ± SEM, $n = 4$. (From Ref. 61.)

These results extend previous studies (63–65) demonstrating that chymase is a proinflammatory agent in the skin. Whether these observations apply to other organs (e.g., the airways) remains to be determined. Additional in vitro studies will be necessary to clarify whether chymase acts directly on endothelial cells in a manner similar to that of the related proteinases cathepsin G and thrombin (66–68). The recent observations that chymase activates precursor IL-1β to an active IL-1 species by limited proteolysis (69) and that keratinocytes synthesize a precursor IL-1β but are unable to produce active IL-1 (70) may suggest a different hypothesis: chymase activates precursor IL-1β present in the skin, which subsequently has multiple effects on endothelial cells, promoting coagulation and inflammatory responses (71).

VI. FIBROBLAST PROLIFERATION

Several lines of evidence linking mast cells to connective tissue turnover have been presented (72–74). In particular:

1. An increased number of mast cells is found in fibrotic diseases such as systemic sclerosis (75,76), pulmonary fibrosis (77,78), and neurofibromatosis (79,80) as well as in experimentally induced pulmonary fibrosis (81,82).
2. Mast cell degranulation is detected in fibrotic diseases by morphological studies and by measurements of granule-associated mediators (76,78,83).

3. Fibroblast proliferation appears to occur following agonist-induced mast cell degranulation (84).

Because tryptase is related to thrombin and other proteinases known to be mitogens for various cell types, including fibroblasts (85,86), Ruoss et al. (87) studied the effect of dog mastocytoma cell tryptase on the growth of fibroblasts in culture. Tryptase in nanomolar concentrations markedly stimulated the proliferation of Chinese hamster lung and rat-1 fibroblasts (87) and was found to be synergistic with other mitogens (i.e., insulin, epidermal growth factor, and basic fibroblast growth factor). Tryptase-induced proliferation was abolished by the tripeptide inhibitor leupeptin, demonstrating that the response involves the active site of the proteinase. However, the smaller inhibitor diisopropyl fluorophosphate only diminished the response, suggesting that it may be partially independent of catalytical activity. Also, proteolytic activity alone does not suffice: trypsin and chymase did not stimulate proliferation of the fibroblast cell lines studied. Fibroblast proliferation induced by tryptase and by thrombin appear to be two distinct phenomena: the growth of some cell lines was stimulated by one but not by the other proteinase. Furthermore, both proteinases utilize different signal transduction pathways: in contrast to thrombin, tryptase does not induce the generation of inositol phosphates and the mobilization of calcium, and the response does not depend on pertussis-sensitive GTP-binding proteins.

These results, which have been extended to human lung tryptase (88), demonstrate that tryptase is a potent mitogen for fibroblasts. Fibroblast proliferation is also stimulated by other mast cell–derived mediators (e.g., histamine, leukotrienes, tumor necrosis factor-α, and transforming growth factor $\beta 1$), by an as-yet-undefined heat-stable 10-kDa mitogen (89–93), and by direct contact with mast cells (94). In addition, degradation of extracellular matrix components induced by chymase and by tryptase (via the activation of stromelysin and subsequently collagenase) may influence connective tissue turnover (95,96). In reverse, fibroblasts affect mast cell proliferation and differentiation (46,97). These multiple interactions between mast cells and fibroblasts indicate that the role of mast cell–derived mediators in the pathogenesis of fibrotic disorders is complex.

VII. CONCLUSIONS AND PERSPECTIVES

The studies reviewed demonstrate the marked effects of mastocytoma proteinases in models of airway secretion, bronchoconstriction, cutaneous vascular permeability, and tissue fibrosis (Fig. 11). The concentrations required to induce effects are low and within the range of those expected after partial degranulation of the large numbers of mast cells present in lung and skin. Furthermore, sufficient levels of tryptase have already been measured in the bronchoalveolar

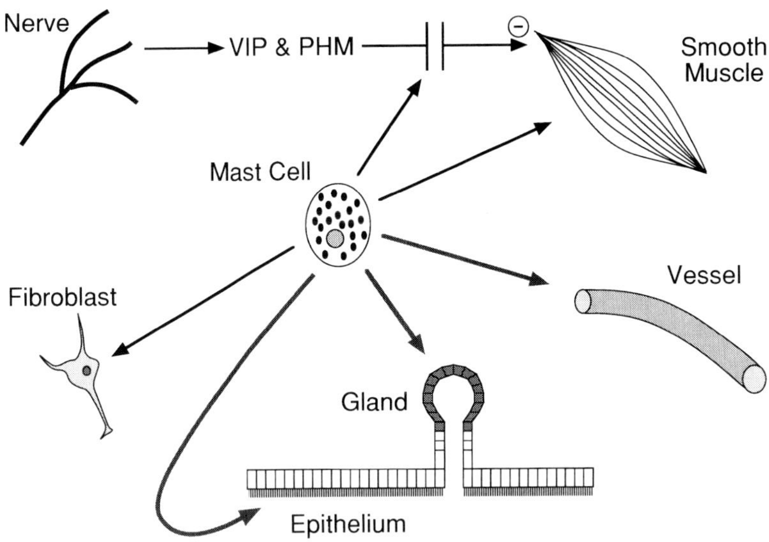

Figure 11 Mast cell proteinases potentially cause characteristic symptoms of asthma, including airway hyperreactivity, hypersecretion, edema, and fibrosis. Studies using dog mastocytoma proteinases have demonstrated that chymase (shaded arrows) may stimulate secretion from submucosal gland serous cells, release glycoconjugates from the cell surface of epithelial cells, and augment histamine-induced vascular permeability. Tryptase (solid arrows) may increase the responsiveness of bronchial smooth muscle, inactivate the bronchodilatatory neuropeptides vasoactive intestinal peptide (VIP) and peptide histidine-methionine (PHM), and stimulate fibroblast proliferation.

lavage fluid of patients with asthma (53–56) and pulmonary fibrosis (78,83) and in the nasal lavage fluid of patients with allergic rhinitis (98,99).

However, the studies reviewed were performed with proteinases derived from dog mastocytoma cells in model systems utilizing tissues or cells of animal species. Therefore, key experiments using human proteinases and human tissues are required before these observations can be applied to human pathophysiology. To date, the effects of dog mastocytoma tryptase on the degradation of vasoactive intestinal peptide (49) and on the proliferation of fibroblasts (87) have been verified with human tryptase (50,88), and the results obtained with the enzymes are similar but not identical.

Studies performed in vitro will also clarify whether the effects of tryptase on bronchial reactivity and of chymase on vascular permeability are due to direct actions of the proteinases on the target cells (i.e., smooth muscle and endothelial cells, respectively) and will elucidate the detailed mechanisms involved. It is conceivable that tryptase and chymase activate cells by a mechanism similar to that recently suggested for the related proteinase thrombin: this enzyme apparently cleaves the extracellular portion of a "thrombin receptor," thereby exposing a new amino terminus which subsequently functions as a tethered ligand to activate the receptor (100,101).

Finally, to validate the roles of tryptase and chymase in vivo, specific and noncytotoxic proteinase inhibitors will be essential. The models described may be used to evaluate promising compounds, both synthetic [e.g., benzamidine derivatives as inhibitors for tryptase (102,103)] and naturally occurring or recombinant inhibitors [e.g., eglin C for chymase (104)]. Suitable inhibitors may be applicable as pharmacological tools to evaluate the role of mast cell proteinases and may evolve as future drugs for the therapy of mast cell–related disorders.

ACKNOWLEDGMENTS

The author wishes to thank the scientists, staff, and fellows of the Cardiovascular Research Institute, University of California, for years of challenging atmosphere, support, and friendship. This work was supported in part by Sonderforschungsbereich 207 of the University of Munich (Grant G-11).

REFERENCES

1. L. B. Schwartz, R. A. Lewis, and K. F. Austen, Tryptase from human pulmonary mast cells: purification and characterization, *J. Biol. Chem. 256*:11939 (1981).
2. S. C. Lazarus, R. DeVinney, L. J. McCabe, W. E. Finkbeiner, D. J. Elias, and W. M. Gold, Isolated canine mastocytoma cells: propagation and characterization of two cell lines, *Am. J. Physiol. 251*:C935 (1986).
3. R. DeVinney and W. M. Gold, Establishment of two dog mastocytoma cell lines in continuous culture, *Am. J. Respir. Cell Mol. Biol. 3*:413 (1990).
4. L. D. Calonico, M. J. Phillips, D. M. McDonald, and W. M. Gold, An ultrastructural analysis of dog mastocytoma cells and normal mast cells, *Anat. Rec. 212*:399 (1985).
5. G. H. Caughey, N. F. Viro, L. D. Calonico, D. M. McDonald, S. C. Lazarus, and W. M. Gold, Chymase and tryptase in dog mastocytoma cells: asynchronous expression as revealed by enzyme cytochemical staining, *J. Histochem. Cytochem. 36*:1053 (1988).
6. G. H. Caughey, S. C. Lazarus, N. F. Viro, W. M. Gold, and J. A. Nadel, Tryptase and chymase: comparison of extraction and release in two dog mastocytoma lines, *Immunology 63*:339 (1988).

7. G. H. Caughey, N. F. Viro, J. Ramachandran, S. C. Lazarus, D. B. Borson, and J. A. Nadel, Dog mastocytoma tryptase: affinity purification, characterization, and amino-terminal sequence, *Arch. Biochem. Biophys. 258*:555 (1987).
8. G. H. Caughey, N. F. Viro, S. C. Lazarus, and J. A. Nadel, Purification and characterization of dog mastocytoma chymase: identification of an octapeptide conserved in chymotryptic leukocyte proteinases, *Biochim. Biophys. Acta 952*:142 (1988).
9. N. M. Schechter, D. Slavin, R. D. Fetter, G. S. Lazarus, and J. E. Fräki, Purification and identification of two serine class proteinases from dog mast cells biochemically and immunologically similar to human proteinases tryptase and chymase, *Arch. Biochem. Biophys. 262*:232 (1988).
10. G. H. Caughey, W. W. Raymond, and P. Vanderslice, Dog mast cell chymase: molecular cloning and characterization, *Biochemistry 29*:5166 (1990).
11. P. Vanderslice, C. S. Craik, J. A. Nadel, and G. H. Caughey, Molecular cloning of dog mast cell tryptase and a related protease: structural evidence of a unique mode of serine protease activation, *Biochemistry 28*:4148 (1989).
12. R. Breuer, T. G. Christensen, E. C. Lucey, P. J. Stone, and G. L. Snider, An ultrastructural morphometric analysis of elastase-treated hamster bronchi shows discharge followed by progressive accumulation of secretory granules, *Am. Rev. Respir. Dis. 136*:698 (1987).
13. T. F. Boat, P. W. Cheng, J. D. Klinger, C. M. Liedtke, and B. Tandler, Proteinases release mucins from airway goblet cells, *Ciba Found. Symp. 109*:72 (1984).
14. S. Varsano, C. B. Basbaum, L. S. Forsberg, D. B. Borson, G. H. Caughey, and J. A. Nadel, Dog tracheal epithelial cells in culture synthesize sulfated macromolecular glycoconjugates and release them from the cell surface upon exposure to extracellular proteinases, *Exp. Lung Res. 13*:157 (1987).
15. K. C. Kim, K. Wasano, R. M. Niles, J. E. Schuster, P. J. Stone, and J. S. Brody, Human neutrophil elastase releases cell surface mucins from primary cultures of hamster tracheal epithelial cells, *Proc. Natl. Acad. Sci. USA 84*:9304 (1987).
16. R. Breuer, T. G. Christensen, R. M. Niles, P. J. Stone, and G. L. Snider, Human neutrophil elastase causes glycoconjugate release from the cell surface of hamster trachea in organ culture, *Am. Rev. Respir. Dis. 139*:779 (1989).
17. L. Reid, Measurement of the bronchial mucous gland layer: a diagnostic yardstick in chronic bronchitis, *Thorax 15*:132 (1960).
18. C. P. Sommerhoff, G. H. Caughey, W. E. Finkbeiner, S. C. Lazarus, C. B. Basbaum, and J. A. Nadel, Mast cell chymase: a potent secretagogue for airway gland serous cells, *J. Immunol. 142*:2450 (1989).
19. C. P. Sommerhoff, J. A. Nadel, C. B. Basbaum, and G. H. Caughey, Neutrophil elastase and cathepsin G stimulate secretion from cultured bovine airway gland serous cells, *J. Clin. Invest. 85*:682 (1990).
20. W. E. Finkbeiner, J. A. Nadel, and C. B. Basbaum, Establishment and characterization of a cell line derived from bovine tracheal glands, *In Vitro Cell. Dev. Biol. 22*:561 (1986).
21. C. P. Sommerhoff, W. E. Finkbeiner, J. A. Nadel, and C. B. Basbaum, Media-

tors of anaphylaxis stimulate ^{35}S-labeled macromolecule secretion from cultured bovine tracheal gland serous cells, *Bull. Eur. Physiopathol. Respir. 23 (Suppl. 12)*:363s (1987).
22. C. P. Sommerhoff, W. E. Finkbeiner, J. A. Nadel, and C. B. Basbaum, Prostaglandin D_2 and prostaglandin E_2 stimulate ^{35}S-labeled macromolecule secretion from cultured bovine tracheal gland serous cells, *Am. Rev. Respir. Dis. 135*:A363 (1987).
23. C. B. Basbaum, J. M. Madison, C. P. Sommerhoff, J. K. Brown, and W. E. Finkbeiner, Receptors on airway gland cells, *Am. Rev. Respir. Dis. 141*:S141 (1990).
24. A. A. Irani, N. M. Schechter, S. S. Craig, G. DeBlois, and L. B. Schwartz, Two types of human mast cells that have distinct neutral protease compositions, *Proc. Natl. Acad. Sci. USA 83*:4464 (1986).
25. R. Matin, E. K. Tam, J. A. Nadel, and G. H. Caughey, Distribution of chymase-containing mast cells in human bronchi, *J. Histochem. Cytochem. 40*:781 (1992).
26. L. B. Schwartz, A. A. Irani, S. Mroueh, and A. Spock, Reversed ratio of T/TC types of human mast cells in cystic fibrosis, *Clin. Res. 35*:255A (1987).
27. R. De Water, L. N. A. Williams, G. N. P. van Muijen, C. Franken, J. A. M. Fransen, J. H. Dijkman, and J. A. Kramps, Ultrastructural localization of bronchial antileukoprotease in central and peripheral airways by a gold-labeling technique using monoclonal antibodies, *Am. Rev. Respir. Dis. 133*:882 (1986).
28. C. H. Lee, Y. Igarashi, R. J. Hohman, H. Kaulbach, M. V. White, and M. A. Kaliner, Distribution of secretory leukoprotease inhibitor in the human nasal airway, *Am. Rev. Respir. Dis. 147*:710 (1993).
29. C. Vogelmeier, R. C. Hubbard, G. A. Fells, H. P. Schnebli, R. C. Thompson, H. Fritz, and R. G. Crystal, Anti-neutrophil elastase defense of the normal human respiratory epithelial surface provided by the secretory leukoprotease inhibitor, *J. Clin. Invest. 87*:481 (1991).
30. C. P. Sommerhoff, S. J. Ruoss, and G. H. Caughey, Mast cell proteoglycans modulate the secretagogue, proteoglycanase, and amidolytic activities of dog mast cell chymase, *J. Immunol. 148*:2859 (1992).
31. B. Uvnäs and J. Antonsson, Triggering action of phosphatidase A and chymotrypsins on degranulation of rat mesentery mast cells, *Biochem. Pharmacol. 12*:867 (1963).
32. D. Lagunoff, E. Y. Chi, and H. Wan, Effects of chymotrypsin and trypsin on rat peritoneal mast cells, *Biochem. Pharmacol. 24*:1573 (1975).
33. B. Schick, K. F. Austen, and L. B. Schwartz, Activation of rat serosal mast cells by chymase, an endogenous secretory granule protease, *J. Immunol. 132*:2571 (1984).
34. C. P. Sommerhoff, G. H. Caughey, and J. A. Nadel, Classical second messengers are not involved in the stimulation–secretion coupling of chymase-induced degranulation of airway gland serous cells, *FASEB J. 4*:A1940 (1990).
35. R. Penner, Multiple signaling pathways control stimulus–secretion coupling in rat peritoneal mast cells, *Proc. Natl. Acad. Sci. USA 85*:9856 (1988).

36. B. D. Gomperts, GE: a GTP-binding protein mediating exocytosis, *Annu. Rev. Physiol. 52*:591 (1990).
37. B. Schick and K. F. Austen, Rat serosal mast cell degranulation mediated by chymase, an endogenous secretory granule protease: active site-dependent initiation at 1 degree C, *J. Immunol. 136*:3812 (1986).
38. S. C. Dietze, C. P. Sommerhoff, and H. Fritz, Inhibition of histamine release from human mast cells ex vivo by natural and synthetic chymase inhibitors, *Biol. Chem. Hoppe-Seyler 371 (Suppl.)*:75 (1990).
39. H. Kido, N. Fukusen, and N. Natanuma, Antibody and inhibitor of chymase inhibit histamine release in immunoglobulin E-activated mast cells, *Biochem. Int. 10*:863 (1985).
40. H. Kido, N. Fukusen, and N. Katunuma, Antibodies and inhibitor of chymase are incorporated into mast cell granules and inhibit histamine release, *Biol. Chem. Hoppe-Seyler 369 (Suppl.)*:95 (1988).
41. T. Hultsch, M. Ennis, and H. H. Heidtmann, The role of chymase in ionophore-induced histamine release from human pulmonary mast cells, *Adv. Exp. Med. Biol. 240*:133 (1988).
42. W. J. Strittmatter, Molecular mechanisms of exocytosis: the adrenal chromaffin cell as a model system, *Cell. Mol. Neurobiol. 8*:19 (1988).
43. D. A. Baxter, D. Johnston, and W. J. Strittmatter, Protease inhibitors implicate metalloendoprotease in synaptic transmission at the mammalian neuromuscular junction, *Proc. Natl. Acad. Sci. USA 80*:4174 (1983).
44. L. B. Schwartz, C. Riedel, J. P. Caulfield, S. I. Wasserman, and K. F. Austen, Cell association of complexes of chymase, heparin proteoglycan, and protein after degranulation by rat mast cells, *J. Immunol. 126*:2071 (1981).
45. R. L. Stevens, M. E. Rothenberg, F. Levi-Schaffer, and K. F. Austen, Ontogeny of in vitro-differentiated mouse mast cells, *Fed. Proc. 46*:1915 (1987).
46. M. E. Rothenberg and K. F. Austen, Influence of the fibroblast environment on the structure of mast cell proteoglycans, *Ann. NY Acad. Sci. 556*:233 (1989).
47. E. Ruoslahti, Structure and biology of proteoglycans, *Annu. Rev. Cell Biol. 4*:229 (1988).
48. K. Sekizawa, G. H. Caughey, S. C. Lazarus, W. M. Gold, and J. A. Nadel, Mast cell tryptase causes airway smooth muscle hyperresponsiveness in dogs, *J. Clin. Invest. 83*:175 (1989).
49. G. H. Caughey, F. Leidig, N. F. Viro, and J. A. Nadel, Substance P and vasoactive intestinal peptide degradation by mast cell tryptase and chymase, *J. Pharmacol. Exp. Ther. 244*:133 (1988).
50. E. K. Tam and G. H. Caughey, Degradation of airway neuropeptides by human lung tryptase, *Am. J. Respir. Cell Mol. Biol. 3*:27 (1990).
51. G. M. Franconi, P. D. Graf, S. C. Lazarus, J. A. Nadel, and G. H. Caughey, Mast cell tryptase and chymase reverse airway smooth muscle relaxation induced by vasoactive intestinal peptide in the ferret, *J. Pharmacol. Exp. Ther. 248*:947 (1989).
52. S. Ollerenshaw, D. Jarvis, A. Woolcock, C. Sullivan, and T. Scheibner, Absence

of immunoreactive vasoactive intestinal polypeptide in tissues from the lungs of patients with asthma, *N. Engl. J. Med. 320*:1244 (1989).
53. S. E. Wenzel, A. A. Fowler 3d, and L. B. Schwartz, Activation of pulmonary mast cells by bronchoalveolar allergen challenge: in vivo release of histamine and tryptase in atopic subjects with and without asthma, *Am. Rev. Respir. Dis. 137*:1002 (1988).
54. D. H. Broide, G. J. Gleich, A. J. Cuomo, D. A. Coburn, E. C. Federman, L. B. Schwartz, and S. I. Wasserman, Evidence of ongoing mast cell and eosinophil degranulation in symptomatic asthma airway, *J. Allergy Clin. Immunol. 88*:637 (1991).
55. J. V. Bosso, L. B. Schwartz, and D. D. Stevenson, Tryptase and histamine release during aspirin-induced respiratory reactions, *J. Allergy Clin. Immunol. 88*:830 (1991).
56. A. C. Ferguson, M. Whitelaw, and H. Brown, Correlation of bronchial eosinophil and mast cell activation with bronchial hyperresponsiveness in children with asthma, *J. Allergy Clin. Immunol. 90*:609 (1992).
57. C. E. Smith, H. S. Koren, D. G. Graham, and D. A. Johnson, Mast cell tryptase is increased in the nasal and bronchoalveolar lavage fluids of humans after ozone exposure, *Inhal. Toxicol. 5*:117 (1993).
58. M. J. Holtzman, Inflammation of the airway epithelium and the development of airway hyperresponsiveness, *Prog. Respir. Res. 19*:165 (1985).
59. W. G. Spector and D. A. Willoughy, The suppression by anti-esterases of increased capillary permeability in acute inflammation, *J. Pathol. Bacteriol. 79*:21 (1960).
60. D. Lagunoff and E. P. Benditt, Proteolytic enzymes of mast cells, *Ann. NY Acad. Sci. 103*:185 (1963).
61. I. Rubinstein, J. A. Nadel, P. D. Graf, and G. H. Caughey, Mast cell chymase potentiates histamine induced wheal formation in the skin of ragweed-allergic dogs, *J. Clin. Invest. 86*:555 (1990).
62. A. B. Becker, K. F. Chung, D. M. McDonald, S. C. Lazarus, O. L. Frick, and W. M. Gold, Mast cell heterogeneity in dog skin, *Anat. Rec. 213*:477 (1985).
63. O. Hagermark, G. Rajka, and U. Berqvist, Experimental itch in human skin elicited by rat mast cell chymase, *Acta Dermato-Venereol. 55*:125 (1972).
64. J. E. Fräki, Human skin proteases: effect of separated proteases on vascular permeability and leukocyte emigration in the skin, *Acta Dermato-Venereol. 57*:393 (1977).
65. H. Seppä, The role of chymotrypsin-like protease of the rat mast cells in inflammatory vasopermeability and fibrinolysis, *Inflammation 4*:1 (1980).
66. M. W. Peterson, P. Stone, and D. M. Shasby, Cationic neutrophil proteins increase transendothelial albumin movement, *J. Appl. Physiol. 62*:1521 (1987).
67. J. G. N. Garcia, A. Siflinger-Birnboim, R. Bizios, P. J. Del Vecchio, J. W. Fenton II, and A. B. Malik, Thrombin-induced increase in albumin permeability across the endothelium, *J. Cell. Physiol. 128*:96 (1986).
68. J. L. Aschner, J. M. Lennon, J. W. Fenton II, M. Aschner, and A. B. Malik, Enzymatic activity is necessary for thrombin-mediated increase in endothelial permeability, *Am. J. Physiol. 259*:L270 (1990).

69. H. Mizutani, N. Schechter, G. Lazarus, R. A. Black, and T. S. Kupper, Rapid and specific conversion of precursor interleukin 1β (IL-1β) to an active IL-1 species by human mast cell chymase, *J. Exp. Med. 174*:821 (1991).
70. H. Mizutani, R. Black, and T. S. Kupper, Human keratinocytes produce but do not process pro-interleukin 1 beta (pro-IL-1β), *J. Clin. Invest. 87*:1066 (1991).
71. J. S. Pober and R. S. Cotran, Cytokines and endothelial cell biology, *Physiol. Rev. 70*:427 (1990).
72. H. N. Claman, New hope for neurofibromatosis? The mast cell connection, *JAMA 258*:823 (1987).
73. H. N. Claman, On scleroderma: mast cells, endothelial cells, and fibroblasts, *JAMA 262*:1206 (1989).
74. M. Jordana, Mast cells and fibrosis: who's on first, *Am. J. Respir. Cell Mol. Biol. 8*:7 (1993).
75. K. Nishioka, Y. Kobayashi, I. Katayama, and C. Takijiri, Mast cell numbers in diffuse sclerodoma, *Arch. Dermatol. 123*:205 (1987).
76. J. R. Seibold, R. C. Giorno, and H. N. Claman, Dermal mast cell degranulation in systemic sclerosis, *Arthritis Rheum. 33*:1702 (1990).
77. A. Pesci, G. Bertorelli, M. Gabrielli, and D. Olivieri, Mast cells in fibrotic lung disorders, *Chest 103*:989 (1993).
78. P. Chanez, J. Y. Lacoste, B. Guillot, J. Giron, G. Barneon, I. Enander, P. Godard, F. B. Michel, and J. Bousquet, Mast cells' contribution to the fibrosing alveolitis of the scleroderma lung, *Am. Rev. Respir. Dis. 147*:1497 (1993).
79. K. Donhuijsen, M. Sastry, B. Volker, and L. D. Leder, Mast cell frequency in soft tissue tumors: relation to type and grade of malignancy, *Pathol. Res. Pract. 188*:61 (1992).
80. M. D. Johnson, J. Kamso-Pratt, C. F. Federspiel, and W. O. Whetsell, Jr., Mast cell and lymphoreticular infiltrates in neurofibromas: comparison with nerve sheath tumors, *Arch. Pathol. Lab. Med. 113*:1263 (1989).
81. S. Watanabe, K. Watanabe, T. Ohishi, M. Aiba, and K. Kageyama, Mast cells in the rat alveolar septa undergoing fibrosis after ionizing irradiation, *Lab. Invest. 31*:555 (1974).
82. T. Goto, D. Befus, R. Low, and J. Bienenstock, Mast cell heterogeneity and hyperplasia in bleomycin-induced pulmonary fibrosis in rats, *Am. Rev. Respir. Dis. 130*:797 (1984).
83. A. F. Walls, A. R. Bennett, R. C. Godfrey, S. T. Holgate, and M. K. Church, Mast cell tryptase and histamine concentrations in bronchoalveolar lavage fluid from patients with interstitial lung disease, *Clin. Sci. 81*:183 (1991).
84. P. Druvefors and K. Norrby, Molecular aspects of mast-cell-mediated mitogenesis in fibroblasts and mesothelial cells in situ, *Virchows Arch. B 55*:187 (1988).
85. M. M. Burger, Proteolytic enzymes initiating cell division and escape from contact inhibition of growth, *Nature 227*:170 (1970).
86. L. B. Chen and J. M. Buchanan, Mitogenic activity of blood components. I. Thrombin and prothrombin, *Proc. Natl. Acad. Sci. USA 72*:131 (1975).
87. S. J. Ruoss, T. Hartmann, and G. H. Caughey, Mast cell tryptase is a mitogen for cultured fibroblasts, *J. Clin. Invest. 88*:493 (1991).

88. T. Hartmann, S. J. Ruoss, W. W. Raymond, K. Seuwen, and G. H. Caughey, Human tryptase as a potent, cell-specific mitogen: role of signaling pathways in synergistic responses, *Am. J. Physiol. 262*:L528 (1992).
89. R. K. Boucek and N. L. Nobel, Histamine, norepinephrine and bradykinin stimulation of fibroblast growth and modification of serotonin response, *Proc. Soc. Exp. Biol. Med. 144*:929 (1973).
90. M. Jordana, A. D. Befus, M. T. Newhouse, J. Bienenstock, and J. Gauldie, Effect of histamine on proliferation of normal human adult lung fibroblasts, *Thorax 43*:552 (1988).
91. L. Baud, J. Perez, M. Denis, and R. Ardaillou, Modulation of fibroblast proliferation by sulfidopeptide leukotrienes: effect of indomethacin, *J. Immunol. 138*:1190 (1987).
92. D. W. Pennington, A. R. Lopez, P. S. Thomas, C. Peck, and W. M. Gold, Dog mastocytoma cells produce transforming growth factor β1, *J. Clin. Invest. 90*:35 (1992).
93. D. W. Pennington, S. J. Ruoss, and W. M. Gold, Dog mastocytoma cells secrete a growth factor for fibroblasts, *Am. J. Respir. Cell Mol. Biol. 6*:625 (1992).
94. E. T. Dayton, J. P. Caulfield, A. Hein, K. F. Austen, and R. L. Stevens, Regulation of the growth rate of mouse fibroblasts by IL-3-activated mouse bone marrow–derived mast cells, *J. Immunol. 142*:4307 (1989).
95. R. A. Briggaman, N. M. Schechter, J. Fräki, and G. S. Lazarus, Degradation of the epidermal–dermal junction by proteolytic enzymes from human skin and human polymorphonuclear leukocytes, *J. Exp. Med. 160*:1027 (1984).
96. B. L. Gruber, M. J. Marchese, K. Suzuki, L. B. Schwartz, Y. Okada, H. Nagase, and N. S. Ramamurthy, Synovial procollagenase activation by human mast cell tryptase: dependence upon matrix metalloproteinase 3 activation, *J. Clin. Invest. 84*:1657 (1989).
97. T. Furitsu, H. Saito, A. M. Dvorak, L. B. Schwartz, A. M. Irani, J. F. Burdick, K. Ishizaka, and T. Ishizaka, Development of human mast cells in vitro, *Proc. Natl. Acad. Sci. USA 86*:10039 (1989).
98. M. Castells and L. B. Schwartz, Tryptase levels in nasal-lavage fluid as an indicator of the immediate allergic response, *J. Allergy Clin. Immunol. 82*:348 (1988).
99. S. Juliusson, K. Holmberg, C. R. Baumgarten, M. Olsson, I. Enander, and U. Pipkorn, Tryptase in nasal lavage fluid after local allergen challenge: relationship to histamine levels and TAME-esterase activity, *Allergy 46*:459 (1991).
100. U. B. Rasmussen, V. Vourset-Craviari, S. Jallat, Y. Schlesinger, G. Pages, A. Pavirani, J. P. Lecocq, J. Pouyssegur, and E. Van Obberghen-Schilling, DNA cloning and expression of a hamster α-thrombin receptor coupled to Ca^{++} mobilization, *FEBS Lett. 288*:123 (1991).
101. T. K. H. Vu, D. T. Hung, V. I. Wheaton, and S. R. Coughlin, Molecular cloning of a functional thrombin receptor reveals a novel proteolytic mechanism of receptor activation, *Cell 64*:1 (1991).
102. J. Stürzebecher, D. Prasa, and C. P. Sommerhoff, Inhibition of human mast cell tryptase by benzamidine derivatives, *Biol. Chem. Hoppe-Seyler 373*:1025 (1992).

103. G. H. Caughey, W. W. Raymond, E. Bacci, R. J. Lombardy, and R. R. Tidwell, Bis(5-amidino-2-benzimidazolyl)methane and related amidines are potent, reversible inhibitors of mast cell tryptases, *J. Pharmacol. Exp. Ther. 264*:676 (1993).
104. E. Fink, R. Nettelbeck, and H. Fritz, Inhibition of mast cell chymase by eglin c and antileukoprotease (HUSI-I): indications for potential biological functions of these inhibitors, *Biol. Chem. Hoppe-Seyler 367*:567 (1986).

9

Secretory Endo- and Exopeptidases of Mouse Mast Cells: Structure, Genetics, and Regulation of Expression

Eric B. Springman and William E. Serafin
Vanderbilt University School of Medicine, Nashville, Tennessee

I. INTRODUCTION

Mouse mast cells are thought to express as many as 12 secretory granule endo- and exopeptidases, although this number depends on what one considers to be a distinct protease versus possible minor sequence differences simply due to interstrain variability. Mouse mast cell (MMC) secretory proteases include as many as eight chymases, two tryptases, a carboxypeptidase, and an aminopeptidase. The chymases and tryptases are serine endopeptidases with predicted chymotrypsin-like activity (cleaving carboxyl to amino acids with hydrophobic side chains) or trypsin-like activity (cleaving carboxyl to amino acids with basic side chains), respectively. Mouse mast cell carboxypeptidase A is designated "A" because it prefers to remove carboxyterminal aromatic and aliphatic amino acids, similar to pancreatic carboxypeptidase A. Mouse mast cell aminopeptidase preferentially removes single aminoterminal hydrophobic amino acids. The cDNA sequences have been determined for most of these enzymes. So far, all have proven to be closely related to, but distinct from, the analogous proteases found in the secretory granules of other hematopoietic cells, which include the granzymes of cytotoxic T lymphocytes and the neutrophil/monocyte secretory proteases elastase and cathepsin G. They are more distantly related to their counterpart digestive enzymes of the pancreas. Despite the large body of research pertaining to the many secretory peptidases of mouse mast cells, knowledge of the protein or peptide targets for these enzymes in vivo has remained elusive.

II. IDENTIFICATION AND NOMENCLATURE OF THE MOUSE MAST CELL PROTEASES

A. Chymases

Originally, the term *chymase* (1) was used to describe what we now call rat mast cell protease I (RMCP-I), because this enzyme has substrate preferences similar to those of pancreatic chymotrypsin. Subsequently, a second chymotrypsin-like protease was identified in rat mast cells (2), and chymotryptic activities were described in mouse (3), dog (4), and human mast cells (5). The story has become significantly more complicated in the mouse, where, depending on how you count, as many as eight chymases may exist. The reader should realize that while these murine proteins have all been labeled as chymases, few of them have had substrate preferences determined experimentally. Instead, these proteases are labeled as chymases based on their predicted substrate-binding pockets being similar to other chymases and to pancreatic chymotrypsin.

Mouse mast cell proteases (MMCP) 1, 2, 4, and 5 are chymases for which aminoterminal protein and full-length cDNA sequences have been determined (6–11). Each was initially identified as a protein and then subsequently sequenced as a cDNA. All four are known to be expressed in vivo in one or more mast cell populations. Four other potential mouse mast cell chymases have been found but are less well characterized. These are termed MMCP-3, MMCP-5′, MMCP-L, and chymase 2.

MMCP-3 was identified as a 28-kDa protein found in the peritoneal mast cells of BALB/c mice (8). The N-terminal sequence of MMCP-3 is distinct from that of other mouse mast cell proteases, but is highly similar to MMCP-4, showing a difference of only 1/21 amino acids. A cDNA for MMCP-3 has not been cloned. MMCP-3 and MMCP-4 do not represent allelic variability, because they were both obtained from highly inbred BALB/c mice. MMCP-5′ was identified in mouse peritoneal mast cells as a 30-kDa protein, distinct in size from the 31-kDa MMCP-5 found in these same cells (8). However, MMCP-5′ and MMCP-5 are identical in amino acid sequence over the 20 N-terminal amino acids that have been determined for MMCP-5′. In our opinion, MMCP-5′ and MMCP-5 probably represent varying degrees of posttranslational processing of the same protein, although they may possibly represent separate, highly conserved proteins.

MMCP-L (mouse mast cell protease-like) is not known definitively to be a mast cell protease or even to be an expressed protein in any cell type. MMCP-L was identified (9) when a BALB/c mouse genomic library was screened under high stringency conditions with an MMCP-4 cDNA probe. In addition to the MMCP-4 gene, the highly conserved but distinct MMCP-L gene also was cloned. The predicted mRNA and protein sequences for MMCP-L indicate a mast cell protease, since MMCP-L is far more similar to mast cell chymases than to other hematopoietic cell proteases. Nevertheless, neither MMCP-L nor its mRNA has been shown conclusively to be present in any mast cell type. Mouse cells that have been examined include Kirsten sarcoma virus–immortalized mast cells (KiSV-MCs), peritoneal mast cells, cultured bone marrow–derived mast cells (BMMCs), and mucosal mast cells elicited by parasite infestation (9). Although it is possible that the MMCP-L genomic DNA represents a pseudogene, we consider this unlikely due to the open reading frame throughout its predicted complete mRNA sequence.

Chu et al. (12) reported cDNA sequences for two mouse mast cell chymases. One of these, termed chymase 1, was identical to MMCP-5 with the exception of a single position over the entire 247-amino acid protein. Thus chymase 1 may actually be an allelic variant of MMCP-5, it may represent MMCP-5′, or it may be an entirely different gene product. Considering that the chymase 1 was cloned from a transformed mast cell line [Abelson murine leukemia virus–transformed mast cells (13)] that was derived from the NSF/N strain mouse, we consider the single amino acid substitution to represent allelic variability. Mouse chymase 2 reported by Chu and co-workers appears to represent an unspliced partial mRNA, lacking exons 1 and 2 and containing a portion of intron 2, by comparison to other mouse mast cell chymase genes. Chymase 2 most closely resembles MMCP-L, with 93% sequence identity for the portions identified. The reported chymase 2 sequence lacks nearly 70 amino acids from the amino terminus by comparison to other mast cell chymases. We consider chymase 2 possibly to be an additional mouse mast cell chymase. However, it remains to be determined if this transcript is actually appropriately spliced and translated in vitro or in vivo.

B. Tryptases

Two tryptases, MMCP-6 and MMCP-7, have been identified in mouse mast cells (14,15). MMCP-6 was initially identified as a 32-kDa protein present in KiSV-MC5 and in peritoneal mast cells (8). The cDNA and gene sequences of MMCP-6 have been determined, and the mRNA has been demonstrated to be expressed by in vitro–cultured BMMCs in addition to peritoneal mast cells and KiSV-MC5 (14). MMCP-7 (15) was initially identified using a strategy similar to that for the identification of MMCP-L. When a genomic DNA library was screened with the MMCP-6 cDNA, a cross-hybridizing but unique clone was identified. The mRNA sequence deduced for this gene indicated that it was a mast cell tryptase. Although MMCP-7 could not be identified in peritoneal mast cells, KiSV-MCs, or mucosal mast cells, it was shown to be present transiently in cultures of BMMCs, indicating that MMCP-7 is selectively expressed early in mast cell differentiation and can be lost with further maturation of either the connective tissue mast cell (CTMC) or MMC phenotype. In certain mouse strains MMCP-

7 is expressed in terminally differentiated perivascular mast cells (see Section V.B).

C. Exopeptidases

Two exopeptidases have been identified in mouse mast cells, a carboxypeptidase (16) and an aminopeptidase (17). The carboxypeptidase has been isolated and characterized (16,18). As with the carboxypeptidase from rat mast cells (19), the mouse mast cell carboxypeptidase has been termed mouse mast cell carboxypeptidase A (MC-CPA) on the basis of its specificity for carboxyterminal hydrophobic amino acid substrates (16). The MC-CPA protein sequence deduced from a cDNA places it in the zinc metallocarboxypeptidase family along with CPAs from rat (20) and human mast cells (21).

Aminopeptidase activity has been identified in murine mast cells and has been partially purified and characterized (17). This mast cell aminopeptidase (MC-AP) has been localized to the secretory granules of mouse BMMCs and appears to be a zinc metalloaminopeptidase with specificity for alanyl and leucyl residues. Upon activation of BMMCs, MC-AP is released in large complexes with chondroitin sulfate E and heparin proteoglycans along with the BMMC serine endopeptidases and MC-CPA. The MC-AP is probably related to the known zinc metalloexopeptidase leucine aminopeptidase, but no protein isolation or sequence information has been reported for the mast cell enzyme.

Interestingly, an aminopeptidase activity has been shown to be associated with leukotriene A_4 (LTA_4) hydrolases from guinea pig small intestine, as well as human erythrocytes and neutrophils (22–25). LTA_4 hydrolase is responsible for converting LTA_4 to LTB_4 in various inflammatory cells. Mouse BMMCs are know to produce LTB_4 upon activation (26), and therefore possess an LTA_4 hydrolase. We consider it possible that the BMMC aminopeptidase and LTA_4 hydrolase activities are due to the same enzyme. LTA_4 hydrolases are believed to be zinc metallohydrolases, and both the aminopeptidase and LTA_4 hydrolase activities are inhibited by bestatin. Human neutrophil LTA_4 hydrolase has specificity for Leu- and Lys-*p*-nitroanilides in the ratio of 10:1 (23). Similarly, mouse MC-AP has specificities for Leu- and Arg-*p*-nitroanilides in the ratio 10:1 (17). Mouse MC-AP and guinea pig small intestine LTA_4 hydrolases are inhibitable by phenanthroline derivatives. Since mouse MC-AP is localized to the secretory granules of BMMCs (17), any LTA_4 hydrolase activity that MC-AP possesses would presumably act upon extracellular LTA_4. There are precedents for this, because transcellular arachidonic acid metabolism does occur in other systems (27). At present MC-AP has only been studied in the mouse, where all mast cell populations examined contain this activity, including BMMCs (17), peritoneal mast cells (17), and KiSV-MCs (W.E. Serafin, unpublished). It remains to be seen whether mast cells from other species contain similar aminopeptidase activity.

III. STRUCTURE

The naturally occurring mature mast cell secretory proteases range in size from 26 to 36 kDa as determined by SDS-PAGE and range in pI value from 5.95 for MMCP-7 to 10.03 for MMCP-L (Table 1). The acidic pI values of the tryptases contrast sharply with the basic pI values of the chymases and carboxypeptidase. This is important, because charge interactions are thought to play a critical role in binding of these secretory proteases with proteoglycans (see Section III.E). All of the secreted peptidases of mouse mast cells for which the primary sequence is known are expressed as three-domain preproenzymes consisting of an N-terminal signal peptide followed by a propeptide and a catalytic domain. The dimensions of the propeptide and catalytic domains vary considerably among the different protease groups but are much less variable among the individual proteases within each group. For instance, the chymases have a propeptide of two amino acids, while both of the tryptases are believed to have propeptides consisting of 10 amino acids, and the carboxypeptidase propeptide is composed of 94 amino acids (Fig. 1). The catalytic domains vary from 226 amino acids for the chymases to 245 for the tryptases and 308 for the carboxypeptidase.

A. Primary Structure and Sequence Similarity

As with all known serine proteases, the mast cell chymases and tryptases contain the characteristic catalytic acid triad of His, Asp, and Ser residues. However, the substrate specificities of the mast cell serine proteases are different from their pancreatic homologs, as shown by their more limited cleavage of protein substrates and different reactivities toward protease inhibitors (38,39).

All murine mast cell chymases for which a full-length amino acid sequence has been determined share ≥50% identity (Table 2). The mouse chymases are closely related to the chymases of rat, dog, and human mast cells. Several of the chymases of different species are quite striking in their similarity. For instance, MMCP-4 shares 89% identity with RMCP-I, and MMCP-5 shares 73% identity with dog chymase and 75% with human chymase. These examples stand out from the typical sequence identities of 50 to 60%, which may indicate that they derived from a common ancestor prior to speciation, whereas others are more recent additions to the family. It is additionally interesting that mouse mast cells appear to have a higher propensity to generate new chymases than do either human or dogs. It may indicate that more chymases are as yet unidentified in the other species or that a selective pressure is acting in the mouse to generate and retain new chymases.

Among the known tryptases, the sequence identities between species are generally higher than for the chymases (Table 2). Several unique features are conserved among the tryptases, including four Trp residues which are predicted

Table 1 Size and Charge Properties of Mouse Mast Cell Secretory Peptidases

Class	Type	Designation	MW observed[a] (kDa)	MW predicted (kDa)	pI predicted[b]	Charge pH 5.5 net	Charge pH 7.4 net	Ref.
Endopeptidases	Chymase	MMCP-1	26	24.9	8.22	11	2	10
		MMCP-2	28	24.7	8.87	13	7	7
		MMCP-3	28	?	?	?	?	8
		MMCP-4	29	25.1	9.66	23	15	9
		MMCP-5	31	25.3	9.44	21	12	11
		MMCP-L	?	24.2	10.03	28	21	9
		Chymase-2[c]	?	?	?	?	?	12
	Tryptase	MMCP-6	32	27.4	6.50	8	-5	14
		MMCP-7	?	27.4	5.95	3	-10	15
Exopeptidases	Carboxypeptidase	MC-CPA	36	35.8	9.54	24	17	18
	Aminopeptidase	MC-AP	237	?	?	?	?	17

[a]All observed molecular weights determined by SDS-PAGE, except MMC-AP determined by gel filtration.
[b]Calculated using DNASTAR software and the reported amino acid sequence for the mature enzyme.
[c]The complete sequence of chymase-2 is not known.

	PROPEPTIDE	:	CATALYTIC DOMAIN
Mouse Mast Cell Protease-1	E E	:	I I G G V E A R + (218)
Mouse Mast Cell Protease-2	E E	:	I I G G V E A K + (216)
Mouse Mast Cell Protease-4	E E	:	I I G G V E S R + (216)
Mouse Mast Cell Protease-5	G E	:	I I G G T E C I + (218)
Mouse Mast Cell Protease-L	E E	:	I I G G V E S E + (214)
Human Chymase	G E	:	I I G G T E C K + (217)
Rat Mast Cell Protease-II	E E	:	I I G G V E S I + (219)
Human Neutrophil Cathepsin G	G E	:	I I G G R E S R + (227)
Human Neutrophil Elastase	S E	:	I V G G R R A R + (231)
Mouse Lymphocyte Granzyme-C	E E	:	I I G G N E I S + (220)
Mouse Lymphocyte Granzyme-D	E E	:	I I G G H V V K + (220)
Rat NK Cell Protease-1	G E	:	I I G G H E A K + (220)
Mouse Mast Cell Carboxypeptidase A	(84) + I E K Q F D V K D E	:	I A G R H S Y A + (300)
Human Mast Cell Carboxypeptidase A	(84) + I E K Q F D V K E D	:	I P G R H S Y A + (300)
Rat Mast Cell Carboxypeptidase A	(?) + E	:	I A G R H S Y A + (300)
Mouse Mast Cell Protease-6	A P R P A N Q R V G	:	I V G G H E A S + (237)
Mouse Mast Cell Protease-7	A P G P A M T R E G	:	I V G G Q E A H + (238)
Human Tryptase α	A P A P V Q A L Q Q A G	:	I V G G Q E A P + (238)
Human Tryptase β	A P A P G Q A L Q R V G	:	I V G G Q E A P + (237)
Dog Tryptase	Q A L Q R V G	:	I V G G R E A P + (237)

Figure 1 Activation sites of mast cell and selected hematopoietic cell secretory granule proteases. The junctions of the propeptide and catalytic domains are shown separated by a colon to mark the site of enzymatic cleavage. Additional amino- and carboxy-terminal residues are noted in parentheses. Human and dog tryptases have an indeterminant cleavage point between the signal sequence and the propeptide. The proposed propeptide of dog tryptase represents our interpretation of the sequence data. The references for the sequences are as follows: MMCP-1 (10), MMCP-2 (7), MMCP-4 (9), MMCP-5 (11), MMCP-L (7), human chymase (28), rat mast cell protease II (67), human neutrophil cathepsin G (30), human neutrophil elastase (31), mouse lymphocyte granzyme-C (32), mouse lymphocyte granzyme-D (33), rat NK cell protease 1 (34), mouse mast cell carboxypeptidase A (18), human mast cell carboxypeptidase A (21), rat mast cell carboxypeptidase (20), MMCP-6 (14), MMCP-7 (15), human tryptase α (35), human tryptase β/I-III (36,100), and dog tryptase (37). The complete aminoterminal sequence of rat MC-proCPA is not known.

to contribute to a hydrophobic patch on the surface of tryptases opposite the active site (40). Also, the tryptases contain an unusual, conserved proline-rich hydrophobic sequence LPPP(F/Y)PL which is thought to lie on the surface (40). The function of this hydrophobic surface feature is not known, but it may be a site of surface contact with other proteins or lipids. Alternatively, human tryptases are known to form tetramers (39), and this hydrophobic patch could serve in this capacity (see further details in Section III.E).

The carboxypeptidases of mouse (28), rat (20), and human (21) mast cells are all strikingly similar, with sequence identities ≥85% (Table 2). In addition to the highly conserved zinc-binding residues found in all zinc carboxypeptidases, all five cysteine residues in the MC-CPAs are also found in pancreatic

In the mouse, the amino acid sequence of MC-CPA is 53% identical to pancreatic carboxypeptidase B with but a single gap of one amino acid (18). Pancreatic carboxypeptidase B and mouse MC-CPA sequences are 63% identical considering conserved amino acid substitutions. Consequently, these two enzymes are likely to be similar in their tertiary structure. The three-dimensional structures of the catalytic portions of pancreatic CPAs and CPBs have been reported and are highly similar. More recently, Coll et al. (53) have reported a refined crystallographic structure (R = 0.169) of porcine pancreatic proCPB, and Guasch et al. (540 have reported a refined crystallographic structure (R = 0.198) of porcine pancreatic proCPA. Both porcine pancreatic procarboxypeptidases consist of two globular domains, a large catalytic domain with a smaller propeptide domain covering the active site. The two domains are connected by a flexible extended α-helical segment. In contrast to the serine proproteases, the active sites of proCPA and proCPB are fully formed and the catalytic domain does not undergo a refolding step after proteolytic removal of the propeptide domain.

In proCPB, the primary site of cleavage, which results in removal of the propeptide and activation of the carboxypeptidase catalytic domain, occurs at residue Arg^{A95} (Arg^{95} of the activation segment), which is the most exposed residue in the connecting loop. The tertiary structure of the CPB propeptide itself is simple and stable, in accord with its high melting temperature, 82°C, and its resistance to proteolysis. In proCPB, the backbone fold of the propeptide effectively excludes entrance of protein substrates into the active site. Additionally, the side chain Asp^{A41} projects into the active site and participates in a salt bridge with Arg^{145}, the residue responsible for stabilizing the carboxy terminus of substrate molecules during catalysis. This results in substrates being excluded from the catalytic site. These structural properties may explain the observations that the CPB propeptide does not act as a competitive inhibitor after proteolytic removal and that proCPB shows no activity even toward small substrates.

The activation mechanism is somewhat different in pancreatic proCPA, which does possess low activity and an unaltered K_m value toward small substrates even when the propeptide remains attached. The Asp–Arg salt bridge between the CPB-propeptide and its active site is not found in proCPA. Instead, side chains of the propeptide interact with residues of the substrate binding pocket via water-mediated hydrogen bonding (54). This may explain why proCPA retains low activity toward small substrates rather than being absolutely inactive. The pancreatic CPA propeptide acts as a competitive inhibitor of CPA activity even after it is proteolytically cleaved from the amino terminus.

The mouse MC-CPA propeptide shares 50% sequence identity with the CPB propeptide and the analogous aspartyl residue is conserved, suggesting that the mechanism of inhibition of pancreatic CPB by its propeptide and the Asp–Arg interaction between the propeptide and catalytic domains may also be pertinent to MC-CPA. Interestingly, the bridging Asp residue of the CPB propeptide is not conserved in human proMC-CPA.

The catalytic domain of porcine pancreatic CPB contains a pentacoordinate zinc ion, as is also seen in pancreatic CPA. This zinc ion is coordinated by a water molecule, His^{69}, His^{196}, and doubly coordinated by Glu^{72}. Although the zinc content of MC-CPA has not been determined directly, the zinc binding residues of pancreatic CPB are conserved in MC-CPA (His^{67}, His^{195}, and Glu^{70}) and in the human and rat MC-CPAs. The substrate specificity pocket of CPB, defined by residues Ser^{207}, Gly^{243}, Ala^{250}, and Asp^{255}, confers the preference for arginyl groups by CPB. Mouse MC-CPA, however, has a substrate specificity more akin to pancreatic CPA in its preference for aromatic and aliphatic side chains. In pancreatic CPA, the substrate-binding pocket is defined by residues Gly^{207}, Ile^{243}, Ala^{250}, and Ile^{255}. The corresponding substrate-binding residues in MC-CPA are Gly^{206}, Ile^{242}, Thr^{249}, and Leu^{254}, indicating substrate preference similar to but not identical to that of pancreatic CPA.

E. Superstructure

While stored in the granules and following exocytosis, murine mast cell secretory proteases have been shown to exist in macromolecular complexes (>20,000,000 Da) with secretory granule proteoglycans (16,17,55). Charge interactions are thought to play a role in formation of the proteoglycan–protease complexes. In murine mast cells, the secretory granules contain highly acidic heparin and chondroitin sulfate E glycosaminoglycans (55,56). The mouse mast cell secretory proteases range from mildly acidic (MMCP-6 and MMCP-7) to highly basic (e.g., MMCP-4, MMCP-5, MMCP-L, and MC-CPA) (Table 1). At the secretory granule pH of approximately 5.5 (57–60), all of the mouse mast cell secretory proteases carry a net positive charge, ranging from +3 for MMCP-7 to +28 for MMCP-L. At the extracellular pH of 7.4, however, the tryptases carry a net negative charge, -5 for MMCP-6 and –10 for MMCP-7. The human tryptases, which exhibit charge properties similar to those of the mouse tryptases, remain bound to proteoglycan at extracellular pH under physiological conditions (61,62). Consequently, binding of the human tryptases, and similarly the mouse tryptases, to proteoglycans is thought to involve more than simple charge interactions (39,40). Human tryptase is known to exhibit a quaternary structure, existing as a tetramer that is stabilized by the presence of heparin (61). Tetramer formation is likely to result from direct tryptase–tryptase contact via conserved surface hydrophobic regions rather than from binding to heparin (39,40). Persistence of the tetramer is required for the tryptase to remain active (61).

Interchain disulfide bonds are frequently observed in multiprotein complexes, acting to maintain the association of certain complex components and often required for activity of the subunits. Two members of the mouse chymase family, MMCP-5 and MMCP-L, as well as human chymase possess unpaired cysteine residues which are potentially available for formation of interchain disulfide bonds. According to their deduced amino acid sequences, these chymases contain seven cysteine residues in their zymogen and mature forms. The odd cysteines are probably those at position 7 in MMCP-5 and human chymase and position 65 in MMCP-L. Similarly, all three known mast cell CPAs, as well as pancreatic CPB, contain five cysteine residues. By analogy to pancreatic CPB, the non-disulfide-bonded cysteine is Cys^{290} of the mast cell CPAs (20). It is not yet known whether these odd cysteine residues actually participate in interchain disulfide bonds. However, modeling of granzymes A and B suggests that the odd cysteine residues of these murine CTL secretory proteases, corresponding, respectively, to residues 80 and 75 of RMCP-II, are accessible for the formation of disulfide-bonded homodimers (46).

Recent evidence suggests that MMC-proCPA may exist in multiprotein complexes independent of proteoglycan, although the nature and number of the components are not yet known (E. B. Springman, M. M. Dikov, and W. E. Serafin, unpublished). There is precedent for superstructure among the pancreatic carboxypeptidases. ProCPAs from many species may form complexes with proproteinase E and/or chymotrypsinogen C, although the proportion of monomeric proCPA to complex proCPA varies in different species. In contrast, pancreatic proCPBs are only known to exist as monomers (63).

IV. GENETICS

A. Chromosomal Localization

The chromosomal locations (47) and genomic organization (9–11,14,15) of several mouse mast cell proteases has been determined. The genes for four chymases, MMCP-1, MMCP-2, MMCP-4, and MMCP-5, are located on chromosome 14, closely linked to a complex encoding four secretory granule serine proteases from cytotoxic T lymphocytes (granzymes). Three of the chymase genes, MMCP-1, MMCP-2, and MMCP-5, are within 850 kb of each other. However, these four chymases are not coordinately expressed despite their chromosomal proximity, as evidenced by the fact that MMCP-1 and MMCP-2 are expressed in mucosal mast cells (MMCs) from helminth-infected mice, whereas MMCP-4 and MMCP-5 are not. Conversely, MMCP-4 and MMCP-5 are expressed in peritoneal connective tissue mast cells (CTMCs), whereas MMCP-1 and MMCP-2 are not.

The gene for the tryptase MMCP-6 is located on chromosome 17 and that for MC-CPA on chromosome 3 (47). Both of these proteases are selectively expressed in mouse mast cell populations that also express the chymases MMCP-

4 and MMCP-5. Clearly, coordinate and differential expression of the various mouse mast cell endo- and exopeptidases is not simply dependent on chromosomal location.

B. Genomic Organization and Evolution

The genes for the chymases MMCP-1 (10), MMCP-2 (47), MMCP-4 (9), and MMCP-5 (11), the putative chymase MMCP-L (9), and the tryptase MMCP-7 (15) are composed of five exons, in contrast to six exons for the gene of the tryptase MMCP-6 (14). In all eukaryotic serine protease genes known to date, including the mast cell chymases and tryptases, each member of the catalytic triad His, Asp, and Ser is encoded on a different exon. The three amino acids of the catalytic triad are located on exons 2, 3, and 5 in the chymases and in MMCP-7 and on exons 3, 4, and 6 in MMCP-6.

The genetic lineage of the mast cell proteases remains unclear. The five-exon structure of the chymases contrasts markedly with the seven-exon structure of pancreatic chymotrypsin and elastase but is comparable to the five-exon structure of the neutrophil cathepsin G and elastase, cytotoxic T-lymphocytes granzymes, and adipsin [a serine protease found in adipocytes that is identical to complement factor D (64,65)]. The five-exon structure and the intron placement found in this group of proteases suggest divergence of the hematopoietic cell proteases from trypsin after divergence of trypsin from chymotrypsin. Nevertheless, the sequence similarities of the chymases with chymotrypsin places the chymases in a branch derived from chymotrypsin, not trypsin.

MMCP-6 is the only serine protease of this group whose gene contains six exons; still, its genomic organization is nearly identical to that of the other mast cell serine proteases. The single difference between the gene for MMCP-6 and the other mast cell serine proteases is that an extra intron (intron 1) is inserted between the 5′ untranslated region and the translation initiation site of the of the MMCP-6 gene. A similar sequence is present in the MMCP-7 gene. However, this portion of the transcript is not removed during splicing, due to a point mutation in the acceptor splice site. Thus MMCP-7 has a long 5′-untranslated region in its transcript. The gene structure and sequences of the tryptases suggest that they diverged from trypsin after divergence of trypsin and chymotrypsin. Obviously, the true genetic lineage of the mammalian mast cell serine proteases remains in doubt; an examination of nonmammalian mast cell serine proteases (if any) might shed more light on the evolutionary heritage of these enzymes.

The mouse MC-CPA gene has not yet been sequenced fully but is expected to be similar to the human MC-CPA gene, which has been sequenced (66). It is composed of 11 exons, similar to the rat pancreatic carboxypeptidases A_2 and B. Recently, an extensive phylogenetic tree of the carboxypeptidases based on minimal nucleotide replacements was proposed by Reynolds et al. (66). The most probable tree placed the mammalian MC-CPAs in a branch derived from a common precursor with the pancreatic CPBs prior to mammalian speciation. The pancreatic CPB/MC-CPA precursor shared a common ancestor with the pancreatic CPAs.

C. Genomic Regulatory Elements

While a substantial amount of work has been devoted to defining the differences in protease expression in various mast cell populations and the particular cytokines that control these differences (see Section V), less work has been peformed at the level of the genetic control mechanisms. Using chimeric constructs with a linked receptor gene (human growth hormone), Sarid and co-workers (67) identified a mast cell–specific enhancer in the 5′-flanking region between bases −1578 and −1387 of the RMCP-II gene (29). This enhancer is similar to the enhancer core consensus sequence in the genes of pancreatic serine proteases (68) and is thought to interact with a mast cell–specific trans-acting factor (67). Two additional regions of similarity with the pancreatic serine protease enhancer core were identified with the 5′ flank of the RMCP-II gene, extending from bases −1431 to −1410 and from −73 to −51, as well as an apparent negative regulatory region between bases −1385 and −1030. A region from bases −1574 to −1492 of the 5′-flanking sequence of the MMCP-4 gene is highly similar to the RMCP-II enhancer region between −1578 and −1387 and the serine protease enhancer core consensus (9). We searched the other murine mast cell serine protease genes for a corresponding sequence but were unable to identify any regions of high sequence similarity with this upstream RMCP-II enhancer element in MMCP-1, MMCP-2, or MMCP-6. The reported sequences of the MMCP-5 and MMCP-7 genes reach only 178 and 80 nucleotides 5′ of the transcription initiation site, respectively, which is not far enough to rule out the presence of this upstream element in these genes.

The region of the RMCP-II gene between bases −73 and −51 is composed of 23 bp, which are 87% identical to the pancreatic serine protease enhancer core as aligned by Sarid et al. (67). A stretch of 23 bp corresponding to the −73 to −51 stretch of RMCP-II is also found in the 5′-flanking sequence of MMCP-1 extending from bases −75 to −53 (10) (Fig. 2). Because of this, we searched for this element in the other murine mast cell serine protease genes and found that it is also present in the identical position in MMCP-2, MMCP-4, and MMCL-L, but not in MMCP-5, MMCP-6, or MMCP-7 (Fig. 2). We also found that the RMCP-II, MMCP-1, MMCP-2, MMCP-4, and MMCP-L genes contain a nearly identical stretch of 19 nucleotides that is highly similar to a putative enhancer element found in the human neutrophil elastase (HNE) gene (Fig. 2) (69). This potential regulatory element is immediately 3′ to the pancreatic serine protease core–like element (Fig. 2). The HNE-like element was not found in MMCP-5, MMCP-6, or MMCP-7. The actual importance of any of these putative enhancer elements in expression of the MMCP-1, MMCP-2, and MMCP-4 genes has not yet been determined experimentally.

Genetic control of expression of MC-CPA in mouse and rat mast cells has been partially characterized. GATA-binding proteins are zinc-finger-type transcription factors that bind enhancers with a conserved motif of (A/T)GATA(A/G). Such enhancers have been found in genes from a number of hematopoietic cells, including erythroblasts, megakaryocytes, and mast cells (70–72). These cells also express one or more of the GATA-binding proteins GATA-1, GATA-2, and GATA-3. Mouse (73) and human (21) MC-CPA genes contain the conserved GATA motif AGATAA starting 42 nucleotides upstream of the MC-CPA gene transcription initiation site. Zon and co-workers (73) investigated expression of mRNAs for the GATA-binding proteins in mouse BMMC, KiSV-MC, P815 mouse mastocytoma cells, and rat basophil leukemia cells (RBL). They found that BMMC, KiSV-MC, and RBL contained GATA-1 mRNA and produced MC-CPA mRNA. P815 cells contained no GATA-1 mRNA and contained minimal MC-CPA mRNA. GATA-2 mRNA was found in BMMC, KiSV-MC, RBL, and P815. BMMC, KiSV-MC, and RBL contained minimal GATA-3 transcript. These authors assessed the effects of GATA-1 by permanently transfecting P815 cells with an expression vector for the GATA-1 cDNA. They then used a human growth hormone reporter gene linked to the putative MC-CPA enhancer to assess the effects of the expressed GATA-1 protein on expression

```
                    PAN-1 (consensus)
                    GTCACCTGTGCTTT-T--CCCTG
                                             CCCCTTCTCCCCCTTTTCA
                                                  HNE (putative)
RMCP-II    -73      CAGACCTGTGGTTTCTAGCCCTTCCTTTCTCCCACTTGTGA     -33
MMCP-1     -75      CAGATCTGTGGTTCCCAGCCCTcCCTTTCTCCCACGTTGGA     -35
MMCP-2     -75      CAGATCTGTGGTTCCCAGCCCTGCCTTTCTCCCACaTTGGA     -35
MMCP-4     -75      CAGATCTGTGGTTCCCAGCCCTGCCTTTCTCCCACGTTGGA     -35
MMCP-L     -75      CAGATCTGTGGTTCCCAGCCCTGCCTTTCTCCCACGTTGGA     -35
MMCP-5              NOT PRESENT                   NOT PRESENT
MMCP-6              NOT PRESENT                   NOT PRESENT
MMCP-7              NOT PRESENT                   NOT PRESENT
```

Figure 2 Putative enhancer region in 5′ sequences of mouse mast cell secretory protease genes: similarities to a pancreatic enhancer core consensus (PAN-1) and a putative enhancer region of human neutrophil elastase (HNE). Note that within the mouse genes nearly 100% sequence identity occurs. Only one nucleotide in MMCP-1 and one nucleotide in MMCP-2 differ from the other sequences. These differing nucleotides are shown in lowercase letters. Sequence data are from: PAN-1 consensus (99); HNE putative enhancer region (69); alignment of PAN-1 and RMCP-II (67); MMCP-1 (10); MMCP-2 (47); MMCP-4 (9); MMCP-5 (11); MMCP-L (9); MMCP-6 (14); MMCP-7 (15).

of the MC-CPA gene. An 18-fold increase in human growth hormone expression was seen with the GATA-1-expressing P815 cells compared to the wild-type P815. Mobility shift assays using wild-type and mutant (AGTTAA) GATA regions in an MC-CPA promoter segment (bases –27 to –62) were used to probe nuclear extracts for GATA activities, providing evidence for GATA-1 binding to the MC-CPA enhancer. A putative GATA-binding sequence, TGATAA, is located beginning at base –68 in the 5′-flanking sequence of the MMCP-5 gene (11). We have examined all other rodent chymase genes for which sequences are available. They all have potential GATA-binding sites. For the MMCP-1, MMCP-2, MMCP-4, and MMCP-L genes the sequence TGATAA is located at bases –103 to –98. For the RMCP-II gene, an identical sequence is located at bases –101 to –96. The functional properties of these elements have not been determined experimentally for any of the chymases.

V. CELLULAR REGULATION OF EXPRESSION

A. Differential Expression of Proteases in Mast Cell Populations

Immature mast cells contain a panel of proteases which is markedly different from that of mature mast cells, and distinct subsets of secretory proteases are expressed in different mature mast cell subsets. Different expression of secretory proteases among different murine mast cell populations suggests distinctive functions for these proteases in physiology and pathology, indicates coordinated regulation of secretory protease gene expression, and underscores differences in pathophysiologic roles between the various mast cell populations. In the rat and mouse, two mast cell subclasses have been described based on their staining properties, anatomical locations, and protease phenotypes. These subclasses are the mucosal mast cells (MMCs) and CTMCs. The original description of these subclasses and their staining properties are reviewed by Galli and Austen (74). With the advent of mouse mast cell protease cDNA probes (7) and antisera (75,76), substantial knowledge has been gained concerning the protease phenotyping of different mast cell populations.

Due to the extreme degree of sequence identities among the various proteases, caution must be used in interpreting results with antisera and cDNA probes unless they are carefully screened for single-protease specificity. For example, MMCP-4 was originally reported to be present in MMCs of helminth-infected mice (9). Subsequently, based on the results of RNAase protection assays, MMCP-4 appeared not to be present in the MMCs. Instead, it appears that the particular cDNA probe initially used for MMCP-4 cross reacts to a limited extent with the highly similar MMCP-1 mRNA. This was not known until the MMCP-1 cDNA was cloned and sequenced (77).

B. Protease Phenotyping of In Vivo–Derived Mast Cells

Several studies have determined the secretory proteases expressed by the MMCs of helminth-infected mice (Table 3). These MMCs are well granulated, contain a moderate amount of histamine compared with connective tissue cells, and stain with Alcian blue but not safranin. Using immunohistochemical methods, Newlands et al. (75) localized a chymotrypic serine protease, later found to be the same as MMCP-1 (6,8,10), to mast cells in the intestinal mucosa of *Trichinella spiralis*–infected mice. MMCs from the small intestines of *Nippostrongylus brasiliensis*–infected mice express MMCP-2 (7) and possibly MMCP-L mRNAs (9) but not mRNAs for MMCP-6 (14) or MC-CPA (9). MMCs from *T. spiralis*–infected mice do not to express MMCP-4 (77), MMCP-5 (11), MMCP-6 (14), or MMCP-7 (15) mRNAs. A more recent study of *T. spiralis*–infected mice has identified MMCP-2 mRNA in total gut RNA by Northern blotting, whereas MMCP-6 and MMCP-7 mRNAs were absent (15). Although it has not been determined explicitly, it is likely that MMCs elicited by *T. spiralis* infection are of the same phenotype as those elicited by *N. brasiliensis* infection. Thus the MMCs of helminth-infected mice are currently thought to have the protease phenotype MMCP-1$^+$, MMCP-2$^+$, MMCP-4$^-$, MMCP-5$^-$, MMCP-6$^-$, MMCP-7$^-$, MC-CPA$^-$ (Table 3).

Much less is known regarding the resident MMCs (i.e., MMCs present in the absence of helminth infection). Newlands et al. (75) localized MMCP-1 to intestinal and gastric MMCs of uninfected mice using immunohistochemistry. Other studies, all examining the total RNA from uninfected gut, have re-

Table 3 Protease Phenotyping of Mast Cells Derived In Vivo

		BALB/c		WBB6F1/J-+/+	
Strain Marker	MMC (helminth infected)	CTMC serosal	CTMC perivascular	CTMC serosal	CTMC perivascular
MMCP-1	+	–	–	–	–
MMCP-2	+	–	–	+	+
MMCP-3	?	+	?	?	?
MMCP-4	–	+	+	+	+
MMCP-5	–	+	+	+	+
MMCP-6	–	+	+	+	+
MMCP-7	–	–	+	+/–	+
MC-CPA	–	+	+	+	+
MC-AP	?	+	?	?	?
Safranin	–	+	+	+	+

ported the presence of mouse chymase-2 (12) and the absence of MMCP-2 (7), MMCP-4 (9), MMCP-5 (11,12), MMCP-6 (14), MMCP-7 (15), and MC-CPA (9), resulting in the phenotype of mouse chymase-2^+, MMCP-1^+, MMCP-2^-, MMCP-4^-, MMCP-5^-, MMCP-6^-, MMCP-7^-, MC-CPA^- for the resident MMCs in uninfected mice. It must be cautioned that the absence of a particular mRNA in the total gut RNA of uninfected mice may be due to the small numbers of MMCs that reside in uninfected gut rather than true phenotypic differences between resident and infection-elicited MMCs. At this point it is not possible to state definitively whether helminth infection alters the protease phenotype of MMC as well as stimulating their marked multiplication.

CTMCs present a histochemical phenotype distinct from resident or infection-elicited MMC (74). They hold numerous, large secretory granules that contain abundant histamine and stain strongly with safranin. Serosal mast cells (SMCs) isolated by peritoneal lavage are the prototypic CTMCs and have been shown to express a different protease phenotype than MMC (Table 3). As determined by RNA blotting, SMCs expression MMCP-4 (9), MMCP-5 (11,76), MMCP-6 (14,15), and MC-CPA (9,18). MMCP-7 transcripts are absent or minimally present in SMCs, depending on the mouse strain examined (15). MMCP-2 transcripts are absent from BALB/c SMC (7,78) but are abundantly present in $WBB6F_1$/J+/+ mice (78). Thus SMCs are currently thought to have the protease phenotype MMCP-1^-, MMCP-2^- (or MMCP-2^+ depending on the strain), MMCP-4^+, MMCP-5^+, MMCP-6^+, MMCP-7^- (or minimally MMCP-7^+, depending on the strain), MC-CPA^+.

Perivascular mast cells (PvMCs) of skin are CTMCs by histochemical characterization, but they may represent a distinct subtype of CTMCs based on protease phenotyping (Table 3). Peritoneal SMCs and PvMCs are similar in containing MMCP-4, MMCP-5, MMCP-6, and MC-PA (78). However, Stevens and co-workers (78) have recently reported differential expression of MMCP-7 between SMCs and PvMCs of BALB/cJ mice. While BALB/cJ SMCs were found to express no MMCP-7 mRNA, PvMCs of the of the same mice were found to express abundant MMCP-7 mRNA. This finding suggests that the SMCs and PvMCs of BALB/cJ mice represent two distinct CTMC populations. In $WBB6F_1$/+ mice there were smaller differences between SMCs and PvMCs in that the SMCs expressed minimal but detectable MMCP-7 transcripts. These findings indicate that it is at least possible for mice to maintain mast cell populations with protease phenotypes beyond the simple either/or model of MMCs vs. CTMCs and suggest caution in making statements about phenotypes of mouse mast cell poulations unless the particular mouse strain is specified.

C. Protease Phenotyping of In Vitro–Derived Mast Cells

The development of murine Kirsten sarcoma virus–immortalized mast cell lines (KiSV-MC) (79) provided a factory for the initial isolation and characterization of most of the murine mast cell secretory granule proteases (8). A particular line, KiSV-MC5, has been examined extensively (Table 4) due to its CTMC-like appearance, high granularity, high histamine content, and Safranin-positive staining. KiSV-MC5 have been shown to express MMCP-2 (8,9,15), MMCP-4 (8,9), MMCP-5 (8,11,76), MMCP-6 (8,14,15), and MC-CPA (9) but not MMCP-7 (15). Thus the KiSV-MC5 protease phenotype is MMCP-2$^+$, MMCP-4$^+$, MMCP-5$^+$, MMCP-6$^+$, MMCP-7$^+$, MC-CPA$^+$. When this phenotype was first determined, the expression of MMCP-2 in a CTMC-like mast cell was considered aberrant and possibly due to an artifact from the retrovirus infection. However, we now know that peritoneal CTMCs from certain mouse strains can indeed express abundant MMCP-2 mRNA, and we consider it likely that KiSV-MC5 truly represents a transformed CTMC containing a normal protease phenotype for that mast cell type. Since protease phenotyping has not been done for peritoneal CTMCs from DBA/2 mice (the stain providing the target cells giving rise to KiSV-MC5) we cannot yet be certain that this is the case.

Immature mast cells can be cultured from bone marrow precursors in vitro in the presence of IL-3 (80). These BMMCs stain Safranin-negative, have small numbers of secretory granules, and contain low amounts of histamine and carboxypeptidase activity (16,81,82). In vitro–differentiated BALB/c BMMCs grown for 3 weeks in 50% WEHI-3B-conditioned medium as a source of IL-3 express MMCP-5, MMCP-6, MMCP-7, and MC-CPA mRNAs (11,14,15,18). However, after 3 weeks in culture the level of MMCP-7 mRNA declines sharply, whereas the levels of MMCP-5 and MC-CPA remain constant and MMCP-6 declines only slightly (15). Thus MMCP-7 mRNA is only transiently expressed at high levels. MMCP-1 and MMCP-2, both found in intestinal MMCs, are not

Table 4 Protease Phenotyping of Mast Cells Derived In Vitro

Marker	KiSV-MC	BMMC WEHI-3B	BMMC rIL-3	BMMC con A	BMMC SCF:IL-9	BMMC IL-3:Il-10	BMMC rSCF
MMCP-1	–	–	–	?	+	+	–
MMCP-2	+	–	–	+	+	+	–
MMCP-3	+	?	?	?	?	?	?
MMCP-4	+	–	–	?	+	–	+
MMCP-5	+	+	+	+	+	+	+
MMCP-6	+	+	+	+	?	+	+
MMCP-7	–	+	?	?	?	?	?
MC-CPA	+	+	+	+	+	+	+
MC-AP	+	+	?	?	?	?	?
Safranin	+	–	–	–	–[a]	?	+

[a]Approximately 8% of the BMMC treated with SCF and IL-9 were Safranin$^+$.

D. M. Rennick, Interleukin 10: a novel stimulatory factor for mast cells and their progenitors, *J. Exp. Med. 173*:507 (1991).

87. N. Ghildyal, D. S. Friend, C. F. Nicodemus, K. F. Austen, and R. L. Stevens, Reversible expression of mouse mast cell protease-2 mRNA and protein in cultured mast cells exposed to IL-10, *J. Immunol. 151*:3206 (1993).
88. K. K. Eklund, N. Ghildyal, K. F. Austen, and R. L. Stevens, Induction by IL-9 and suppression by IL-3 and IL-4 of the levels of chromosome 14–derived transcripts that encode late-expressed mouse mast cell proteases, *J. Immunol. 151*:4266 (1993).
89. O. N. White, Steel locus defines new multipotent growth factor, *Cell 63*:5 (1990).
90. D. E. Williams, J. Eisenman, A. Baird, C. Rauch, K. Van Ness, C. J. March, L. S. Park, U. Martin, D. Y. Mochizuki, H. S. Boswell, G. S. Burgess, D. Cosman, and S. D. Lyman, Identification of a ligand for the c-*kit* proto-oncogene, *Cell 63*:167 (1990).
91. N. G. Copeland, D. J. Gilbert, B. C. Cho, P. J. Donovan, N. A. Jenkins, D. Cosman, D. Anderson, S. D. Lyman, and D. E. Williams, Mast cell growth factor maps near the steel locus on mouse chromosome 10 and is deleted in a number of steel alleles, *Cell 63*:175 (1990).
92. J. G. Flanagan and P. Leder, The *kit* ligand: a cell surface molecule altered in steel mutant fibroblasts, *Cell 63*:185 (1990).
93. K. M. Zsebo, J. Wypych, I. K. McNiece, H. S. Lu, K. A. Smith, S. B. Karkare, R. K. Sachdev, V. N. Yuschenkoff, N. C. Birkett, L. R. Williams, V. N. Satyagal, W. Tung, R. A. Bosselman, E. A. Mendiaz, and K. E. Langley, Identification, purification, and biological characterization of hematopoietic stem cell factor from Buffalo rat liver–conditioned medium, *Cell 63*:195 (1990).
94. F. H. Martin, S. V. Suggs, K. E. Langley, H. S. Lu, J. Ting, K. H. Okino, C. F. Morris, I. K. McNiece, F. W. Jacobsen, E. A. Mendiaz, N. C. Birkett, K. A. Smith, M. J. Johnson, V. P. Parker, J. C. Flores, A. C. Patel, E. F. Fisher, H. O. Erjavec, C. J. Herrera, J. Wypych, R. K. Sachdev, J. A. Pope, I. Leslie, D. Wen, C. H. Lin, R. L. Cupples, and K. M. Zsebo, Primary structure and functional expression of rat and human stem cell factor DNAs, *Cell 63*:203 (1990).
95. K. M. Zsebo, D. A. Williams, E. N. Geissler, V. C. Broudy, F. H. Martin, H. L. Atkins, R.-Y. Hsu, N. C. Birkett, K. H. Okino, D. C. Murdock, F. W. Jacobsen, K. E. Langley, K. A. Smith, T. Takeishi, B. M. Cattanach, S. J. Galli, and S. V. Suggs, Stem cell factor is encoded at the *SI* locus of the mouse and is the ligand for the c-*kit* tyrosine kinase receptor, *Cell 63*:213 (1990).
96. E. Huang, K. Nocka, D. R. Beier, T.-Y. Chu, J. Buck, H.-W. Lahm, D. Wellner, P. Leder, and P. Besmer, The hematopoietic growth factor KL is encoded by the *SI* locus and is the ligand of the c-*kit* receptor, the gene product of the *W* locus, Cell 63:225 (1990).
97. D. M. Anderson, S. D. Lyman, A. Baird, J. M. Wignall, J. Eisenman, C. Rauch, C. J. March, H. S. Boswell, S. D. Gimpel, D. Cosman, and D. E. Williams, Molecular cloning of mast cell growth factor, a hematopoietic that is active in both membrane bound and soluble forms, *Cell 63*:235 (1990).
98. M. F. Gurish, N. Ghildyal, H. P. McNeil, K. F. Austen, S. Gillis, and R. L.

Stevens, Differential expression of secretory granule proteases in mouse mast cells exposed to interleukin 3 and c-*kit* ligand, *J. Exp. Med. 175*:1003 (1992).

99. A. M. Boulet, C. R. Erwin, and W. J. Rutter, Cell-specific enhancers in the rat exocrine pancreas, *Proc. Natl. Acad. Sci. USA 83*:3599 (1986).

100. P. Vanderslice, S. M. Ballinger, E. K. Tam, S. M. Goldstein, C. S. Craik, and G. H. Caughey, Human mast cell tryptase: multiple cDNAs and genes reveal a multigene serine protease family, *Proc. Natl. Acad. Sci. USA 87*:3811 (1990).

101. M. M. Dikov, E. B. Springman, S. Yeola, and W. E. Serafin, Processing of procarboxy-peptidase A and other zymogens in murine mast cells, *J. Biol. Chem. 269*:25897 (1994).

10

Mast Cell Chymases in Helminthosis and Hypersensitivity

Hugh R. P. Miller
The University of Edinburgh, Edinburgh, Scotland

John F. Huntley and George F. J. Newlands
Moredun Research Institute, Edinburgh, Scotland

I. INTRODUCTION

Recruitment of mast cells is a feature of intestinal nematode infection and has been described in many species, including humans (1), but the functional role of the recruited cells has not been defined. Because the production of parasite-specific IgE is a hallmark of helminthosis and a large number of studies have implicated immediate hypersensitivity reactions in the rejection of parasites (2,3), the role of mast cells in the protective response has been investigated intensively. The debate as to whether mast cells participate in immune responses against helminths is based largely on (a) enumerating cells in the lesion, (b) comparing responses in mast cell deficient/sufficient mice, (c) monitoring the accumulation or release of putative mast cell–derived mediators, or (d) comparing the effects of various mast cell–blocking or mast cell–stabilizing drugs on worm expulsion.

While it is now generally agreed that mast cell accumulation is common to most helminth infections (1) and that it is highly regulated by T cells (1–4), there is debate about the precise timing of mastocytosis in relation to worm expulsion. Furthermore, systemic infection with some tissue-dwelling helminths, where the parasites can survive for long periods, can be associated with mastocytosis (5,6). Under these circumstances mast cells presumably have no protective role. Conversely, the immune elimination of nematodes from mast

highly soluble chymase MMCP-1 is expressed predominantly (>99%) in the small intestine of the mouse (27,30). The intraepithelial murine mast cell is sparsely granulated but fluoresces strongly and rather diffusely when detected with FITC-labeled sheep anti-MMCP-1, whereas CTMCs are not labeled (Fig. 4) (27). The immunofluorescent observations, together with the quantitative data on the distribution of MMCP-1 in normal and parasitized mice (Table 5), show that this protease is expressed almost exclusively in enteric MMCs. Recent analyses of the distribution of MMCP-1 mRNA in normal and *T. spiralis*–infected murine intestine confirmed that there is abundant transcription of this gene in parasitized mice (43). However, only weak hybridization is detected with the MMCP-1 probe in normal mice (43). Similarly, Northern blots demonstrated abundant transcripts for MMCP-2 in parasitized mouse intestine but only weak hybridization in controls (44). As yet no translated product for the MMCP-2 gene has been identified in either normal or parasitized mice. Compared with the immunohistochemical and ELISA data (Table 5), the Northern hybridizations apparently do not adequately detect low levels of expression of MMCP-1.

There is also expression of an RMCP I-like granule protease in murine MMC which is in the form of discrete granules (9). As mentioned earlier, the anti-RMCP-I antibody absorbed against MMCP-1 identifies MMCP-4 but not MMCP-1 on Western blots (27). Murine MMC may, therefore, contain a very insoluble form of MMCP-4 because attempts to detect MMCP-4 in Western blots after homogenization and extraction of parasitized intestine in 2 *M* NaCl were not successful (Newlands and Miller unpublished observation). An alternative but less likely possibility, because of the poor amino acid sequence homologies and predicted high solubility of MMCP-2 (25), is that the anti-RMCP-I antibody cross-reacts with MMCP-2. The most recent studies using gene probes to iden-

Table 5 Concentrations of MMCP-1 and Number of Mast Cells in *N. brasiliensis*–infected and Normal Mice[a]

	Control		Infected	
	Jejunum	Tongue	Jejunum	Tongue
MMCP-1 (μg/g)	13 + 3.6	0	183 + 23.7	0.5 + 0.12
Mast cell/vcu				
Toluidine blue	1 + 0.2	A*	7 + 0.7	A*
Fluorescence	1 + 0.3	0	6 + 0.3	0

Source: Data from Ref. 27.
[a]Jejunal mast cell counts are expressed as cells per villus/crypt unit (vcu). A*, mast cells were abundant but were not counted.

tify MMCP transcripts suggest that MMCP-4 is not expressed by enteric MMC (43); therefore, the nature of RMCP-I-like granule protein in murine MMC requires further analysis.

The distribution of tryptase and chymase in human mast cells is discussed in detail elsewhere (see Chapter 7 and Ref. 45), but it is important to note that the enteric and pulmonary mast cells contain tryptase but little chymase, whereas the intestinal submucosal and dermal mast cells contain both tryptase and chymase. This contrasts with the mouse and rat, where tryptase is apparently not expressed by MMCs in the gut (25,46) but is present in some connective tissue mast cells with, in the rat, 7% of dermal mast cells and 63% of peribronchial mast cells containing this serine protease (46).

3. *Sheep*

As already mentioned, SMCP is present in gastric and enteric mast cells, but it is not yet known whether this protease or an antigenically similar enzyme is located in nonmucosal sites. For example, some reactivity has also been observed in CTMCs and basophils in sections of ovine skin, although these immunoreactions required the mild pretreatment of sections with trypsin (J. F. Huntley, H. R. P. Miller, and Jenkinson, unpublished observations). Partial proteolytic digestion may unmask cross-reactive epitopes, and these preliminary findings suggest that ovine CTMCs and basophils contain antigenically similar proteases that have yet to be characterized. Importantly, ovine BMMCs contain a 31,000-Da putative protease which, antigenically, is similar to SMCP (33). Purified MMCs isolated from parasitized sheep contain only the 28,000-Da chymase, SMCP (33). This would suggest that ovine BMMCs are heterogeneous in their expression of granule chymases. Cross-species reactivity also occurs, with strong labeling of MMCs in bovine and caprine gastrointestinal tissue sections using anti-SMCP antibodies (J. F. Huntley and H. R. P. Miller, unpublished observations).

B. Parasitized Host

As discussed in Section I, mast cell recruitment is a consistent feature of gastrointestinal helminth infection in rodents and ruminants. In addition, mast cell recruitment occurs in the gut and at other sites during infection with some larval cestodes (*Taenia taeniaeformis, Mesocestoides corti*) (5,6) and with trematodes such as *Fasciola hepatica* and *Schistosoma mansoni* (1,47). However, there are major species differences. For example, rats infected with *S. mansoni* develop a pronounced hepatic mastocytosis, whereas mice do not (47). Similarly, infection of rats with the hookworm-like migratory nematode *Nippostrongylus brasiliensis* results in pulmonary mastocytosis (48), whereas no such reaction occurs in *Nippostrongylus*-infected mice (H. R. P. Miller and G. F. J. Newlands, unpublished). In humans there have been relatively few reports of intestinal mastocytosis associated with helminth infections (49,50), but this probably reflects a lack of detailed investigations rather than an intrinsically different biological response.

1. *Gastrointestinal and Pulmonary Helminthosis*

a. Rat. More than 99% of the mast cells recruited to the gastrointestinal mucosa of parasitized rats have the RMCP-II^{+}/RMCP-I^{-} phenotype (36–38) and the concentrations of RMCP-II are highly correlated with the density of mast cells in the mucosa (51,52). Mast cells in gastric submucosa and mesenteric lymph node are heterogeneous and three protease phenotypes can be identified: -RMCP-I^{+}/RMCP-II^{-}, RMCP-I^{-}–/RMCP-II^{+}, and RMCP-I^{+}/RMCP II^{+} (38). Infection with *N. brasiliensis* substantially increases the proportion of RMCP-I^{+}/RMCP-II^{+} mast cells in the node, at the apparent expense of RMCP-I^{+}/RMCP-II^{-} mast cells (Table 3) (53). Similarly, an increased proportion of RMCP-I^{+}/RMCP-II^{+} mast cells and a reduced proportion of of RMCP-I^{+}/RMCP-II^{-} cells are found in lung (53), presumably in response to the early transpulmonary migration of the parasite. These results suggest that protease expression by mature mast cells might be modified by infection, and since the concentration of RMCP-II in lung homogenates increases almost tenfold whereas the RMCP-I content is virtually unaltered, there is a possibility that mature RMCP-I^{+}/RMCP-II^{-} cells can be induced to express RMCP-II. Similar changes were noted in peritoneal mast cells, where on day 10 of infection with *N. brasiliensis,* 100% of peritoneal mast cells were RMCP-I^{+}/RMCP-II^{+} (Fig. 6), and this contrasted with peritoneal mast cells from normal rats, which were 100% RMCP-I^{+}/RMCP-II^{-} (H. R. P. Miller and G. F. J. Newlands, unpublished). The significance of these phenotypic changes are discussed later in this chapter.

Infection of rats with *N. brasiliensis* is also associated with altered and/or increased expression of mast cell granule proteases in liver, thymus, uterus, brain, and other tissues (38). Most of these changes are relatively trivial compared with the levels of RMCP-II in the gut. Nevertheless, they suggest that a parasite with a hookworm-like migratory life cycle can influence the expression of mast cell granule proteases in a wide variety of tissues. As yet no comparable study has been carried out on the effect of nonmigratory intestinal nematodes.

b. Mouse. Infection with intestinal nematodes is associated with substantial recruitment of intraepithelial mast cells in the murine gut together with increased intestinal concentrations of MMCP-1 (30,54). Infection with *T. spiralis* resulted in intraepithelial MMCs in the gut which were strongly, although diffusely, labeled with anti-MMCP-1 antibody conjugates, whereas RMCP-I–like immunoreactivity was discrete and often confined to only a few granules within the cell (9). Mast cell numbers increased tenfold and the concentrations of

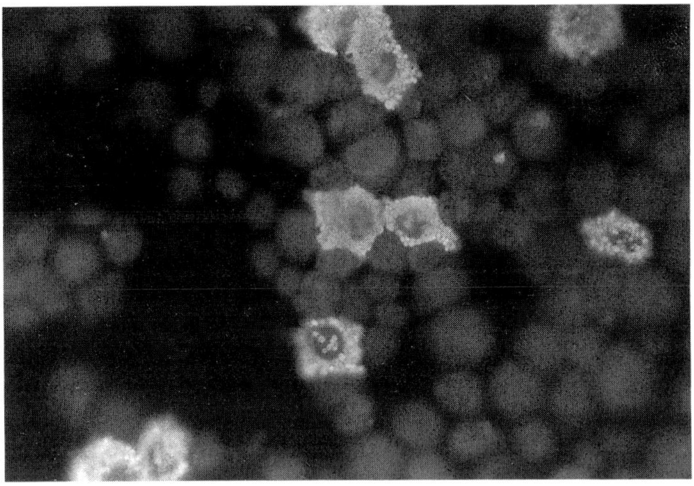

(a)

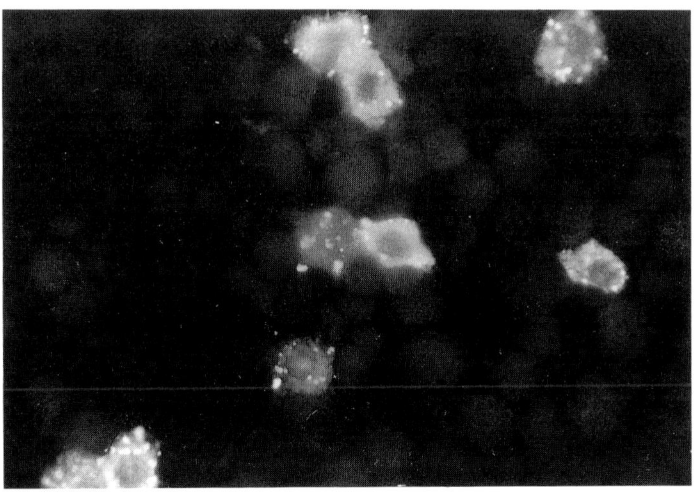

(b)

Figure 6 Localization of RMCP-I and RMCP-II by paired immunofluorescence in cytosmear preparations of rat peritoneal mast cells isolated 10 days after infection with 3000 *N. brasiliensis* L_3 (a and b), or after 7 days of stimulated growth in vitro with rrSCF (100 ng/mL) and rrIL-3 (10 U/mL; c and d). (a) and (c), RMCP-II (FITC) and (b) and (d), RMCP-I (TRITC) and shows CTMCs expressing both RMCP-I and RMCP-II. Some depletion of RMCP-I is evident in cultured PMCs.

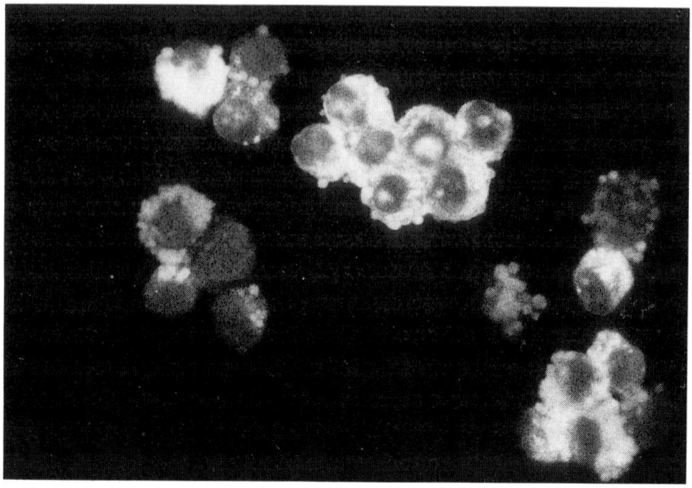

(c)

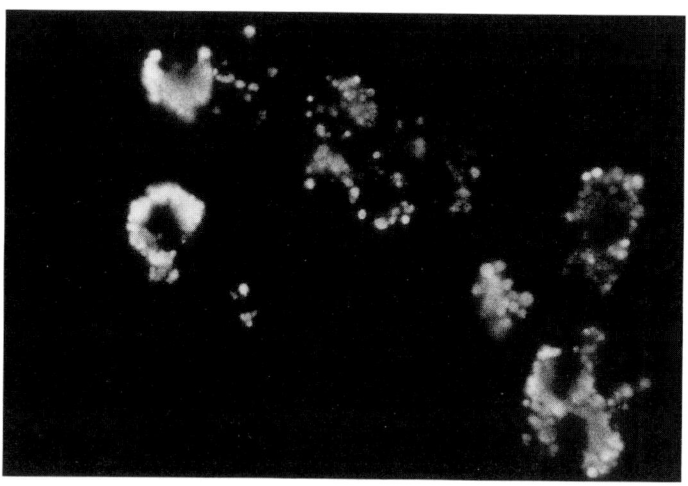

(d)

Figure 6 Continued.

MMCP-1 increased up to 6000-fold (30), suggesting that the concentration of MMCP-1/cell was greatly increased during infection with *T. spiralis*. These results are consistent with Northern hybridization data (43).

c. Sheep. Recruitment of ovine mast cells occurs more slowly than in rodents and requires repeated oral challenge with infective larvae over a period of 10 to 12 weeks (55). Daily infection with the gastric nematode *Haemonchus contortus* resulted in a sevenfold increase in mast cells in the gastric mucosa and a 570-fold increase in the concentration of sheep mast cell protease (SMCP) (35,56). This may reflect a greatly increased concentration of enzyme per cell or may be an artifact associated with the difficulty, in normal sheep, of detecting low concentrations of SMCP in the presence of plasma proteins (35). Immunohistochemistry confirms that SMCP is a product of ovine MMCs (32).

In summary, very substantial increases in the MMC-specific chymases RMCP-II, MMCP-1, and SMCP were detected during gastrointestinal helminthosis. In the rat and sheep, RMCP-II and SMCP appear to be the only chymases expressed by the recruited cells, whereas murine MMCs expressed an additional RMCP-I-like granule protein. It will be interesting to determine whether comparable responses will eventually be observed in parasitized human subjects.

2. *Hepatic/Pancreatic Helminthosis*

Infection of laboratory rats with the trematode *S. mansoni* results in attrition and death of the parasite 21 to 28 days later during the hepatic phase of infection before egg laying commences (57). The demise of the parasite is associated with substantial recruitment of hepatic mast cells (Table 6) (47). The protease phenotype of the recruited cells is predominantly RMCP-I$^-$/RMCP-II$^+$, although a scattering of RMCP-I$^+$/RMCP-II$^+$ mast cells is also present. Despite the migration of schistosomulae through the lungs, pulmonary mastocytosis is not pronounced and is associated with a relatively slight increase in concentration of RMCP-II (47). The hepatic phase of infection is also associated with intestinal mastocytosis (100% RMCP-I$^-$/RMCP-II$^+$) (47).

Infection of mice with *S. mansoni* produces a patent infection and eggs are sequestered into the liver and intestinal tract. Hepatic mastocytosis does not occur over the first 49 days (Table 6) and the few mast cells that are present lack MMCP-1. However, 35 and 49 days after infection and in association with egg deposition in the gut wall, there is an intestinal mucosal mastocytosis together with a substantial increase in the tissue content of MMCP-1 (Table 6) (47). The contrasting responses in the two species are probably a function of distinctive T-cell responses and emphasize that closely related species can mount profoundly different mast cell responses against the same parasite.

In the mouse, responses against *S. mansoni* are initially regulated by the T_H1 subset of helper T cells (58). The cytokines released by T_H1 cells include

Table 6 Distribution of Mast Cells and of MMCP-1 and RMCP-II in Mice and Rats, Respectively, Infected with *S. mansoni*[a]

	Rat				
	Gut		Liver		
	Mast cells/ vcu[b]	RMCP-II (μg/g)[c]	Mast cells/ field[d]	RMCP-II (μg/g)	Serum RMCP-II/ (ng/mL)
Control	6.8 ± 0.6	140 ± 5.8	2.6 ± 0.4	0.9 ± 0.3	157 ± 12
Day 28	13.6 ± 0.5[e]	222 ± 31.9	26.2 ± 3.5[e]	126 ± 32.5[e]	1076 ± 153[e]
Day 35	14.3 ± 0.9[e]	284 ± 20.9[e]	147.1 ± 18.6[e]	297 ± 61.4[e]	827 ± 12.0

	Mouse				
	Mast cells/ vcu	MMCP-1 (μg/g)	Mast cells/ field	MMCP-1 (μg/g)	MMCP-1 (ng/mL)
Control	0.02 ± 0.1	0.5 ± 0.1	0.0	0.0	0.01 ± 0.006
Day 42	2.8 ± 0.08[e]	86.6 ± 59[e]	0.0	0.05 ± 0.01	0.05 ± 0.01
Day 49	17.0 ± 1.7[e]	1083 ± 208[e]	0.26 ± 0.04	1.4 ± 0.2[e]	5.4 ± 0.76[e]

Source: Adapted from Ref. 47.
[a]Fischer F 344 inbred females (6 to 8 weeks old) and C57/BL6 female mice (6 weeks old) were infected with doses of 2500 and 200 Cercariae, respectively. RMCP-II and MMCP-1 were quantified by ELISA.
[b]Mast cells counted per villus/crypt unit.
[c]μg RMCP-II/g wet weight of tissues.
[d]Mast cells counted at a magnification of ×10.
[e]Significantly different ($p < 0.005$ to < 0.001) from control value by Student's *t* test.

interferon-γ and IL-2 (59). When egg laying commences, T_H2 cells respond to egg antigens (58) and the cytokines secreted by this T-cell subset include IL-4, IL-5, IL-9, and IL-10 (4,59). Both subsets produce IL-3 following antigenic stimulation and there are both in vivo and in vitro data which suggest that T_H2 cells promote eosinophilia and mast cell responses (4,59). It is significant, therefore, that the intestinal mucosal mastocytosis and expression of MMCP-1 are associated with the deposition of eggs in the murine gut (4,7). Comparable data are not available in the rat, but it is probable, in view of the hepatic mastocytosis, that there are T_H2-like responses in this species which are directed against worm rather than egg antigens.

Infection of rats with the larval metacestode stage of the tapeworm *Mesocestoides corti,* which parasitizes the liver and also causes peripancreatic granuloma formation, is associated with mastocytosis at both sites of infection (6). Phenotyping studies using immunoperoxidase rather than paired fluorescence suggested that RMCP-I and RMCP-II were expressed in a mutually exclusive fashion in both liver and pancreatic granulomata, with 35% of cells in the pancreas and 68% in the liver expressing RMCP-II. Concentrations of RMCP-II in the liver and pancreas increased from undetectable in normal rats to, respectively, 200 and 350 μg/per gram of soluble protein 20 weeks after infection. When the metacestodes were inoculated subcutaneously, none of the mast cells recruited to skin were RMCP-II$^+$. No RMCP-II was detected in the dermis of subcutaneously infected rats (6). These data suggest that mast cells with distinct protease phenotypes can coexist in some tissues but that RMCP-II is not expressed by mast cells in the dermis, regardless of the stimulus.

IV. REGULATION OF THE EXPRESSION OF GRANULE CHYMASES

Two main hypotheses can be advanced to explain the presence of distinct mast cell phenotypes in humans and rodents. The first, which is supported by data from a number of in vivo and in vitro experiments, is that, during differentiation, the mast cell precursor has the potential to express a variety of granule products. In this case, expression is regulated by signals in the immediate vicinity of the precursor cell (1,14). The second hypothesis is that the progenitors develop along lineage-specific pathways when, after leaving the bone marrow, they lodge in the appropriate tissue (1,14). For example, progenitors of MMC would differentiate only if they were seeded into mucosal tissue.

Much of the argument in favor of the first hypothesis is based on an analysis of the proteoglycan content of bone marrow–derived mast cells (BMMCs) grown in the presence or absence of fibroblasts and on elegant in vivo studies by Kitamura and colleagues (60,61). The latter showed that the transfer of peritoneal mast cells from +/+ littermates into the gastric wall of mast cell–deficient W/W^v mice results in the subsequent proliferation of MMC-like cells in the gastric mucosa (60,61). For reasons discussed below, the converse experiment, in which MMC-like BMMCs developed into CTMC-like cells after intraperitoneal injection into W/W^v mice (60,61), is more difficult to interpret.

The first hypothesis is also supported, to some extent, by recent studies on the role of cytokines with, for example, different combinations of cytokines promoting the in vitro development of populations of BMMCs with distinctive patterns of protease transcription (43,44). However, as discussed below, this approach is open to the criticism that there may be substantial heterogeneity in the starting population (28) and the addition of different cytokines may select for different precursor populations.

A. Regulation of Mast Cell Granule Chymase Expression In Vitro

The earliest in vitro studies where mast cells were co-cultured with fibroblasts and which indicated that there were distinct populations of mast cells were those of Ginsburg et al., who described a "fibroblast-derived" CTMC-like cell and a "lymph node–derived" MMC-like cell (62). Subsequently, it was shown that bone marrow was the source of mast cell precursors, and in the presence of the appropriate conditioned medium, derived from activated T cells, BMMCs grew readily in the absence of fibroblasts (63). Comparison of the development of murine and rat BMMCs quickly established that both contained neutral serine proteases (64–66). The pattern in rat BMMCs was relatively simple in that substantial quantities of RMCP-II were expressed in a uniform fashion throughout the BMMC population (65). Little or no RMCP-I was detected in rat BMMCs and because the cells were phenotypically RMCP-I$^-$/RMCP-II$^+$, they were considered to be the in vitro equivalent of rat MMCs (65).

Neutral protease expression in murine BMMCs was complex, with at least four serine esterases identified by labeling with the inhibitor [^{3}H]DFP (66). Western blotting with rabbit anti-MMCP-1 revealed four immunoreactive proteins in BMMCs but only one in CTMCs (28). The latter, with an M_r value of 28,000, and a polypeptide with an identical M_r value from BMMCs were the sole proteins detected with anti-RMCP-I (absorbed against MMCP-1 and specific for MMCP-4). Therefore, the BMMCs grown in the presence of conditioned medium probably contained MMCP-4 and, by ELISA, contained significant quantities of MMCP-1. Paired immunofluorescence revealed that protease expression was heterogeneous, some cells containing no immunoreactive granule proteases, others having a MMCP-1$^+$/RMCP-I$^-$ phenotype, and a third population was MMCP-1$^+$/RMCP-I$^+$ (28). The content of MMCP-1 in murine BMMCs varied according to the culture conditions, being less than 25 ng per 10^6 BMMCs grown in IL-3 derived from WEH1-3B cells and 52-118 ng/10^6 BMMCs grown in spleen cell–conditioned medium (28). These results suggested that conditioned medium contained additional factors that promoted the expression of MMCP-1, whereas IL-3 alone was a poor stimulus for MMCP-1 expression, in agreement with studies on the transcription of MMCP-1 mRNA (43).

The results quantifying translated proteins are at variance with the extensive RNA blotting studies in which MMCP transcripts were analyzed. For example, neither MMCP-1 nor MMCP-4 mRNA transcripts are detectable in murine BMMCs (43), and it has been suggested that both are expressed late in the differentiation of maturing mast cells. The data derived by immunofluorescence and Western blotting show that murine BMMCs are a mixed population and can express MMCP-1 and, probably, MMCP-4 in the early stages of differentiation. Addition of the *kit* ligand, stem cell factor (SCF), to murine BMMCs promotes the transcription of MMCP-4 mRNA in vitro (67), whereas the combination of IL-4 and IL-10, in the absence of IL-3, induces the expression of MMCP-1 and MMCP-2 mRNAs (43,44). However, as mentioned earlier, it is also possible that these cytokines promote the recruitment and growth of subpopulations of BMMC. The same reservation must also apply to the in vitro transfer of BMMCs grown from +/+ mouse bone marrow and transferred intraperitoneally into W/W^v mast cell–deficient littermates (60,61).

The expression of RMCP-II by rat BMMCs occurs at the earliest stage of mast cell development. The RMCP-II content of BMMCs rose from less than 1 pg per cell on day 0 to 7, 15, and 22 pg per cell on days 2, 4, and 6, respectively, of culture in the presence of T cell–conditioned medium (68). Studies have yet to be carried out in the rat to show, conclusively, that IL-3 is alone responsible for the development of mast cells in rat bone marrow and for the expression of RMCP-II. Nevertheless, this cytokine does support the survival and growth of differentiated BMMCs (69).

B. In Vitro Secretion of Granule Chymases

An additional complication in the analysis of granule proteases in vitro is the observation that some chymases are apparently secreted constitutively. The original work of McMenamin and Haig (68,70) is supported by a more recent study, comparing the effect on rat BMMCs of recombinant rat SCF (rrSCF) and rrIL-3 with that of T cell–conditioned medium (71). The secretion of RMCP-II into culture supernatants was modified by the presence of these two cytokines. The further growth of already mature BMMCs (>98% mast cells) increased approximately 10- and 100-fold 7 and 14 days, respectively, after addition of 200 ng/mL of rrSCF. However, within 7 days these cells were substantially depleted of RMCP-II, there being microgram quantities of this protease in the supernatant (71). In the presence of rrIL-3, cell growth was minimal but the BMMCs retained or slightly increased their content of RMCP-II over the first 7 days of culture. Compared with rrSCF, rrIL-3 induced relatively little secretion of RMCP-II (71). In conclusion, therefore, rrSCF promotes growth of already mature BMMC (71) and depletion of granule RMCP-II, whereas IL-3 maintains cell viability and retention of RMCP-II (69,71).

When purified (>99%) rat peritoneal mast cells were subjected to comparable in vitro treatment with rrSCF (200 ng/mL) there was a fivefold increase in cell growth, but the concentration of RMCP-I measured by ELISA 10 days later was constant at approximately 12 to 15 μg per 10^6 cells (71). The cells did not survive until day 20 unless rrSCF was supplemented with rrIL-3, and in this instance the content of RMCP-I was depleted by 50% on day 5, by 60% on day 10, and 90% on day 20. Interestingly, the peritoneal mast cells maintained in the presence of rrSCF and rrIL-3 expressed significant quantities of intracellular RMCP-II (7 μg per 10^6 cells on day 10) and constitutively released RMCP-II but not RMCP-I into the supernatant. Approximately 80% of the mast cells were RMCP-I$^+$/RMCP-II$^+$ by paired fluorescence (Fig. 6) (71).

These in vitro results suggest that rrSCF promotes the growth of CTMC over the short term but that growth and survival are improved in the combined presence of rrSCF and rrIL-3. For both cell types rrIL-3 maintained, or increased, the concentrations of intracellular RMCP-II. The latter result is at variance with the Northern hybridization studies of murine BMMCs where IL-3 apparently suppressed the transcription of MMCP-1 mRNA (43). One of the major features in this, and in previous studies of rat mast cells, was the secretion by BMMCs and peritoneal cells of the soluble chymase RMCP-II. This in vitro secretory response may have in vivo parallels (see below).

While the data above suggest that cytokines modulate the spontaneous secretion of RMCP-II in rat BMMC, IgE-mediated release of this chymase is also readily demonstrated. Between 13 and 18% release of RMCP-II occurred over a 15-min time course when BMMC, sensitized with IgE, were challenged with anti-IgE. Release of RMCP-II was highly correlated with that of another granule constitutent, β-hexosaminidase (72). As yet, the influence of cytokines, such as SCF, on the magnitude and time course of secretion has not been determined.

C. Regulation of Mast Cell Granule Chymases in Vivo

Relatively little is known about the in vivo regulation of mast cell chymases except that they appear to be expressed in a tissue-specific fashion (1). Studies by Kitamura and colleagues (60,61) indicate that the local environment plays a crucial role. The transfer, mentioned earlier, of CTMCs from the peritoneal cavity of +/+ mice into the gastric wall of their mutant, mast cell–deficient W/W^v littermates resulted in the appearance of typical MMCs in the mucosa. Even when a single peritoneal mast cell was transferred, a cluster of gastric MMCs was evident in recipient W/W^v mice (61). Since there is normal expression of SCF in W/W^v mice, expansion of the transferred mast cell population may be regulated via the c-*kit* receptor, a tyrosine kinase, which is fully functional on +/+ mast cells but not on W/W^v haemopoietic cells (73). Recent evidence also suggests that the c-*kit* receptor is central to the development of mucosal mastocytosis in *Trichinella*-infected mice. Treatment of mice with anti-c-*kit* antibodies ablates mucosal mastocytosis and reduces substantially the levels of intestinal and plasma MMCP-1 (74). Coincidentally, this treatment abrogates the development of resistance to infection.

Intravenous injection of rrSCF (25 μg/kg per day) into normal rats for 14 days augmented the numbers of MMCs in gastric and ileal mucosae of recipients and increased the concentration of RMCP-II in these two tissues. Similarly, the numbers of CTMCs in different sites were increased, as was the tissue content of RMCP-I (75). These findings are consistent with in vitro data on mast cell growth but conflict with observations on the depletion of RMCP-II from BMMCs supplemented with SCF (71). In this instance it is possible that local regulatory stimuli in the gut maintained the RMCP-II content of MMC in vivo.

The role of T cell–derived IL-3 and IL-4 in the recruitment of MMCs during infection is documented extensively (4), but despite the experiment describing the effect of anti-c-*kit* (74), it is not clear at what stage SCF is involved in mucosal mastocytosis. Treatment of mice with anti-IL-3 or with anti-IL-4 depresses *Nippostrongylus*-induced intestinal mastocytosis by 40% and 50%, respectively. Mastocytosis was reduced by 90% by combined treatment with the two anti-cytokine antibodies (76). Antibody treatment for this and for the c-*kit* experiment (74) was given at the time of infection, and it is possible that the consequent downregulation of IL-3 and IL-4 and of c-*kit*–expressing cells prevented the early recruitment of mast cell precursors. Since, in vitro, either fibroblasts (62,63) or cytokines are essential for the survival of BMMCs, it is likely that, in vivo, the mast cell population is maintained by one or several cytokines. This hypothesis is supported by studies on the effects of immunosuppressants on mast cell population in vivo (52,56).

Experiments in rats and sheep have established that corticosteroid decimates the mucosal mast cell population within 24 h of treatment (52,56). In vitro studies show that the effect of steroid is not directly on rat BMMCs but on the T cells from which conditioned medium is harvested, presumably by downregulating cytokine production (77). This is further supported by the in vivo effects of cyclosporin A, a drug known to suppress cytokine production by T cells, which also depletes MMCs in the rat (39). Unlike the responses in rats (52) and sheep (56), steroid treatment of mice has no effect on MMCs, nor is the content of MMCP-1 in the gastrointestinal mucosa altered by this treatment (78). This surprising difference in corticosteroid sensitivity between two closely related species is difficult to explain unless murine T cells in the gut are steroid resistant. Given the unique location of murine MMCs within the epithelium, it is possible that intraepithelial T cells provide the necessary local source of cytokines. Extrapolation from results of in vitro studies (43,44), suggest that IL-4 and IL-10 are the most likely candidates.

D. Secretion of Mast Cell Chymases In Vivo

The solubility of RMCP-II and MMCP-1 is such that if either protease were to be released during intestinal immunological reactions, it should, theoretically, be detectable in the blood. The advantage of detecting RMCP-II and MMCP-1 systemically is that since both proteases are expressed by MMCs, gut hypersensitivity responses to parasitic infection should be monitored selectively in vivo.

1. *Gastrointestinal Nematode-Induced Secretion of Chymases*

The systemic release of RMCP-II was initially described in *N. brasiliensis*–primed rats challenged intraduodenally with live worms or with soluble worm antigen (42). Maximal release (1150 ng of RMCP-II/mL serum) occurred 1 h after challenge with 1200 live worms. In comparison with live worm challenge, soluble worm antigen increased the systemic secretion of RMCP-II tenfold (10,500 ng/mL) when given intraduodenally and 100-fold (122,000 ng/mL) when given intravenously (42). These data are entirely consistent with a worm-induced local immediate hypersensitivity reaction in the gut and of an anaphylactic reaction involving an expanded population of MMC and parasite-specific IgE. Quantification of RMCP-II and mast cells at different sites in the gut as well as lung, mesenteric lymph node, trachea, and thymus established that the jejunum and stomach were the only tissues that were depleted significantly following the anaphylactic reaction (51). The presence of 1 to 2 mg of RMCP-II in the gut lumen within 5 min of challenge adds further support to the view that the gut is likely to be the major source of systemically secreted RMCP-II after challenge in previously sensitized rats (79).

Naive rats infected for the first time with *N. brasiliensis* or *T. spiralis* do not develop a significant mastocytosis until day 10 or 12 after infection, yet maximal systemic release of RMCP-II precedes intestinal mastocytosis (Fig. 5) by several days (80). The stimulus for secretion has not been identified, but it may not be mediated via IgE since parasite-specific IgE is not usually detected systemically until after the worms have been expelled (2).

Infection of mice with *T. spiralis* is associated with the systemic release of MMCP-1, and the kinetics are similar to that of RMCP-II in *Trichinella*-infected rats (30,54). In this instance, however, maximal release of protease coincides with the peak of mastocytosis (Table 7). In both species the maximal

Table 7 Mast Cell Counts in the Small Intestine and Concentrations of MMCP-1 in Sera from *T. spiralis*-Infected Mice[a]

Day	MMC/vcu[b]	Serum MMCP-1 (μg/mL)
0	2.2	0.02
6	6.6	1.64
9	20.8	9.48
12	11.6	8.89
15	15.5	7.12
20	14.1	2.43
32	4.8	0.64

Source: Data from Ref. 30.
[a]Results are expressed as mean of three animals per group.
[b]vcu, Villus/crypt unit.

systemic secretion of MMC-derived proteases coincides with worm expulsion (30,54,80). Comparison of NIH and B10 mice showed that the former, which expel *T. spiralis* rapidly, mounted an early and substantial (>15 μg MMCP-1/mL serum) secretory response, whereas B10 mice, which reject *T. spiralis* slowly, responded late, with a relatively low systemic release of MMCP-1 (<3 μg/mL) (54). These differences between the two strains were also reflected in amounts of MMCP-1 released into the gut lumen (54). The question, in both rats and mice, is whether the secretion of chymases during primary intestinal infection is mediated by mast cell–sensitizing IgE or IgG2a and worm antigen, or by activated parasite-specific T cells.

Systemic release of SMCP during infection of the ovine gastric mucosa with the nematodes *H. contortus* and *O. circumcincta* was demonstrable only in primed sheep challenged with a minimum of 50,000 larvae (35). Given that priming involved repeated infection over 10 to 12 weeks, this is likely to be an IgE-mediated response. Again, in sheep, systemic release of SMCP coincides with rapid elimination of the challenge infection (35). The larvae are excluded from the mucosa, being unable to penetrate the gastric pits within the first 4 h of challenge (81). The concentrations of SMCP in blood and in gastric lymph (<2 ng/mL) are apparently very low, and this is almost certainly due to the presence of inhibitors in ovine lymph plasma and serum. Dilution of serum to 1/145 to 1/676 and of lymph plasma to 1/32 to 1/100 was required before addition of 20 ng SMCP/mL could be detected (50% endpoint) by ELISA (35). Fractionation of ovine serum revealed both high (60,000 to >440,000 Da) and low (60,000 to 20,000 Da) inhibitors (82). However, measurable quantities of SMCP were discovered in the superficial mucus of immune sheep and the amount was significantly increased in immune, challenged animals (56).

In summary, the immune expulsion of gastric and intestinal nematodes is associated with the local release from the gut of MMC-specific granule proteases in rat, mouse, and sheep. The probable functional significance of this secretory response is discussed briefly later in the chapter.

2. *Secretion of Chymases After Hepatic Infection*

The recruitment of mast cells to the livers of *S. mansoni*–infected rats is accompanied by systemic secretion of RMCP-II (47). Concentrations of RMCP-II in serum, like those in *Nippostrongylus*-infected rats, are maximal before mast cell recruitment at the site of infection reaches a peak (Table 6). There is, in this instance, the added complication that recruitment of mast cells to, and increased concentrations of RMCP-II in the intestine accompany the systemic secretory response (Table 6) (47). It is possible, therefore, that worm antigens, transported via the bile duct, stimulate intestinal MMCs to secrete RMCP-II during the early stages of infection. These findings may be paralleled in infected mice, where mast cell recruitment to the gut is accompanied late in infection by a small but significant increase in the concentration of MMCP-1 in serum (Table 6) (47).

It is, perhaps, intriguing to note that hepatic and pancreatic infection with *M. corti* is not associated with detectable levels of RMCP-II in blood (6). Unfortunately, samples from the latter experiments were assayed by a less sensitive competitive ELISA and it is possible that a low-grade systemic secretion of RMCP-II occurred. However, this would have been at a substantially reduced rate compared with the systemic responses measured by the same ELISA in *T. spiralis*– and *N. brasiliensis*–infected rats (80). This would suggest that unlike nematode infections, *M. corti* does not activate the locally recruited mast cells.

V. FUNCTIONAL SIGNIFICANCE OF GRANULE CHYMASES IN HELMINTHIASIS

A variety of functions have been proposed for granule chymases. These include physiological regulation of the extracellular environment, modulation of inflammation, and secretagogue activity (1). Some or all of these functions may be important in the response to helminths. As yet, however, most of the functions can only be guessed at. Anaphylactic release of RMCP-II is associated with shedding of enteric epithelium, increased enteric vascular permeability (79), and focal disruption of the epithelial basement membrane (83). Our hypothesis is that RMCP-II and MMCP-1, when released, may, together with other mediators, alter mucosal permeability, increase the translocation of anti-worm antibodies, and contribute to an increasingly hostile environment that is inimical to worm survival (1,79). However, this hypothesis is based on experimental in vivo data showing correlations between the distribution of secreted proteases in blood and in the intestinal lumen. More definitive evidence is required to distinguish the effects of chymases from those of PAF, LTC_4, histamine, and free radicals in the development of enteric pathology.

Interpretation of the functional significance of secreted chymases is complicated by the fact that both RMCP-II (84) and MMCP-1 (27,85) are rapidly and virtually irreversibly inhibited by α_1-proteinase inhibitor (α_1-PI). The k_{ass} value for RMCP-II against rat α_1-PI is $1.65 \times 10^5\ M^{-1}\ s^{-1}$ and the k_i value is 0.36 p*M* (84). These figures are in good agreement with those derived for MMCP-1 against human α_1-PI (Table 2) (27). The physiologically significant activity of MMCP-1 and RMCP-II once secreted is, therefore, likely to be highly localized and transient unless sufficient protease is released to overwhelm the α_1-PI in the surrounding tissues. The concentration of α_1-PI would, in addition, be subject to variation depending on whether an acute-phase response is mounted during infection.

Other approaches are now being employed to explore the functions of granule chymases. The first, an ex vivo perfusion system, is being developed to monitor the IgE-mediated release of RMCP-II into the vasculature and intestinal lumen. Perfusion reduces the concentrations of α_1-PI and other plasma proteins in the mucosa and permits their replacement with known concentrations of synthetic and native protease inhibitors. Similarly, RMCP-II or MMCP-1 can be infused directly into the mucosa via the vasculature. These techniques, allied to measurements of vascular and mucosal permeability and to analyses of basement membrane and matrix proteins, may allow a more precise definition of the targets of proteolytic attack. The second approach, that of incapacitating the gene encoding MMCP-1 by homologous targeting, is now a feasible proposition. Experiments on mutant mice, unable to express MMCP-1, may eventually reveal the functional significance of this chymase, which is so abundant during intestinal helminthiasis.

VI. CONCLUSIONS

The study of mast cell secretory granule chymases in rodents and sheep has clearly established that enteric mucosal mast cells are activated during intestinal nematode infection and that the local and systemic release of chymases is usually associated with the immune elimination of the parasites. The release of chymase during hepatic infection with *S. mansoni* in the rat again suggests that mast cells are active at a time when the parasites are under maximum immunological attack. These data generally support the view that mast cells are involved in the immune rejection of helminths.

T cell–derived cytokines appear to be the stimulus for mast cell recruitment and for the expression of granule chymases in helminth infection. From in vitro studies (71) it would appear that SCF and IL-3 regulate the growth of rat BMMC and the expression/secretion of RMCP-II. Similarly, both SCF and IL-3 are implicated in intestinal mastocytosis in mice, although these two cytokines apparently have antagonistic effects on chymase expression in vitro (43). The in vivo regulation of the tissue-specific expression of granule chymases has not been fully resolved. Nevertheless, the substantial accumulation and release of soluble granule chymases in the gut may be central to the development of mucosal permeability changes known to occur during infection. Although the emphasis in this chapter has been on chymase expression and function during helminth infection, there is a body of data which suggests that mast cell granule chymases in the gut may be involved in a variety of allergic and other immunological conditions. Food antigen–induced hypersensitivity reactions, graft-versus-host disease, irradiation-induced damage, psychoneuroimmunological responses, and weaning are all associated with activation of MMCs and release of RMCP-II in the rat (1). While there are many similarities between mice and other species, there are enough differences between mice and rats to suggest that MMCs and their granule chymases will have to be analyzed on a species-by-species basis. The signals that regulate the expression and secretion of a given chymase also appear to vary, even between closely related species, and this may be of considerable functional importance in determining therapeutic approaches to allergic disease involving mast cells.

REFERENCES

1. H. R. Miller, Mast cells: their function and heterogeneity, in *Allergy and Immunity to Helminths: Common Mechanisms or Divergent Pathways?* (R. Moqbel, ed.), Taylor & Francis, Bristol, Pa., 1992, p. 228.
2. E. E. Jarrett and H. R. Miller, Production and activities of IgE in helminth infection, *Prog. Allergy 31*:178 (1982).
3. T. L. Rothwell, Immune expulsion of parasitic nematodes from the alimentary tract, *Int. J. Parasitol. 19*:139 (1989).
4. J. F. Urban, K. B. Madden, A. Svetic, P. P. Cheever, W. C. Trotta, I. M. Gause, I. M. Katona, and F. D. Finkelman, The importance of T_H2 cytokines in protective immunity to nematodes, *Immunol. Rev. 127*:205 (1992).
5. M. C. Lindsay and J. F. Williams, Hepatic mast cells in rats infected with *Taenia taeniaeformis, Int. J. Parasitol. 15*:201 (1985).
6. J. Chernin, H. R. Miller, G. F. Newlands, and D. J. McLaren, Proteinase phenotypes and fixation properties of rat mast cells in parasitic lesions caused by *Mesocestoides corti*: selective and site-specific recruitment of mast cell subsets, *Parasite Immunol. 10*:433 (1988).
7. N. D. Reed, Function and regulation of mast cells in parasite infections, in *Mast Cell and Basophil Differentiation and Function in Health and Disease* (S. J. Galli and K. F. Austen, eds.), Raven Press, New York, 1985, p. 205.
8. N. Arizono, T. Kasugai, M. Yamada, M. Okada, M. Morimoto, H. Tei, G. F. Newlands, H. R. Miller, and Y. Kitamura, Infection of *Nippostrongylus brasiliensis* induces development of mucosal-type but not connective tissue–type mast cells in genetically mast cell-deficient W^s/W^s rats, *Blood 81*:2572 (1993).
9. H. R. Miller, J. F. Huntley, G. F. Newlands, A. Mackellar, D. A. Lammas, and D. Wakelin, Granule proteinases define mast cell heterogeneity in the serosa and the gastrointestinal mucosa of the mouse, *Immunology 65*:559 (1988).
10. H. Alizadeh and D. Wakelin, Comparison of rapid expulsion of *Trichinella spiralis* in mice and rats, *Int. J. Parasitol. 12*:65 (1982).
11. A. D. Befus, F. L. Pearse, J. Gauldie, P. Horsewood, and J. Bienenstock, Mucosal mast cells. I. Isolation and functional characteristics of rat intestinal mast cells, *J. Immunol. 128*:2475 (1982).
12. F. L. Pearce, A. D. Befus, J. Gauldie, and J. Bienenstock, Mucosal mast cells. II. Effects of anti-allergic compounds on histamine secretion by isolated intestinal mast cells, *J. Immunol. 128*:2481 (1982).
13. L. Enerback, The gut mucosal mast cell, *Monogr. Allergy 17*:222 (1981).
14. H. R. Miller, J. F. Huntley, G. F. Newlands, and J. Irvine, Granule chymases and the characterization of mast cell phenotype and function in rat and mouse, *Monogr. Allergy 27*:1 (1990).

15. E. P. Benditt and M. Arase, An enzyme in mast cells with properties like chymotrypsin, *J. Exp. Med. 110*:451 (1959).
16. D. Lagunoff and P. Pritzl, Characterization of rat mast cell granule proteins, *Arch. Biochem. Biophys. 173*:554 (1976).
17. N. Katunuma, E. Kominiami, K. Kobayashi, Y. Banno, K. Suzuki, K. Chichibu, Y. Hamaguchi, and T. Katsunama, Studies on new intracellular proteases in various organs of rat. I. Purification and comparison of their properties, *Eur. J. Biochem. 52*:37 (1975).
18. R. G. Woodbury, G. M. Gruzensky, and D. Lagunoff, Immunofluorescent localization of a serine protease in rat small intestine, *Proc. Natl. Acad. Sci. USA 75*:2785 (1978).
19. R. G. Woodbury and H. Neurath, Purification of an atypical mast cell protease and its levels in developing rats, *Biochemistry 17*:4298 (1978).
20. R. G. Woodbury and H. R. P. Miller, Quantitative analysis of mucosal mast cell protease in the intestines of *Nippostrongylus*-infected rats, *Immunology 46*:487 (1982).
21. R. G. Woodbury and H. Neurath, Structure and localization of the serine proteases of connective tissue, *FEBS Lett. 144*:189 (1980).
22. R. G. Woodbury, H. Le Trong, K. Cole, H. Neurath, and H. R. Miller, Rat mast cell proteases, in *Mast Cell and Basophil Differentiation and Function in Health and Disease* (S. J. Galli and K. F. Austen, eds.), Raven Press, New York, 1989, p. 71.
23. G. F. Newlands, S. Gibson, D. P. Knox, R. Grencis, D. Wakelin, and H. R. Miller, Characterization and mast cell origin of a chymotrypsin-like proteinase isolated from intestines of mice infected with *Trichinella spiralis, Immunology 62*:629 (1987).
24. H. Le Trong, G. F. Newlands, H. R. Miller, H. Charbonneau, H. Neurath, and R. G. Woodbury, Amino acid sequence of a mouse mucosal mast cell protease, *Biochemistry 28*:391 (1989)
25. D. Reynolds, K. F. Austen, and W. E. Serafin, Molecular characterization of mouse mast cell secretory granule proteases, in *The Mast Cell in Health and Disease* (M. Kaliner and D. Metcalf, eds.), Marcel Dekker, New York, 1992, p. 247.
26. H. P. McNeil, D. S. Reynolds, V. Schiller, N. Ghildyal, D. S. Gurley, K. F. Austen, and R. L. Stevens, Isolation, characterization, and transcription of the gene encoding mouse mast cell protease 7, *Proc. Natl. Acad. Sci. USA 89*:11174 (1992).
27. G. F. Newlands, D. P. Knox, S. R. Pirie-Shepherd, and H. R. Miller, Biochemical and immunological characterization of multiple glycoforms of mouse mast cell protease 1: comparison with an isolated murine serosal mast cell protease (MMCP-4), *Biochem. J. 294*:127 (1993).
28. G. F. Newlands, D. A. Lammas, J. F. Huntley, A. MacKellar, D. Wakelin, and H. R. Miller, Heterogeneity of murine bone marrow–derived mast cells: analysis of their proteinase content, *Immunology 72*:434 (1991).
29. R. Y. Huang, T. Blom, and L. Hellman, Cloning and structural analysis of MMCP-1, MMCP-4 and MMCP-5, three mouse mast cell–specific serine proteases, *Eur. J. Immunol. 21*:1611 (1991).
30. J. F. Huntley, C. Gooden, G. F. Newlands, A. Mackellar, D. A. Lammas, D.

Wakelin, M. Tuohy, R. G. Woodbury, and H. R. Miller, Distribution of intestinal mast cell proteinase in blood and tissues of normal and *Trichinella*-infected mice, *Parasite Immunol. 12*:85 (1990).
31. J. F. Huntley, G. F. Newlands, and H. R. Miller, The isolation and characterization of globule leucocytes: their derivation from mucosal mast cells in parasitized sheep, *Parasite Immunol. 6*:371 (1984).
32. J. F. Huntley, S. Gibson, D. P. Knox, and H. R. Miller, The isolation and purification of a proteinase with chymotrypsin-like properties from ovine mucosal mast cells, *Int. J. Biochem. 18*:673 (1986).
33. J. F. Huntley, D. M. Haig, J. Irvine, L. Inglis, A. MacDonald, A. Rance, and R. Moqbel, Characterisation of ovine mast cells derived from in vitro culture of haemopoietic tissue, *Vet. Immunol. Immunopathol. 32*:47 (1992).
34. D. P. Knox and J. F. Huntley, Classification of sheep abomasal mucosal mast cell proteinase as a serine endopeptidase (EC 3.4.21), *Int. J. Biochem. 20*:193 (1988).
35. J. F. Huntley, S. Gibson, D. Brown, W. D. Smith, F. Jackson, and H. R. Miller, Systemic release of a mast cell proteinase following nematode infections in sheep, *Parasite Immunol. 9*:603 (1987).
36. S. Gibson and H. R. Miller, Mast cell subsets in the rat distinguished immunohistochemically by their content of serine proteinases, *Immunology 58*:101 (1986).
37. S. Gibson, A. Mackellar, G. F. Newlands, and H. R. Miller, Phenotypic expression of mast cell granule proteinases: distribution of mast cell proteinases I and II in the rat digestive system, *Immunology 62*:621 (1987).
38. J. F. Huntley, A. Mackellar, G. F. Newlands, J. Irvine, and H. R. Miller, Mapping of the rat mast cell granule proteinases RMCPI- and -II by enzyme-linked immunosorbent assay and paired immunofluorescence, *Acta Pathol. Microbiol. Immunol. Scand. 98*:933 (1990).
39. A. G. Cummins, G. H. Munro, and A. Ferguson, Effect of cyclosporin A on rat mucosal mast cells and the associated protease RMCP-II, *Clin. Exp. Immunol. 72*:136 (1988).
40. G. F. Newlands, A. M. Mowat, M. V. Felstein, and H. R. Miller, Role of mucosal mast cells in intestinal graft-versus-host reaction in the mouse, *Int. Arch. Allergy. Appl. Immunol. 93*:308 (1990).
41. L. B. Schwartz, Tryptase from human mast cells: biochemistry, biology and clinical utility, *Monogr. Allergy 27*:90 (1990).
42. H. R. Miller, R. G. Woodbury, J. F. Huntley, and G. F. Newlands, Systemic release of mucosal mast cell protease in primed rats challenged with *Nippostrongylus brasiliensis, Immunology 49*:471 (1983).
43. N. Ghildyal, H. P. McNeil, S. Stechschulte, K. F. Austen, D. Silberstein, M. F. Gurish, L. L. Somerville, and R. L. Stevens, IL-10 induces transcription of the gene for mouse mast cell protease-1, a serine protease preferentially expressed in mucosal mast cells of *Trichinella spiralis*-infected mice, *J. Immunol. 149*:2123 (1992).
44. N. Ghildyal, H. P. McNeil, M. F. Gurish, K. F. Austen, and R. L. Stevens, Transcriptional regulation of the mucosal mast cell-specific protease gene, MMCP-2, by interleukin 10 and interleukin 3, *J. Biol. Chem. 267*:8473 (1992).

45. A. M. Irani and L. B. Schwartz, Neutral proteases as indicators of human mast cell heterogeneity, *Monogr. Allergy 27*:146 (1990).
46. Z. Chen, A. M. Irani, T. R. Bradford, S. S. Craig, G. F. Newlands, H. R. Miller, T. Huff, W. H. Simmons, and L. B. Schwartz, Localization of rat tryptase to a subset of the connective tissue type of mast cell, *J. Histochem. Cytochem. 41*:961 (1993).
47. H. R. Miller, G. F. Newlands, A. MacKellar, L. Inglis, P. Coulson, and R. A. Wilson, Hepatic recruitment of mast cells occurs in rats but not mice infected with *Schistosoma mansoni, Parasite Immunol. 16*:145 (1994).
48. P. D. Wells, *Nippostrongylus brasiliensis*: mast cell populations in rats, *Exp. Parasitol. 30*:30 (1971).
49. L. Gutowska, E. J. Ruitenberg, A. Elgersma, and W. Kociecka, Increase of mucosal mast cells in the jejunum of patients infected with *Trichinella spiralis, Int. Arch. Allergy Appl. Immunol. 71*:304 (1983).
50. E. S. Cooper, J. Spencer, C. A. Whyte-Alleng, O. Cromwell, P. Whitney, S. Venugopal, D. A. Bundy, B. Haynes, and T. T. MacDonald, Immediate hypersensitivity in colon of children with chronic *Trichuris trichuria* dystrophy, *Lancet 338*:1104 (1991).
51. S. J. King, H. R. Miller, R. G. Woodbury, and G. F. Newlands, Gut mucosal mast cells in *Nippostrongylus*-primed rats are the major source of secreted rat mast cell protease II following systemic anaphylaxis, *Eur. J. Immunol. 16*:151 (1986).
52. S. J. King, H. R. Miller, G. F. Newlands, and R. G. Woodbury, Depletion of mucosal mast cell protease by corticosteroids: effect on intestinal anaphylaxis in the rat, *Proc. Natl. Acad. Sci. USA 82*:1214 (1985).
53. J. F. Huntley, A. MacKellar, and H. R. Miller, Altered expression of mast cell proteases in the rat: quantitative and immunohistochemical analysis of the distribution of rat mast cell proteases I and II during helminth infection, *Atca. Pathol. Microbiol. Immunol. Scand. 101*:953 (1993).
54. M. Tuohy, D. A. Lammas, D. Wakelin, J. F. Huntley, G. F. Newlands, and H. R. Miller, Functional correlations between mucosal mast cell activity and immunity to *Trichinella spiralis* in high and low responder mice, *Parasite Immunol. 12*:675 (1990).
55. J. F. Huntley, G. R. Wallace, and H. R. Miller, Quantitative recovery of isolated mucosal mast cells and globule leucocytes from parasitized sheep, *Res. Vet. Sci. 33*:58 (1982).
56. J. F. Huntley, G. F. Newlands, F. Jackson, and H. R. Miller, The influence of challenge dose, duration of immunity, or steroid treatment on mucosal mast cells and on the distribution of sheep mast cell proteinase in *Haemonchus*-infected sheep, *Parasite Immunol. 14*:429 (1992)
57. S. R. Smithers and R. J. Terry, Acquired resistance to experimental infections of *Schistosoma mansoni* in the albino rat, *Parasitology 55*:711 (1965).
58. P. Scott, E. Pearce, A. W. Cheever, R. L. Coffman, and A. Sher, Role of cytokines and $CD4^+$ T cell subsets in the regulation of parasite immunity and disease, *Immunol. Rev. 112*:161 (1989).
59. T. R. Mosmann and R. L. Coffman, TH1 and TH2 cells: different patterns of lym-

phokine secretion lead to different functional properties, *Annu. Rev. Immunol. 7*:145 (1989).
60. Y. Kitamura, Y. Kanakura, S. Sonoda, H. Asai, and T. Nakano, Mutual phenotypic changes between connective tissue type and mucosal mast cells, *Int. Arch. Allergy App. Immunol. 82*:244 (1987).
61. Y. Kitamura, T. Nakano, and Y. Kanakura, Transdifferentiation between mast cell subpopulations, *Dev. Growth Differ. 28*:321 (1986).
62. H. Ginsburg, D. Ben-Shahar, and E. Ben-David, Mast cell growth on fibroblast monolayers: two cell entities, *Immunology 45*:371 (1982).
63. S. J. Galli, New insights into "the riddle of the mast cells": microenvironmental regulation of mast cell development and phenotypic heterogeneity, *Lab. Invest. 62*:5 (1990).
64. D. M. Haig, T. A. McKee, E. E. Jarrett, R. G. Woodbury, and H. R. Miller, Generation of mucosal mast cells is stimulated in vitro by factors derived from T cells of helminth-infected rats, *Nature 300*:188 (1982).
65. C. McMenamin, D. M. Haig, S. Gibson, G. F. Newlands, and H. R. Miller, Phenotypic analysis of mast cell granule proteinases in normal rat bone marrow cultures, *Immunology 60*:147 (1987).
66. L. DuBuske, K. F. Austen, J. Czop, and R. Stevens, Granule associated serine neutral proteases of mouse bone marrow–derived mast cells that degrade fibronectin: their increase after sodium butyrate treatment of the cells, *J. Immunol. 133*:1535 (1984).
67. M. F. Gurish, N. Ghildyal, H. P. McNeil, K. F. Austen, S. Gillis, and R. L. Stevens, Differential expression of secretory granule proteases in mouse mast cells exposed to interleukin 3 and c-*kit* ligand, *J. Exp. Med. 175*:1003 (1992).
68. C. McMenamin, Ph.D. thesis, University of Glasgow, 1986.
69. D. M. Haig, C. McMenamin, J. Redmond, D. Brown, I. G. Young, D. R. Cohen, and A. J. Hapel, Rat IL-3 stimulates the growth of rat mucosal mast cells in culture, *Immunology 65*:205 (1988).
70. D. M. Haig, C. McMenamin, C. Gunneberg, R. G. Woodbury, and E. E. Jarrett, Stimulation of mucosal mast cell growth in normal and nude rat bone marrow cultures, *Proc. Natl. Acad. Sci. USA 80*:4499 (1983).
71. D. M. Haig, J. F. Huntley, A. MacKellar, G. F. Newlands, L. Inglis, R. Sangha, D. Cohen, A. J. Hapel, K. Zsebo, S. J. Galli, and H. R. Miller, Effects of stem cell factor (*kit*-ligand) and interleukin-3 on the growth and serine proteinase expression of rat bone marrow-derived or serosal mast cells, *Blood 83*:72 (1994).
72. A. J. MacDonald, D. M. Haig, H. Bazin, A. C. McGuigan, R. Moqbel, and H. R. Miller, IgE-mediated release of rat mast cell protease II, β-hexosaminidase and leukotriene C4 from cultured bone marrow–derived rat mast cells, *Immunology 67*:414 (1989).
73. S. J. Galli, M. Tsai, and B. K. Wershil, The c-*kit* receptor, stem cell factor, and mast cells: what each is teaching us about the others, *Am. J. Pathol. 142*:965 (1993).
74. R. K. Grencis, K. J. Else, J. F. Huntley, and S. I. Nishikawa, The in vivo role of stem cell factor (c-*kit* ligand) on mastocytosis and host protective immunity to the intestinal nematode *Trichinella spiralis* in mice, *Parasite Immunol. 15*:55 (1993).

75. M. Tsai, L. S. Shih, G. F. Newlands, T. Takeishi, K. E. Langley, K. M. Zsebo, H. R. Miller, E. N. Geissler, and S. J. Galli, The rat c-*kit* ligand, stem cell factor, induces the development of connective tissue-type and mucosal mast cells in vivo: analysis by anatomical distribution, histochemistry, and protease phenotype, *J. Exp. Med. 174*:125 (1991).
76. K. B. Madden, J. F. Urban, Jr., H. J. Ziltener, J. W. Schrader, F. D. Finkelman, and I. M. Katona, Antibodies to IL-3 and IL-4 suppress helminth-induced intestinal mastocytosis, *J. Immunol. 147*:1387 (1991).
77. C. C. McMenamin, E. A. Gault, and D. M. Haig, The effect of dexamethasone on growth and differentiation of bone-marrow derived mucosal mast cells in vitro, *Immunology 62*:29 (1987).
78. G. F. Newlands, A. MacKellar, and H. R. Miller, Intestinal mucosal mast cells in *Nippostrongylus*-infected mice: lack of sensitivity to corticosteroids, *Int. J. Parasitol. 20*:669 (1990).
79. S. J. King and H. R. Miller, Anaphylactic release of musocal mast cell protease and its relationship to gut permeability in *Nippostrongylus*-primed rats, *Immunology 51*:653 (1984).
80. R. G. Woodbury, H. R. Miller, J. F. Huntley, G. F. Newlands, A. C. Paliser, and D. Wakelin, Mucosal mast cells are functionally active during the spontaneous expulsion of primary intestinal nematode infections in the rat, *Nature 312*:450 (1984).
81. H. R. Miller, F. Jackson, G. F. Newlands, and W. T. Appleyard, Immune exclusion, a mechanism of protection against the ovine nematode *Haemonchus contortus, Res. Vet. Sci. 35*:357 (1983).
82. J. F. Huntley, Ph.D. thesis, University of Edinburgh (1991).
83. M. K. Patrick, I. J. Dunn, A. Buret, H. R. Miller, J. F. Huntley, S. Gibson, and D. G. Gall, Mast cell protease release and mucosal ultrastructure during intestinal anaphylaxis in the rat, *Gastroenterology 94*:1 (1988).
84. S. R. Pirie-Shepherd, H. R. Miller, and A. Ryle, Differential inhibition of rat mast cell proteinase I and II by members of the α-1-proteinase inhibitor family of serine proteinase inhibitors, *J. Biol. Chem. 266*:17314 (1991).
85. J. Irvine, G. F. Newlands, J. F. Huntley, and H. R. Miller, Interaction of murine intestinal mast cell proteinase with inhibitors (serpins) in blood: analysis by SDS-PAGE and Western blotting, *Immunology 69*:139 (1990).

11

Control of Tryptase and Chymase Activity by Protease Inhibitors

Hiroshi Kido
Institute for Enzyme Research, The University of Tokushima, Tokushima, Japan

Nobuhiko Katunuma
Institute for Health Science, Tokushima Bunri University, Tokushima, Japan

I. INTRODUCTION

Two types of serine protease are localized in the granules of mast cells (1–3), which are thought to be involved in immunoglobulin E (IgE)–dependent immediate hypersensitivity reactions (4,5) and pulmonary fibrosis (6). These are chymases, chymotrypsin-like proteases, and tryptases, trypsin-like proteases. These enzymes convert proproteins or propeptides into biologically active forms and also degrade various protein substrates intracellularly or extracellularly. The proteolytic activity of these enzymes is controlled in vivo by protease inhibitors and/or modulators. However, regulation of their catalytic activity may differ among species, since the properties of these enzymes differ among humans, rodents, and dogs. In this chapter we review the properties of inhibitors and modulators of chymases and tryptases and discuss their control of these enzymes in vivo.

II. PROPERTIES OF INHIBITORS OF MAST CELL TRYPTASES AND CHYMASES

A. Inhibitors and Modulators of Tryptases

The biochemical properties of mast cell tryptase in humans (7–9), rats (10), and dogs (11) differ, including susceptibility to inhibitors. In addition, mast cell tryptase from humans is distributed in both connective tissue and mucosal mast cells, whereas that from rats is found only in connective tissue (H. Kido, unpublished data).

Table 1 summarizes the effects of various synthetic and natural inhibitors on the activities of mast cell tryptase from various species. Trypstatin, which is an endogenous inhibitor of rat tryptase (12), has been purified as a complex with tryptase (10). However, human tryptase isolated from the lung (8), pituitary (9), skin (13), and uterine cervix (14) is not complexed with an inhibitor. In addition, natural endogenous low-molecular-weight peptide inhibitors of tryptase are not present in the human lung (H. Kido, unpublished data).

Rat mast cell tryptase free of trypstatin is sensitive to all synthetic and natural trypsin inhibitors tested, including high-molecular-weight serum inhibitors, although tryptase associated with trypstatin is inhibited only by diisopropyl fluorophosphate (DFP), antipain, leupeptin, and α_1-proteinase inhibitor. These inhibitors of tryptase complexed with trypstatin may have higher affinity than trypstatin (10). On the contrary, human mast cell tryptase resists a number of natural high-molecular-weight trypsin inhibitors, such as Kunitz soybean trypsin inhibitor, aprotinin, Bowmann-Birk soybean trypsin inhibitor (BBI), α_1-proteinase inhibitor, and α_2-macroglobulin, and is inhibited only by DFP and several low-molecular-weight microbial inhibitors, such as antipain and leupeptin (2). Table 1 shows that the inhibitor specificity of human mast cell tryptase is similar to that of rat mast cell tryptase associated with trypstatin, except for the effects of NaCl, $CaCl_2$, and α_1-proteinase inhibitor.

Human mast cell tryptase is reversibly inhibited by increasing concentrations of NaCl (2,8), but rat mast cell tryptase is not. Heparin, which is a major component of the secretory granules in connective tissue mast cells, binds and stabilizes but does not inhibit tryptase from human mast cells (2), whereas it has no affinity for tryptase from rat mast cells. The mechanism of heparin-dependent stabilization of human mast cell tryptase involves prevention of spontaneous subunit dissociation, resulting in conversion of the active tetramer to inactive monomers (15).

In Fig. 1A, the amino acid sequence of trypstatin is compared with those of other known Kunitz protease inhibitors, such as bovine inter-α-trypsin inhibitor (16), human inter-α-trypsin inhibitor (17), the inhibitor domain of the human $A4_{751}$ amyloid protein precursor of Alzheimer's disease (A4 inhibitor) (18), and basic pancreatic trypsin inhibitor (BPTI) (19). Trypstatin is a single polypeptide consisting of 61 amino acid residues, and the spacing of the six cysteines is consistent with that found in the minikringle structure common to this family. The presumed covalent structures of trypstatin and BPTI are shown in Fig. 1B. Regions of extensive congruence with highly conserved regions of this family are underlined in Fig. 1A. The reactive site of trypstatin may be Arg^{14} $-Ala^{15}$, judging from the sequence homology. Trypstatin shows 35 to 70%

Table 1 Effect of Synthetic and Natural Inhibitors on Tryptase

Inhibitor		Relative activity[a] (%)			
		Rat tryptase complexed with trypstatin	Rat tryptase	Human lung tryptase	Human pituitary tryptase
None		100	100	100	100
DFP	1 μ*M*	12.5	17.0	—	—
	10 μ*M*	0	0	—	0
	1 m*M*	—	—	—	0
	2 m*M*	0	0	0	—
Soybean trypsin inhibitor	2.4 μ*M*	—	—	—	100
	4.8 μ*M*	—	—	100.5	—
	10 μ*M*	90.9	28.9	—	—
Aprotinin	10 μ*M*	90.4	28.9	—	—
	15 μ*M*	—	—	84.9	—
	30 μ*M*	—	—	—	91
Trypstatin	10 μ*M*	—	0	—	—
A_4 inhibitor	10 μ*M*	—	2.3	—	—
α_1-Proteinase inhibitor	10 μ*M*	22.2	33.7	—	93
	18 μ*M*	—	—	100.5	—
BBI	10 μ*M*	—	60.5	—	—
Antipain	10 μ*M*	5.3	1.8	—	6
	65 μ*M*	—	—	0	—
Leupeptin	10 μ*M*	2.7	0	—	5
	235 μ*M*	—	—	0.9	—
Chymostatin	10 μ*M*	88.2	81.3	—	—
	165 μ*M*	—	—	93.7	—
Elastatinal	10 μ*M*	100	75.3	—	—
	196 μ*M*	—	—	97.1	—
E-64	1 μ*M*	—	—	—	100
	10 μ*M*	100	75.3	—	—
Pepstatin	10 μ*M*	96.0	99.6	—	97.0
	146 μ*M*	—	—	96.4	—
Phosphoramidon	10 μ*M*	86.6	80.9	—	—
	184 μ*M*	—	—	93.2	—
EDTA	2 m*M*	—	—	87.5	90
	10 m*M*	101.7	100	—	—
$CaCl_2$	2 m*M*	—	—	—	95
	10 m*M*	100	218	—	—
	20 m*M*	—	—	35.1	15
NaCl	0.6 *M*	100	—	18	28.2

[a]Values are activities as percentage of rat tryptase with *N-tert*-butoxycarbonyl (Boc)–Phe–Ser–Arg[7]–amino-4-methylcoumarin amide (MCA) (10), human lung tryptase with *N*-benzoyl (Bz)–DL-Arg-*p*-nitroanilide (8), and human pituitary tryptase with benzyloxylcarbonyl (Z)–Ala–Lys–Arg–MCA (9) without inhibitors.

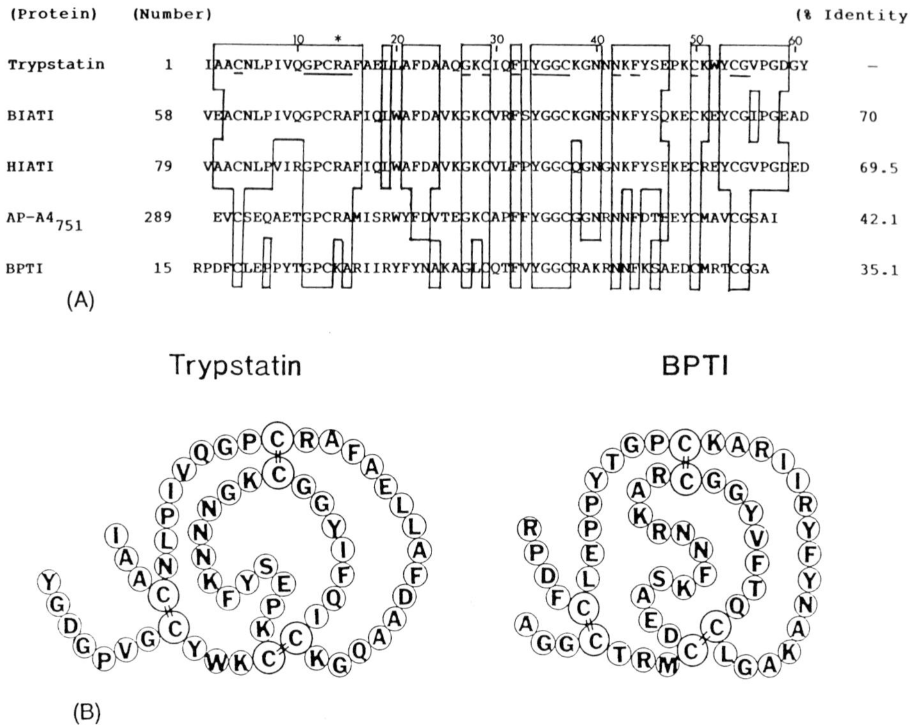

Figure 1 (A) Comparison of amino acid sequences of trypstatin and other known Kunitz protease inhibitors; (B) presumed covalent structures of trypstatin and BPTI. The residues homologous to those in the highly conserved sequences of this family are underlined. The basic amino acid residue (Arg) in the active site is indicated by an asterisk. Residues identical to those in trypstatin are boxed. Residue numbers from the first aligned amino acid are shown on the left. Inhibitors: BIATI, bovine inter-α-trypsin inhibitor; HIATI, human inter-α-trypsin inhibitor; AP-$A4_{751}$, human $A4_{751}$ amyloid protein precursor; BPTI, basic pancreatic trypsin inhibitor.

sequence homology with these inhibitors, being highest with fragments of bovine and human inter-α-trypsin inhibitors. It is not yet known whether trypstatin is a proteolytic product of a precursor of high molecular weight, such as inter-α-trypsin inhibitor.

Figure 2 shows plots of the inhibition of trypsin and rat mast cell tryptase by trypstatin. These plots show that one molecule of trypstatin reacts with one molecule of either a tryptase subunit or of trypsin to form an enzyme-inhibitor complex. Rat mast cell tryptase consists of four identical subunits, each with a

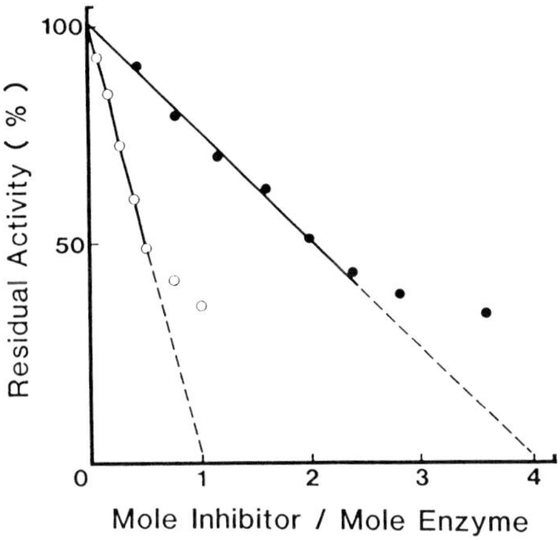

Figure 2 Inhibition of trypsin and mast cell tryptase by trypstatin. Trypsin (30 p*M*) (○) and tryptase (8 p*M*) (●) were preincubated with various concentrations (3 to 30 p*M*) of purified trypstatin in 0.1 *M* Tris-HCl, pH 7.5, and 0.1 *M* Tris-HCl, pH 8.5, containing bovine serum albumin (0.1 mg/mL) at 25°C for 5 min, respectively. Residual activities of these enzymes were measured using Boc–Phe–Ser–Arg–MCA as the substrate.

molecular weight of 35,000 Da (10), and thus 4 mol of trypstatin binds to 1 mol of the tryptase.

The effects of trypstatin on various proteases are shown in Table 2. Trypstatin inhibits the amidase activities of rat mast cell tryptase and bovine blood coagulation factor Xa about 100-fold more rapidly than those of pancreatic trypsin and rat mast cell chymase. It does not inhibit bovine α-thrombin, porcine pancreatic elastase, or papain. The inhibition constants of trypstatin for factor Xa and rat mast cell tryptase are very low, being 1.2 and 3.6×10^{-10} *M*, respectively (12).

Trypstatin that is copurified with tryptase from rat mast cells as an associated protein acts not only as an inhibitor of tryptase amidolytic activity but also as a modulator of substrate specificity and of the optimum pH of the enzyme. Rat tryptase free of trypstatin has maximal activity at pH 8.5 as determined using Boc–Phe–Ser–Arg–MCA as a substrate, as shown in Fig. 3. The optimum pH of tryptase associated with trypstatin, however, shifts to pH 7.5 and no activity is evident above pH 8.5. Table 3 shows trypstatin's effects on the substrate specificity of rat tryptase. Tryptase free of trypstatin rapidly hydrolyzes the substrate for factor Xa, Boc–Ile–Glu–Gly–Arg–MCA, although this substrate is scarcely hydrolyzed by the enzyme associated with trypstatin. Try-

Table 2 Inhibitory Effects of Trypstatin upon Various Proteases

Protease	Substrate[a]	pH	K_i value[b] (*M*)
Porcine pancreatic trypsin	Boc–Phe–Ser–Arg–MCA	7.5	1.4×10^{-8}
Rat mast cell tryptase	Boc–Phe–Ser–Arg–MCA	8.5	3.6×10^{-10}
Bovine factor Xa	Boc–Ile–Glu–Gly–Arg–MCA	8.0	1.2×10^{-10}
Bovine α_1-thrombin	Boc–Val–Pro–Arg–MCA	8.0	NI
Rat chymase (RMCP-I)	Suc–Leu–Leu–Val–Tyr–MCA	8.0	2.4×10^{-8}
Porcine pancreatic elastase	Suc–Ala–Pro–Ala–MCA	7.0	NI
Papain	Z–Phe–Arg–MCA	6.0	NI

[a]Suc-, succinyl-; Z-, benzyloxycarbonyl-.
[b]NI, no inhibition at a concentration of 1 μ*M* trypstatin.

ptase complexed with trypstatin hydrolyzed Boc–Phe–Ser–Arg–MCA most efficiently among the synthetic substrates tested. These results suggest that trypstatin binds near the substrate binding site and modulates both the substrate and inhibitor specificities of the enzyme.

B. Inhibitors of Chymases

Chymase has been found in connective tissue mast cells in rodents, humans, and dogs and in mucosal mast cells in rodents. In rats, connective tissue mast cells

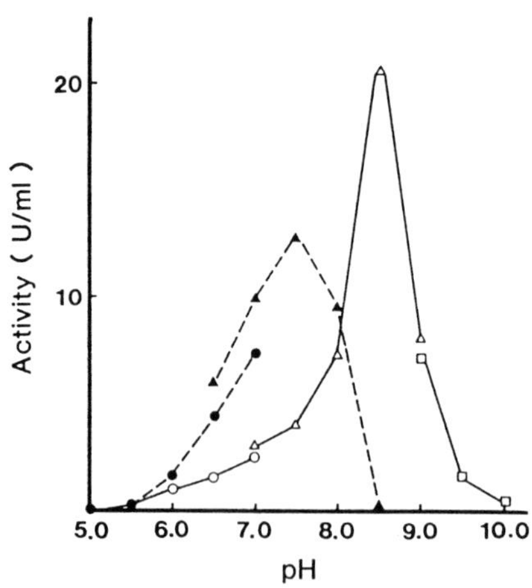

Figure 3 Effects of pH on the activities of tryptase with (---) or without (—) trypstatin. The activity of tryptase was measured by adding Boc–Phe–Ser–Arg–MCA after preincubation with or without trypstatin for 5 min. The buffers used at final concentrations of 0.2 *M* were sodium cacodylate-HCl, pH 5.0 to 7.0 (○, ●), Tris-HCl, pH 6.0 to 9.5 (Δ, ▲); and triethanolamine-HCl, pH 9.0 to 10.0 (□).

Table 3 Trypstatin Modulates the Substrate Specificity of Rat Tryptase

Substrates[a] (10 μ*M*)	Relative activity[b] (%) — Tryptase and trypstatin complex	Relative activity[b] (%) — Tryptase
Boc–Phe–Ser–Arg–MCA (trypsin)	100	100
Boc–Val–Pro–Arg–MCA (α-thrombin)	80.3	50.6
Boc–Val–Leu–Lys–MCA (plasmin)	4.7	6.1
Boc–Ile–Glu–Gly–Arg–MCA (factor Xa)	4.1	149.0
Glt–Gly–Arg–MCA (urokinase)	3.5	4.6
Pro–Phe–Arg–MCA (kallikrein)	0	4.6
Bz–Arg–MCA (trypsin)	1.2	0
Suc–Ala–Ala–Pro–Phe–MCA (α-chymotrypsin)	0	2.0
Suc–Leu–Leu–Val–Tyr–MCA (α-chymotrypsin)	0	12.3

[a]The enzymes for which the compounds are substrates are shown in parentheses.
[b]Values are activities of tryptase as a percentage of that with Boc–Phe–Ser–Arg–MCA.

contain one type of chymase [rat mast cell protease I (RMCP-I)], whereas mucosal mast cells contain a different type [rat mast cell protease II (RMCP-II)]. The primary structures of chymases in these species are homologous (20) and the inhibition properties of these enzymes are similar, with a few exceptions, as shown in Table 4. Although the inhibitor spectrum of RMCP-II is similar to that of other chymases from humans, rats, and dogs, it is less sensitive. In addition, human skin chymase, dog chymase, and RMCP-I are more susceptible to BBI, soybean trypsin inhibitor, and BPTI, respectively, than RMCP-II (Table 4). Among those tested, BBI is the strongest inhibitor of chymases in various species, having K_i values of 10 n*M*, 13 n*M*, and 41 μ*M* for human skin chymase (3), RMCP-I, and RMCP-II (22), respectively. An endogenous inhibitor, trypstatin, which is the most powerful inhibitor of rat tryptase, also inhibits rat chymase with a K_i value of 2.4×10^{-8} *M* (12). Protein inhibitors that do not inhibit chymases but do inhibit pancreatic α-chymotrypsin are hen ovoinhibitor and the A4 inhibitor of Alzheimer's amyloid protein precursor (23).

Chymases from humans and rats are inhibited by plasma protease inhibitors, such as α_1-antichymotrypsin and α_1-proteinase inhibitor (Table 4), although both are better substrates than inhibitors of chymase (24). The rate of human chymase inhibition by each inhibitor, appearing as pseudo-first-order and apparent second-order rate constants, are very slow, as shown in Table 4: $K_a = 2.1 \times 10^4\ M^{-1}\ s^{-1}$ for α_1-antichymotrypsin and $k_a = 7.5 \times 10^3\ M^{-1}\ s^{-1}$ for α_1-proteinase inhibitor (23). Thus plasma inhibitors do not appear to have an important role in the physiological regulation of mast cell chymase in vivo.

Table 4 Effects of Synthetic and Natural Inhibitors on Chymases

Inhibitor		Relative activity[a] (%)			
		Human skin chymase	Rat chymase (RMCP-I)	Rat chymase (RMCP-II)	Dog chymase
None		100	100	100	100
DFP	10 μ*M*	0	—	—	—
	2.0 μ*M*	—	—	—	8.6
Soybean	4.8 μ*M*	—	—	—	2.9
trypsin inhibitor	10 μ*M*	—	57	100	—
	15 μ*M*	50	—	—	—
		—	$K_i = 8.3 \times 10^{-8}$ *M*	—	—
Bovine pancreatic	1.5 μ*M*	—	—	—	97.7
trypsin inhibitor	10 μ*M*	—	57	100	—
(BPTI)	15 μ*M*	100	—	—	—
Kazal-type pancreatic secretory trypsin inhibitor		—	$K_i = 8.1 \times 10^{-8}$ *M*	—	—

BBI		$K_i = 1 \times 10^{-8}$ *M*	$K_i = 1.3 \times 10^{-8}$ *M*	$K_i = 4.1 \times 10^{-7}$ *M*	—
Eglin C		$K_i = 2.3 \times 10^{-8}$ *M*	$K_i = 9.5 \times 10^{-7}$ *M*	—	—
Hen ovoinhibitor	2 μ*M*	—	—	—	100.9
	6 μ*M*	100	—	—	—
Trypstatin		—	$K_i = 2.4 \times 10^{-8}$ *M*	—	—
A_4 inhibitor	1 μ*M*	—	100	—	—
α_1-Antichymotrypsin	10 μ*M*	—	0	—	—
		$k_a = 2.1 \times 10^{4}\ M^{-1}\ s^{-1}$	—	—	—
α_1-Proteinase	1.8 μ*M*	—	—	—	103.5
inhibitor	10 μM	—	32.0	—	—
		$k_a = 7.5 \times 10^{3}\ M^{-1}\ s^{-1}$	—	—	—
Chymostatin		$K_i = 2 \times 10^{-8}$ *M*	$K_i = 6.2 \times 10^{-8}$ *M*	$K_i = 6.2 \times 10^{-6}$ *M*	—
Leupeptin	10 μ*M*	—	101.5	—	—
E-64	10 μ*M*	—	96.2	—	—
Elastatinal	10 μ*M*	—	101.8	—	—
Phosphoramidon	10 μ*M*	—	100.7	—	—

[a]Values are activities as a percentage of those of human skin chymase with Bz-L-tyrosine ethyl ester (23), of rat chymases with Suc–Leu–Leu–Val–Tyr–MCA (10,21) and of dog chymase with Suc–Phe–Pro–Phe–*p*-nitroanilide (21).

Rat chymase is markedly inhibited by phosphoglycerides (25) and long-chain fatty acids (26). As shown in Table 5, phosphoglycerides, such as phosphatidic acid, phosphatidylserine, and phosphatidylinositol, inhibit the activity, but acylglycerides, phosphoglyceroserine, serine, inositol, and glycerol do not. The binding of a primary amine to the anionic polar head of a phosphatidic acid, such as in phosphatidylserine and phosphatidylethanolamine, decreases the inhibitory effect of phosphatidic acid, and conversely, binding of a strong cation to the head, such as in phosphatidylcholine, results in chymase activation. Rat chymase is also inhibited by fatty acids with carbon chain lengths of 14 to 22 at concentrations above 0.02 μ*M* (26). However, a fatty acid with a terminal functional amide, oleamide, has little effect, and the trans isomer of oleic acid,

Table 5 Effects of Phosphoglycerides: Acylglycerols, and Fatty Acids on Chymase Activity

Compound	Inhibition [ID_{50} (μ*M*)][a]
Phosphatidylserine	0.29
Phosphoglyceroserine	NI
Serine	NI
Phosphatidylinositol	0.14
Inositol	NI
Phosphatidylethanolamine	5.6
Ethanolamine	NI
Phosphatidylcholine	Activation (V_a/V_a = 1.5)[b]
Phosphoglycerocholine	NI
Phosphatidic acid from bovine brain	0.06
1-Monostearin	NI
1,2-Disterin	NI
Tristearin	NI
Glycerol	NI
Stearic acid (18:0)	0.14
Oleic acid (18:1 *cis*-9)	0.12
Elaidic acid (18:1 *trans*-9)	7.60
Linoleic acid (18:2 *cis*-9, *cis*-12)	0.13
Linolenic acid (18:3 *cis*-9, *cis*-12, *cis*-15)	0.14
Oleamide	200

[a]ID_{50}, concentration inhibiting 50% of the enzyme activity with Suc–Ala–Ala–Pro–Phe–MCA as the substrate. NI, no inhibition at a concentration of 200 μ*M* was considered ineffective.

[b]V_a/V_a, ratio of *p*-amino-4-methylcoumarin release in the presence of 1 μ*M* phosphatidylcholine to that in the absence of the compound.

elaidic acid, shows a slight inhibitory effect. These results suggest that both a nonpolar hydrophobic hydrocarbon chain and a polar hydrophilic anionic head are essential for the inhibitory effects of phosphoglycerides and fatty acids. Both inhibitions are noncompetitive; the inhibition by fatty acids is reversible but that by phosphoglycerides is pseudoirreversible. Structural studies of RMCP-I indicate there are two clusters of lysine on its surface (27), which may be responsible for binding to the anionic polar heads of phosphoglycerides.

Chymases from various species bind to heparin in mast cell granules at physiological ionic strength (28) and are secreted with heparin from mast cells after stimulation. Heparin and heparan sulfate may provide sites for the accumulation of chymase on basement membranes and/or the surface of various cells near their sites of release.

Serotonin (5-hydroxytryptamine), but not histamine, is a competitive inhibitor of RMCP-I (29), although human mast cells do not contain serotonin. It may limit RMCP-I's proteolytic activity.

III. BIOLOGICAL FUNCTIONS OF INHIBITORS OF TRYPTASE

Differences in the substrate and inhibitor specificities of tryptase have been found among species. Although many functions of tryptase have been reported in vitro, its biological functions and control of its activity by inhibitors in vivo have not been convincingly demonstrated because no genetic disease involving a dysfunction or deficiency of tryptase and its inhibitors has been identified. Nevertheless, tryptase is released from activated mast cells with various vasoactive substances, such as histamine, prostaglandin D_2, leukotrienes C_4 and D_4, and heparin. Human tryptase complexed with heparin has low solubility in the extracellular milieu and binds to tissues near its site of release. It rapidly inactivates human fibrinogen and high-molecular-weight kininogen and thus prevents blood coagulation (2). It also proteolytically activates connective tissue stromelysin, which is an activator of procollagenase (30). However, it has no effect on low-molecular-weight kininogen, bradykinin, plasma prekallikrein, or plasminogen. In contrast, tryptase from rats, which complexes with trypstatin and does not bind to heparin, is highly soluble, although the mechanisms of dissociation and association of trypstatin to tryptase remain unknown. Rat tryptase that is free of trypstatin hydrolyzes the specific substrate of factor Xa, Boc–Ile–Glu–Gly–Arg–MCA, most efficiently at neutral pH as shown in Table 3. The coefficients of proteolytic efficiency (k_{cat}/K_m) of rat tryptase and factor Xa for activation of prothrombin in the absence of phospholipid and factor Va are 20.1 and 13.9 s^{-1} μM^{-1}, respectively (31) (Table 6), suggesting that rat tryptase contributes to prothrombin activation in tissue. The time courses of prothrombin activation by the tryptase and by factor Xa are shown in Fig. 4. The process of activation of prothrombin by rat tryptase is similar to that by Xa except

Table 6 Kinetic Constants for Activation of Prothrombin by Tryptase and Factor Xa[a]

Kinetic constant	Tryptase	Xa
K_m (μM)	2.3	0.74
k_{cat} (s^{-1})	46.3	10.3
k_{cat}/K_m (s^{-1} μM^{-1})	20.1	13.9

[a]The rate of conversion of prothrombin to α-thrombin was measured with Boc–Val–Pro–Arg–MCA as an α-thrombin substrate (31).

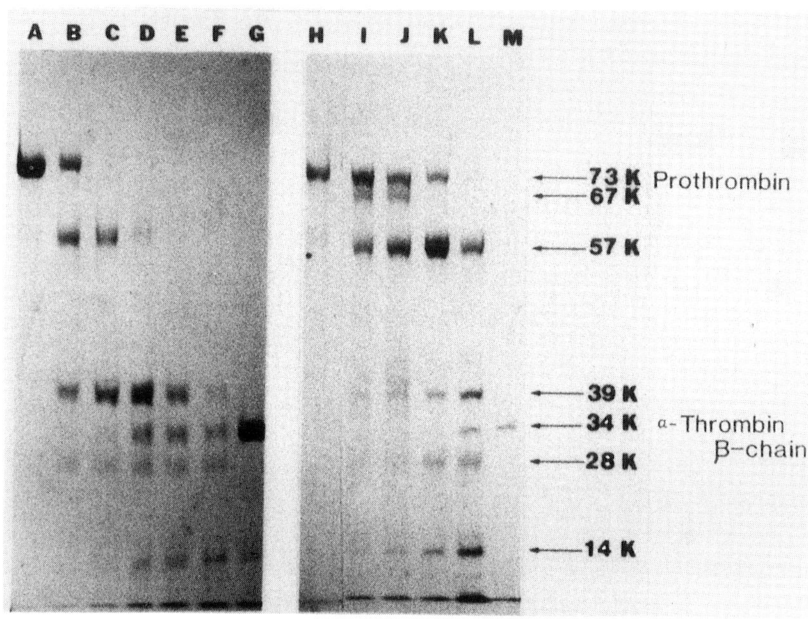

Figure 4 SDS-PAGE of products at various stages in the reaction of bovine prothrombin with tryptase or with factor Xa. Prothrombin (4 μg) and factor Xa (40 ng) were incubated in 20 μL of 50 mM Tris-HCl buffer, pH 7.5, containing 100 mM NaCl and 10 mM $CaCl_2$ for 0 min (A), 45 min (B), 2 h (C), 4 h (D), 6 h (E), and 8 h (F). Prothrombin (4 μg) and tryptase (1.5 ng) were incubated in 20 μL of buffer containing 100 mM NaCl and 10 mM $CaCl_2$ for 0 h (H), 4 h (I), 8 h (J), 12 h (K), and 16 h (L). The reaction was stopped by boiling the mixtures, which were then resolved by electrophoresis in a 6% polyacrylamide gel under reducing conditions. Lanes G and M, 6 μg and 2 μg of purified bovine thrombin, respectively.

that an intermediate peptide of 67,000 Da is formed. Trypstatin inhibits both the activities of rat tryptase and of factor Xa (12). Therefore, trypstatin secreted from rat mast cells may act as an anticoagulant by inhibiting prothrombin activation in vivo (Fig. 5).

IV. BIOLOGICAL FUNCTIONS OF INHIBITORS OF CHYMASE

Chymase is present as a complex with heparin in the granules of mast cells in various species, is released with heparin, and remains near its site of release in tissues because of its low solubility. Phosphatidylinositol, phosphatidylserine, and phosphatidic acid, which are biologically active metabolites in the plasma membrane and membranes of secretory granules of mast cells, are potent inhibitors of chymase, and highly methylated phospholipids, such as phosphatidylcholine, are activators of chymase (25). Recent studies have shown that stimulation of phospholipid methylation in mast cell plasma membranes (32) and phosphatidylinositol turnover in mast cell granules (33,34) are evident after bridging the IgE receptor and that phosphatidylinositol kinase is localized in mast cell granules (34). These facts suggest that a change in the phospholipid composition in the membranes of granules modulates the chymase activity in the granules, although the physiological meaning of the modulation in vivo remains uncertain.

Serine protease inhibitors and substrates inhibit histamine release (35). Inhibitors of chymase, but not of tryptase, inhibit histamine release in a dose-

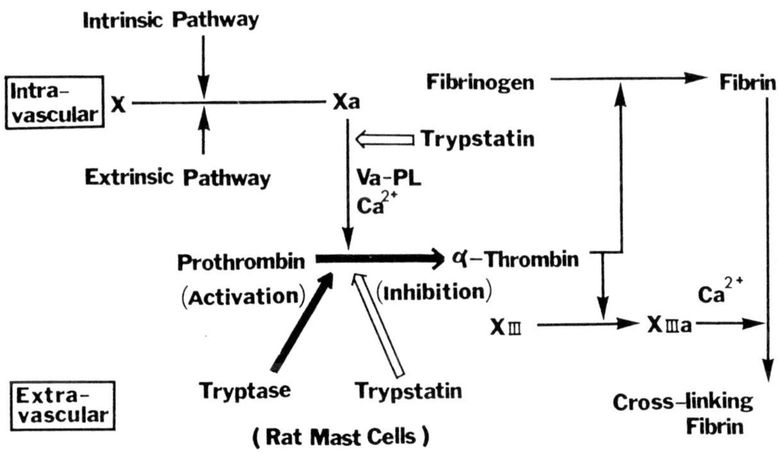

Figure 5 Possible role of rat mast cell tryptase and trypstatin in the process of fibrin formation. Va, activated factor V; PL, phospholipid.

dependent manner from rat and human mast cells (22,36,37). These findings are also confirmed by the effect of $F(ab')_2$ fragments of anti-chymase on histamine release from activated rat mast cells (36,38). Table 7 shows the effects of inhibitors of chymase and tryptase, substrate analogs, and $F(ab')_2$ fragments of anti-chymase on histamine release from activated rat mast cells. Mast cells were activated by anti-rat IgE, compound 48/80, and the Ca ionophore A23187, respectively, which are thought to act at different steps in the process of degranulation. Low-molecular-weight inhibitors of chymase, such as chymostatin and peptide boronic acids, which are substrate analogs, markedly inhibit the histamine release (22) from mast cells induced by anti-rat IgE, but inhibitors of

Table 7 Effects of Inhibitors of Mast Cell Proteases on Histamine Release

		Percent of maximum histamine release[b]		
Inhibitor or antibody[a]		Anti-IgE (150 μg/mL)	Ca ionophore A23187 (0.5 μg/mL)	Compound 48/80 (5 μg/mL)
None		100	100	100
Aprotinin	25 μ*M*	121	92	86
Leupeptin	25	80	106	94
E-64	25	100	86	90
Chymostatin	25	27	62	100
Meo-Suc-Ala-Ala-Pro-(L)boro-Phe-OH	50	30		
Meo-Suc-Ala-Ala-Pro-(L)boro-Phe-pinacol	50	25		
BBI	25	74		
	100	52		
Eglin C	25	80		
	100	62		
Antichymase $F(ab')_2$	25 μg	97		
	50	73		
	100	45	100	120
	150	0		115
Control serum $F(ab')_2$	50	115		
	100	91	100	100
	150	95		105

[a]Meo, methoxy.
[b]Mast cells were preincubated with protease inhibitors, $F(ab)_2$ of anti-chymase, or nonimmunized $F(ab')_2$ for 10 min at 37°C, and histamine release was induced by anti-rat IgE, Ca ionophore A23187, or compound 48/80. Values are averages for triplicate samples.

tryptase, such as aprotinin and leupeptin, cause little inhibition (36). Inhibitors of chymase with a molecular mass of more than 6000 Da, such as BBI and eglin C, also inhibit histamine release dose dependently as shown in Fig. 6, although they are less inhibitory than chymostatin and the substrate analogs. $F(ab')_2$ of anti-chymase, which specifically inhibits proteolytic activity of chymase, inhibits the histamine release induced by anti-rat IgE dose dependently, inhibition being complete with 150 μg/mL of the antibody (36). However, it does not inhibit the histamine release induced by A23187 or compound 48/80. Nonimmunized $F(ab')_2$ and E-64, an inhibitor of thiol protease, have no effect on histamine release. These results suggest that chymase in mast cell granules is involved in the process of IgE-mediated degranulation before calcium entry, and that tryptase is not involved. Secreted endogenous inhibitors of chymase in the granules, such as trypstatin, phosphoglycerides, and fatty acids, may have a role in feedback inhibition of the secretion to prevent excessive degranulation. Under physiological conditions, up to 20 to 40% of the histamine granules in mast cells are secreted by antigen stimulation, and further degranulation is not effected by further stimulation. A similar process is evident during inhibition of the acetylcholine receptor–mediated process of catecholamine secretion from adrenal medullary cells (39) and during inhibition of eosinophilic peroxidase release from

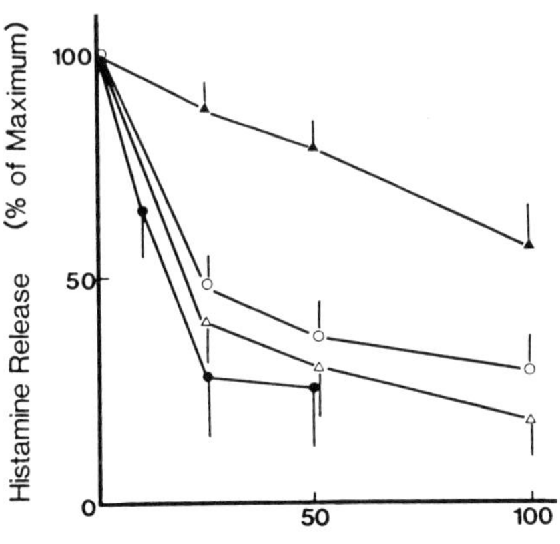

Figure 6 Effects of peptide boronic acids, chymostatin, and BBI on the release of histamine from rat peritoneal mast cells. Purified mast cells were preincubated with MeO–Suc–Ala–Ala–Pro–(L)boro-Phe–OH (○), MeO–Suc–Ala–Ala–Pro–(L)boro-Phe–pinacol (△), chymostatin (●), or BBI (▲) for 20 min at 37°C. The inhibition of histamine release was assayed after further incubation with anti-rat IgE for 10 min at 37°C. Anti-rat IgE induced a maximum release of 30 to 45% of the total histamine at a concentration of 150 μg/mL. Values are means for three separate experiments. Bars, SD.

human eosinophils (40). Therefore, the chymotrypsin-like protease in secretory granules of various cells seems to play an essential role in exocytosis. The main questions that need to be answered are how chymase is activated by receptor bridging, what kind of biologically active peptides chymase produces, and which process of degranulation the proteolytic products stimulate.

Experiments in vitro show that secreted human chymase efficiently converts angiotensin I to angiotensin II (20) and that it degrades bradykinin, although similar hydrolytic activities are not observed for rat chymase (3). On the other hand, rat chymase converts bigendothelin to endothelin (41) and degrades fibronectin and type IV collagen (42). Endogenous inhibitors of chymase may also control extracellular proteolysis.

V. CONCLUSIONS

Species differences in catalytic properties and inhibitor specificities of chymase and tryptase are evident, suggesting that these enzymes and their inhibitors have evolved different biological functions. Inhibitors in serum do not appear to have a role in controlling the activities of secreted tryptase complexed with heparin or trypstatin and of chymase associated with heparin. Trypstatin, an endogenous inhibitor isolated from rat mast cells but not from humans, inhibits the activities of rat tryptase and chymase, and modulates the substrate specificity of tryptase and its optimal pH. Phosphoglycerides and fatty acids, which are biologically important metabolites in plasma membranes and membranes of secretory granules of mast cells, may control chymase activity intracellularly and extracellularly. Further studies to understand the biological functions of chymase and tryptase and control of these enzymes by inhibitors in vivo are necessary to solve the many remaining questions.

REFERENCES

1. N. Katunuma and H. Kido, Biological functions of serine proteases in mast cells in allergic inflammation, *J. Cell. Biochem. 38*:291 (1988).
2. L. B. Schwartz, Tryptase from human mast cells: biochemistry, biology and clinical utility, in *Neutral Proteases of Mast Cells* (L. B. Schwartz, ed.), S. Karger, Basel, 1990, p. 90.
3. N. M. Schechter, Human chymase, in *Neutral Proteases of Mast Cells* (L. B. Schwartz, ed.), S. Karger, Basel, 1990, p. 114.
4. S. I. Wasserman, Mediators of immediate hypersensitivity, *J. Allergy Clin. Immunol. 72*:101 (1983).
5. B. Samuelsson, Leukotrienes: mediators of immediate hypersensitivity reactions and inflammation, *Science 220*:568 (1983).
6. O. Kawanami, V. J. Ferrans, J. D. Fulmer, and R. G. Crystal, Ultrastructure of

pulmonary mast cells in patients with fibrotic lung disorders, *Lab. Invest. 40*:717 (1979).
7. L. B. Schwartz, R. A. Lewis, and K. F. Austen, Tryptase from human pulmonary mast cells: purification and characterization, *J. Biol. Chem. 256*:11939 (1981).
8. T. J. Smith, M. W. Hougland, and D. A. Johnson, Human lung tryptase: purification and characterization, *J. Biol. Chem. 259*:11046 (1984).
9. J. A. Cromlish, N. G. Seidah, M. Marcinkiewicz, J. Hamelin, D. A. Johnson, and M. Chretien, Human pituitary tryptase: molecular forms, NH_2-terminal sequence, immuno-cytochemical localization, and specificity with prohormone and fluorogenic substrates, *J. Biol. Chem. 262*:1363 (1987).
10. H. Kido, N. Fukusen, and N. Katunuma, Chymotrypsin- and trypsin-type serine proteases in rat mast cells: properties and functions, *Arch. Biochem. Biophys. 239*:436 (1985).
11. G. H. Caughey, N. F. Viro, J. Ramachandran, S. C. Lazarus, D. B. Borson, and J. A. Nadel, Dog mastocytoma tryptase: affinity purification, characterization, and amino-terminal sequence, *Arch. Biochem. Biophys. 258*:555 (1987).
12. H. Kido, Y. Yokogoshi, and N. Katunuma, Kunitz-type protease inhibitor found in rat mast cells: purification, properties, and amino acid sequence, *J. Biol. Chem. 263*:18104 (1988).
13. I. T. Harvima, N. M. Schechter, R. J. Harvima, and J. E. Fräki, Human skin tryptase: purification, partial characterization and comparison with human lung tryptase, *Biochim. Biophys. Acta 697*:416 (1988).
14. A. Ito, H. Ihara, and Y. Mori, Partial purification and characterization of a novel neutral proteinase from human uterine cervix, *Biochem. J. 185*:443 (1980).
15. L. B. Schwartz and T. R. Bradford, Regulation of tryptase from human lung mast cells by heparin: stabilization of the active tetramer, *J. Biol. Chem. 261*:7372 (1986).
16. K. Hochstrasser and E. Wachter, Kunitz-type proteinase inhibitors derived by limited proteolysis of the inter-α-trypsin inhibitor. VII. Determination of the amino-acid sequence of the trypsin-released inhibitor from bovine inter-α-trypsin inhibitor, *Hoppe-Seyler's Z. Physiol. Chem. 364*:1679 (1983).
17. E. Wachter and K. Hochstrasser, Kunitz-type proteinase inhibitors derived by limited proteolysis of the inter-α-trypsin inhibitor, IV, *Hoppe-Seyler's Z. Physiol. Chem. 362*:1351 (1981).
18. P. Ponte, P. Gonzulez-Dewhitt, J. Schilling, J. Miller, D. Hsu, B. Greenberg, K. Davis, W. Wallance, I. Lieberburg, F. Fuller, and B. Cordell, A new A4 amyloid mRNA contains a domain homologous to serine protease inhibitor, *Nature 344*:525 (1988).
19. B. Kassell and M. Laskowski, The basic trypsin inhibitor of bovine pancreas. V. The disulfide linkages, *Biochem. Biophys. Res. Commun. 20*:46 (1965).
20. H. Urata, A. Kinoshita, D. M. Perez, K. S. Misono, F. M. Bumpus R. H. Graham, and A. Husain, Cloning of the gene and cDNA for human heart chymase, *J. Biol. Chem. 266*:17173 (1991).
21. G. H. Caughey, N. F. Viro, S. C. Lazarus, and J. A. Nadel, Purification and characterization of dog mastocytoma chymase: identification of an octapeptide conserved in chymotryptic leukocyte proteinases, *Biochim. Biophys. Acta 952*:142 (1988).

22. Y. Kato, H. Kido, N. Fukusen, and N. Katunuma, Peptide boronic acids, substrate analogs, inhibit chymase, and histamine release from rat mast cells, *J. Biochem. 103*:820 (1988).
23. H. Kido, A. Fukutomi, J. Schilling, Y. Wang, B. Cordell, and N. Katunuma, Proteases-specificity of Kunitz inhibitor domain of Alzheimer's disease amyloid protein precursor, *Biochem. Biophys. Res. Commun. 167*:716 (1990).
24. N. M. Schechter, J. L. Sprows, O. L. Schoenberger, G. S. Lazarus, B. S. Cooperman, and H. Rubin, Reaction of human skin chymotrypsin-like proteinase chymase with plasma proteinase inhibitors, *J. Biol. Chem. 264*:21308 (1989).
25. N. Fukusen, H. Kido, and N. Katunuma, Inhibition of chymase activity by phosphoglycerides, *Arch. Biochem. Biophys. 237*:118 (1985).
26. H. Kido, N. Fukusen, and N. Katunuma, Inhibition of chymase activity by long chain fatty acids, *Arch. Biochem. Biophys. 230*:610 (1984).
27. H. Le Trong, D. C. Parmelee, K. A. Walsh, H. Neurath, and R. G. Woodbury, Amino acid sequence of rat mast cell protease I (chymase), *Biochemistry 26*:6988 (1987).
28. S. Sayama, R. V. Iozzo, G. S. Lazarus, and N. M. Schechter, Human skin chymotrypsin-like proteinase chymase: subcellular localization to mast cell granules and interaction with heparin and other glycosaminoglycans, *J. Biol. Chem. 262*:6808 (1987).
29. R. Yurt and K. F. Austen, Preparative purification of the rat mast cell chymase, *J. Exp. Med. 146*:1405 (1977).
30. B. L. Gruber, M. J. Marchese, K. Suzuki, L. B. Schwartz, Y. Okada, H. Nagase, and N. S. Ramamurthy, Synovial procollagenase activation by human mast cell tryptase: dependence upon matrix metalloproteinase 3 activation, *J. Clin. Invest. 84*:1657 (1989).
31. H. Kido, H. Fukusen, N. Katunuma, T. Morita, and S. Iwanaga, Tryptase from rat mast cells converts bovine prothrombin to thrombin, *Biochem. Biophys. Res. Commun. 132*:613 (1985).
32. T. Ishizaka, F. Hirata, A. R. Sterk, K. Ishizaka, and J. A. Axelrod, Bridging of IgE receptors activates phospholipid methylation and adenylate cyclase in mast cell plasma membranes, *Proc. Natl. Acad. Sci. USA 78*:6812 (1981).
33. M. Kurosawa, Protein phosphorylation and inositol phospholipid metabolism in activated mast cells, *Clin. Exp. Allergy 20*:7 (1990).
34. M. Kurosawa and C. Parker, A phosphatidyl-inositol kinase in rat mast cell granules, *J. Immunol. 136*:616 (1986).
35. T. Ishizaka, Analysis of triggering events in mast cells or immunoglobulin E–mediated histamine release, *J. Allergy Clin. Immunol. 67*:90 (1981).
36. H. Kido, N. Fukusen, and N. Katunuma, Antibody and inhibitor of chymase inhibit histamine release in immunoglobulin E–activated mast cells, *Biochem. Int. 10*:863 (1985).
37. S. C. Dietze, C. P. Sommerhoff, and H. Fritz, Inhibition of histamine release from human mast cell ex vivo by natural and synthetic chymase inhibitors, *Biol. Chem. Hoppe-Seyler 371*:75 (1990).
38. H. Kido, N. Fukusen, and N. Katunuma, Antibodies and inhibitor of chymase are

incorporated into mast cell granule and inhibit histamine release, *Biol. Chem. Hoppe-Seyler 369*:95 (1988).
39. A. Nakanishi, K. Morita, M. Oka, and N. Katunuma, Evidence against a possible involvement of the serine and thiol proteases in the exocytotic mechanism of catecholamine secretion in cultured bovine adrenal medullary cells, *Biochem. Int. 13*:799 (1988).
40. Y. Matsunaga, H. Kido, K. Kawaji, K. Kamoshita, N. Katunuma, and T. Ogura, Inhibition of chymotrypsin-like proteases inhibit easinophil peroxidase release from activated human eosinophils, *Arch. Biochem. Biophys. 312*:67 (1994).
41. C. A. Maggi, S. Giuliani, R. Patacchini, P. Rovero, A. Giachetti, and A. Meli, The activity of peptides of the endothelin family in various mammalian smooth muscle preparations, *Eur. J. Pharmacol. 174*:23 (1987).
42. H. Sage, R. G. Woodbury, and P. Bornstein, Structural studies on human type IV collagen, *J. Biol. Chem. 254*:9893 (1979).

12

Role of Mast Cell Proteases and Proteoglycans in Lipoprotein Metabolism

Jorma O. Kokkonen, Ken A. Lindstedt, and Petri T. Kovanen
Wihuri Research Institute, Helsinki, Finland

I. INTRODUCTION

Rat serosal mast cells (i.e., mast cells from rat peritoneal and thoracic cavities) were used to obtain the results presented in this chapter. They were used because of their excellent availability and because they are easily purified to a nearly homogeneous cell population.

A. Mast Cell Degranulation

The most important functional feature of mast cells is their capacity to degranulate upon stimulation. A wide variety of agents can stimulate mast cells, including antigens through the IgE-dependent pathway, various small peptides (e.g., substance P), vasoactive intestinal peptide (VIP), neurotensin, somatostatin, and the complement fragments C3a and C5a (1). In addition, several cells, including macrophages (2), T lymphocytes (3), neutrophils (4), and eosinophils (5), secrete histamine-releasing factors, which act on mast cells as stimulants.

On stimulation of rat serosal mast cells, a fraction of their secretory granules swell and individual granule membranes fuse to form tubes known as degranulation channels, in which the granules lie in chains (6). The degranulation channels then open to the exterior, and the soluble mediators of the rat serosal mast cells, such as histamine, about 40% of the heparin proteoglycans, and all of the chondroitin sulfate proteoglycans are released from the granules into the incubation medium (7). The remaining components of the granules (i.e., all of the neutral proteases and about 60% of the heparin proteoglycans) remain tightly bound to each other and are known as granule remnants (7,8). Our experiments did not reveal the exact mechanism by which the soluble heparin proteoglycan fraction is released from the granules. However, neither chymase nor carboxypeptidase A is released along with this proteoglycan fraction, which strongly suggests that the soluble fraction already exists as a separate pool within the intact membrane-covered granules (7).

After stimulation of rat serosal mast cells, there are two pools of extracellularly located granule remnants. Figure 1 is a scanning electron micrograph of unstimulated (Fig. 1A) and stimulated (Fig. 1B) rat serosal mast cells. Some of the granule remnants have escaped through the openings of the degranulation channels at the cell surface into the "free" extracellular space (Fig. 1B). These granule remnants are seen attached to the coverslip. In rat serosal mast cells, about 25% of the granule remnants that come in contact with the extracellular fluid lose contact with the cell (9). These expelled granule remnants are ultimately phagocytosed and degraded by scavenger cells, such as macrophages, neutrophils, and eosinophils, located in the microenvironment of the stimulated mast cells (10,11). However, as Fig. 1B shows, most of the granule remnants remain cell associated within the degranulation channels and are rapidly (within 10 to 20 min of stimulation) internalized by the recovering mast cells (9,12). During the recovery period, the channel openings close, and each granule remnant ultimately becomes bounded by a membrane, reappearing as a secretory granule in the mast cell cytoplasm (13).

B. Accumulation of LDL-Derived Cholesterol in Atherosclerosis

Low-density lipoproteins (LDL) form the major cholesterol-carrying lipoprotein fraction in human plasma. LDL occurs as spherical particles (diameter about 22 nm) with an M_r value of about 2.5×10^6 (14). Each LDL particle contains cholesteryl esters in a lipid core, which is shielded from the aqueous plasma by a hydrophilic shell composed of unesterified cholesterol, phospholipids, and a single molecule of a 513-kDa protein called apolipoprotein B (apoB) (15,16).

The physiologic role of LDL is to provide the cells of extrahepatic tissues with cholesterol (17). They do this by binding via the apoB component to specific cell surface receptors, the LDL receptors. The cells take up the receptor-bound LDL particles, hydrolyze their cholesteryl esters in lysosomes, and use the unesterified cholesterol that has been liberated for membrane synthesis.

The pathophysiologic effect of LDL in atherosclerosis and its clinical complications are well established. The cholesterol deposited in the arterial intima is generally held to be derived from LDL that has entered the arterial intima from the bloodstream (18). The earliest recognizable gross lesion in

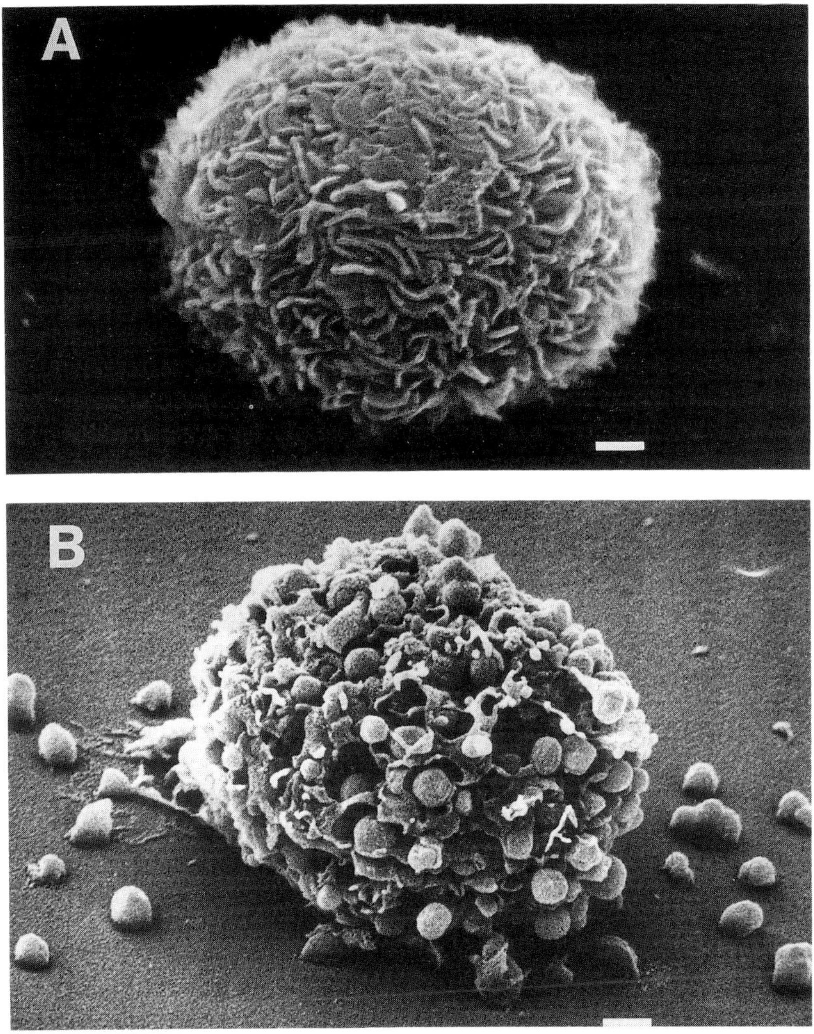

Figure 1 Scanning electron micrographs of an unstimulated (A) and a stimulated (B) rat serosal mast cell. (A) Unstimulated mast cell attached to a polylysine-coated coverslip; (B) mast cell stimulated with compound 48/80 (1 μg/mL) and allowed to recover for 2 min. Most of the spherical granule remnants are cell associated. Some of them are protruding from the degranulation channels. Some granule remnants have lost contact with the cell and are attached to the coverslip. Bar, 1 μm. (From Ref. 39.)

c. Tandem activity of chymase and carboxypeptidase A
d. Release of small peptides (M_r < 5000) and free amino acids (aromatic, branched-chain aliphatic)
3. Features of LDL modification by granule remnants
a. Requirement for binding and extensive proteolysis (>20%) of LDL
b. Fusion of bound LDL particles on the surface of granule remnants
c. Increased LDL particle size up to 100 nm
d. No loss of LDL lipid
e. Decreased LDL density
f. Stronger ionic interaction between apoB and heparin proteoglycan

A. Binding of Lipoproteins to Granule Remnants

Binding of lipoproteins to granule remnants is apolipoproteins (apo) B/E–specific (43). Therefore, only LDL, very low density lipoproteins (VLDL), and HDL subfractions containing apoE are bound to granule remnants. In contrast, a subfraction of human HDL (HDL_3) containing only apoA is not bound to granule remnants. The binding process is rapid, being completed in less than 1 min. At saturation, approximately 100 ng of LDL protein is bound per microgram of granule protein. The granule remnants contain a single class of high-affinity binding sites, being capable of binding 5000 to 10,000 LDL particles per granule remnant. The binding sites are located in the heparin proteoglycan component of the granule remnants (43). LDL particles are bound to the surface of granule remnants only, as indicated by electron microscopic studies (see Figs. 6 to 8). Moreover, the binding of LDL to granule remnants is mediated by ionic interactions between the positively charged domains of LDL (arginine and lysine residues) and the negatively charged sulfate groups of the heparin proteoglycan component of the granule remnants (43).

1. *Binding of LDL to Soluble Heparin Proteoglycans*

We have shown that on stimulation of rat serosal mast cells, the secretory granules involved in degranulation lose about 40% of their heparin proteoglycans and all of the chondroitin sulfate proteoglycans which they initially contained (7). At physiologic ionic strength and calcium ion concentration, the solubilized heparin proteoglycans, but not the chondroitin sulfate proteoglycans, form insoluble complexes with the LDL present. The heparin proteoglycans can bind approximately seven times their own mass (M_r about 1×10^6) of LDL cholesterol. Of the total capacity of the material exocytosed by rat serosal mast cells to sediment LDL, half can be attributed to the granule remnants and half to the soluble heparin proteoglycans (7).

B. Proteolytic Degradation of Lipoproteins by Granule Remnants

All lipoprotein classes (VLDL, LDL, HDL) studies so far in our laboratory are effectively degraded proteolytically by mast cell granule remnants irrespective of whether or not they are able to bind to granule remnants (40,41,46,51). Indeed, the net apoB degradation of a chemically modified LDL (acetylated LDL), which is unable to bind to granule remnants, practically equals that of the native LDL (46). However, if binding takes place as in the case of native LDL, it is only the bound LDL fraction that is degraded (46). On the other hand, the bound LDL particles form only a fraction (10 to 30% in a standard incubation) of the total amount of the LDL in the incubation system (46). Therefore, proteolytic degradation of individual native LDL particles bound to granule remnants proceeds faster, the end result being more extensive degradation per particle after a given incubation time. The data do not explain why only the bound LDL fraction is degraded. The granule remnant-bound LDL particles appear to form a barrier shielding the enzyme molecules from contact with the apoB of the unbound LDL particles (52).

1. *Sequential Degradation of LDL by Chymase and Carboxypeptidase A*

Proteolytic degradation of lipoproteins by mast cell granule remnants results from the coordinated action of the two granule remnant–bound enzymes, chymase and carboxypeptidase A (41). The two neutral proteases act in sequence, the chymase first cleaving peptides from the apoB of LDL, and the carboxypeptidase A then cleaving amino acids from the peptides formed.

Figures 2 and 3 show the results of critical experiments revealing the respective roles of chymase and carboxypeptidase A in the degradation of LDL (41). For these experiments the tyrosine residues of LDL were radiolabeled with ^{125}I. [^{125}I]LDL was incubated with granule remnants in the presence of a selective inhibitor of one enzyme or the other. The formation of radiolabeled low-molecular-weight degradation products was analyzed by sequential fractionation on columns of Bio-Gel A-5m (Fig. 2) and Sephadex G-10 (Fig. 3). Fractionation on Bio-Gel A-5m showed that selective inhibition of carboxypeptidase A did not affect the formation of degradation products (Fig. 2B), whereas selective inhibition of granule chymase totally abolished the degradation of LDL (Fig. 2C). An elution profile identical to that observed after chymase inhibition was obtained in a control incubation in which LDL was incubated in the absence of granule remnants (Fig. 2D). Further analysis of the degradation products on a Sephadex G-10 column disclosed that the inhibition of granule carboxypeptidase A had blocked the production of [^{125}I]tyrosine but not that of the ^{125}I-labeled peptides (Fig. 3B). If granule chymase was inhibited, the only radioactive material present in the eluate was free radioactive iodine which had not been removed from the LDL preparation despite extensive dialysis (Fig. 3C). An identical elution profile was obtained if [^{125}I]LDL was incubated in the absence

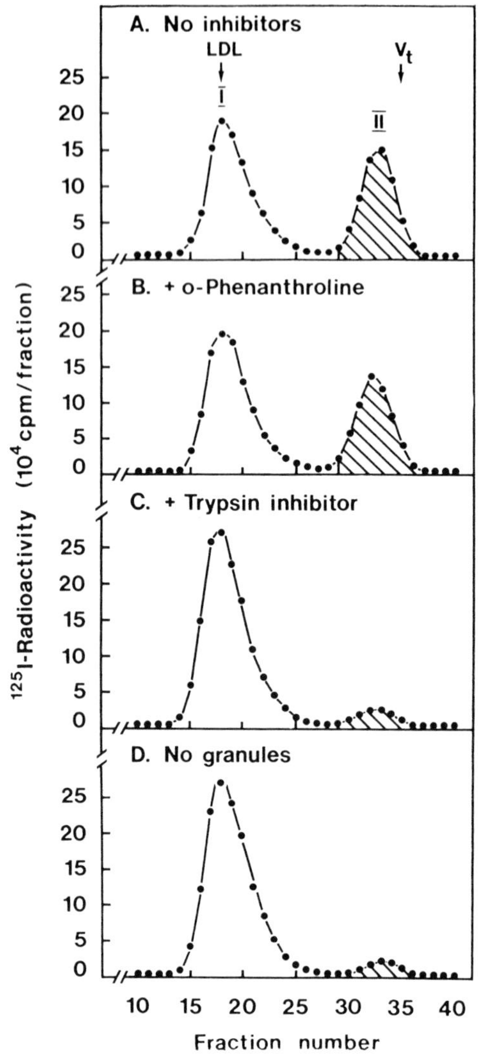

Figure 2 Effect of selective inhibitor of chymase and carboxypeptidase A on LDL degradation by mast cell granule remnants. Granule remnants were preincubated with soybean trypsin inhibitor, an inhibitor of chymase or with *o*-phenanthroline, an inhibitor of carboxypeptidase A, as described in more detail in Ref. 41. The degradation assay contained 100 μg of [^{125}I]LDL and 20 μg of control (A), *o*-phenanthroline–treated (B), or trypsin inhibitor–treated (C) granules. In an additional experiment no granules were added (D). After incubation at 37°C for 3 h the granules were removed by centrifugation. Each supernatant was applied to a Bio-Gel A-5m column (1 × 13 cm). The column was eluted with a buffer (100 m*M* NaCl, 50 m*M* Tris-Cl, pH 7.4) at a flow rate of 3.4 mL/h at 4°C. Fractions of 280 μL were measured for their ^{125}I radioactivity. Fractions 29 to 39 (the shaded area of peak II) were pooled and further analyzed on a Sephadex G-10 column (Fig. 3). The total volume of the column (V_t) and the elution position of native LDL are indicated by arrows. (From Ref. 41.)

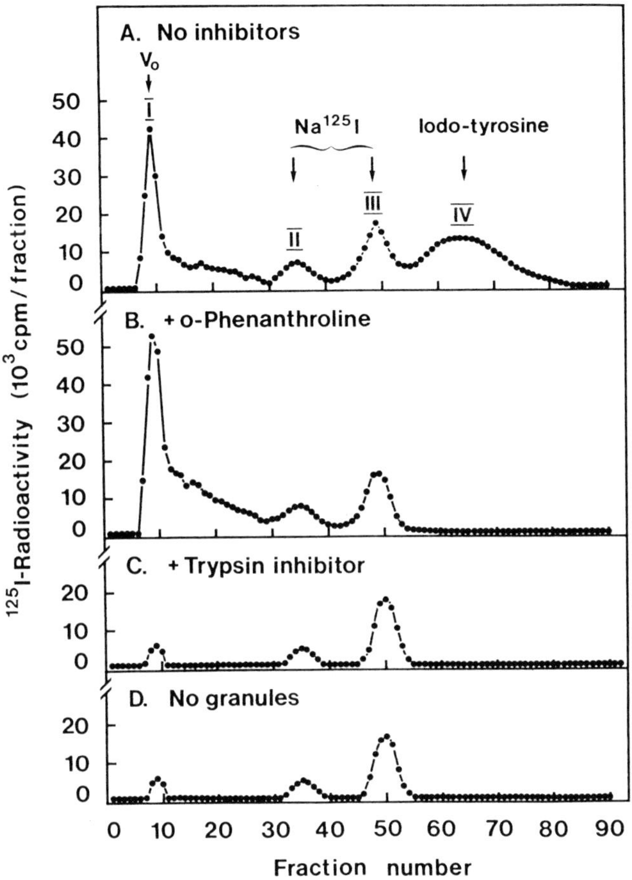

Figure 3 Sephadex G-10 analysis of the degradation products of LDL. The peak II material obtained from the Bio-Gel A-5m analysis (Fig. 2) was applied to a Sephadex G-10 column. Panels A to D show the corresponding analyses of the peak II material represented in Fig. 2 A, B, C, and D, respectively. The column (1 × 30 cm) was eluted with the same buffer as described in Fig. 2 at a flow rate of 5 mL/h at 4°C. Fractions of 370 μL were measured for their ^{125}I radioactivity. The void volume of the column (V_0) and the elution position of commercial $Na^{125}I$ and iodo-L-tyrosine are indicated by arrows. (From Ref. 41.)

of granule remnants (Fig. 3D). Thus for the formation of ^{125}I-labeled peptides only chymase activity was needed, whereas for the formation of [^{125}I]tyrosine both chymase and carboxypeptidase activities were necessary.

Figure 4 shows a model of LDL degradation by these two granule enzymes. Chymase, an endopeptidase, can hydrolyze the peptide bond provided

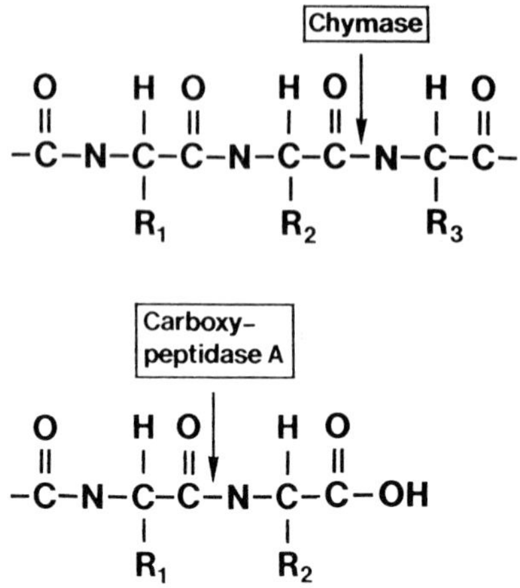

Figure 4 Sequential degradation of apoB of LDL by mast cell granule chymase and carboxypeptidase A. Chymase, an endopeptidase, first cleaves the peptide bond, provided that the amino acid in the R_2 position is aromatic or a branched-chain aliphatic. The peptide formed serves as a substrate for carboxypeptidase A, an exopeptidase. The result of the coordinated action of these two enzymes is liberation of small peptides and aromatic and branched-chain aliphatic amino acids. (From Ref. 50.)

that there is an aromatic (Phe, Tyr, Trp) or a branched-chain aliphatic (Leu, Ile, Val) amino acid in the R_2 position. Analysis of the breakdown products released by chymase from the apoB of LDL revealed the presence of numerous small peptides ($M_r < 5000$) of varying sizes (41). Carboxypeptidase A then acts by cleaving the carboxy-terminal amino acid R_2 from the peptides formed. The soluble heparin proteoglycan fraction contains no measurable chymase or carboxypeptidase A activity (7).

2. *Proteolytic Degradation of LDL by Human Mast Cell Granules*

Human lung T mast cells, which contain tryptase as their main proteolytic enzyme, appear to be unable to degrade LDL (K. A. Lindstedt, J. O. Kokkonen, and P. T. Kovanen, unpublished observations). No experiments have yet been made with human TC mast cells.

C. Modification of LDL by Granule Remnants

Binding and proteolytic degradation of LDL by mast cell granule remnants are closely related events. Indeed, as described in Section II.B, in the presence of excess LDL, only the apoB moieties of those LDL particles that have been bound to the granule remnants are degraded by the granule enzymes. Moreover, we have found that this proteolysis of LDL increases the binding capacity of the granule remnants for LDL. When the binding of LDL to mast cell granule remnants is followed at 37°C for various lengths of time (from 5 min to 8 h), LDL is found to be progressively proteolyzed by the granule remnants. Furthermore, the amount of LDL bound to the granule remnants increases progressively, being about fivefold higher after incubation for 8 h. The increase in LDL binding occurs even when saturating LDL concentrations are used to ensure that all LDL-binding sites are occupied during incubation (44).

The mechanism behind the increased binding of LDL to the proteolytically active granule remnants is fusion of several LDL molecules to form larger LDL particles on the granule surface (44). Figure 5 shows negatively stained preparations of granule remnant–bound LDL after incubation with granule remnants for 8 h. Panel A shows an electron micrograph of a negatively stained preparation of native LDL. The mean diameter of this particle population is 21.6 ± 1.4 nm (±SD). Panel B shows LDL particles released from proteolytically inactive granule remnants [treated with phenylmethanesulfonyl fluoride (PMSF), an inhibitor of chymase]. They are of the same size as native LDL particles, the mean diameter being 21.3 ± 1.1 nm (± SD). Panel C shows the particles released from the proteolytically active (i.e., native) granule remnants. They consist of a heterogeneous population of spherical particles which are, on the

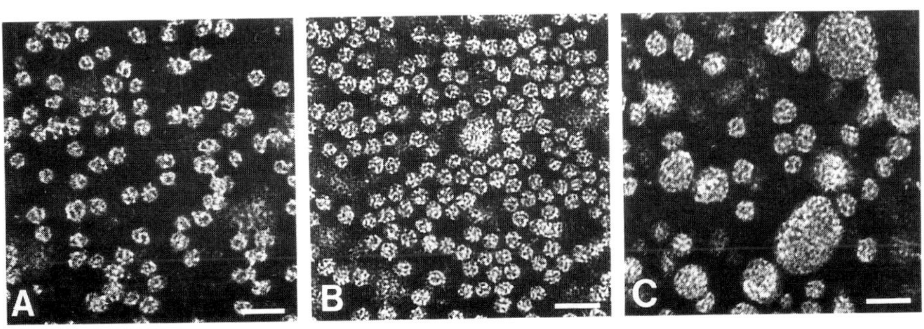

Figure 5 Electron micrographs of negatively stained LDL fractions released from proteolytically inactive and active granule remnants. Twenty micrograms of proteolytically active or inactive (PMSF-treated) granule remnants were incubated at 37°C with 100 μg of LDL. After incubation for 8 h the granule remnants were sedimented to remove unbound LDL and the granule remnant–bound LDL was released by incubation in 0.5 *M* NaCl. Portions of native LDL and released LDL fractions were negatively stained with 1% sodium phosphotungstate as described in more detail in Ref. 44. (A) Native LDL; (B) LDL released from PMSF-treated granule remnants; (C) LDL released from proteolytically active granule remnants. Bar, 50 nm. (From Ref. 44.)

whole, larger than native LDL particles, their mean diameter being 32.8 ± 15.5 nm (± SD). The largest particles have diameters of about 100 nm. It can be calculated that if the diameter of a spherical LDL particle increases from 22 nm to 100 nm, its volume increases 90-fold. Thus the largest particles were formed by fusion of about 90 native LDL particles.

In Fig. 6 the morphology of the LDL particles bound to the granule remnants is shown by transmission electron microscopy. To preserve the associated lipid structures, the granule remnants were postfixed with osmium–thiocarbohydrazine–osmium (OTO) (46). Electron micrographs of the proteolytically active granule remnants show large fused particles of LDL on the granule surface (Fig. 6B). Black rims, typical of OTO-stained lipid droplets, encircle these fused particles. The particles have diameters ranging from about 30 to 80 nm, with a mean diameter of 55 ± 12 nm (± SD). A granule remnant incubated without LDL is shown in Fig. 6A.

The ionic interaction between apoB and heparin proteoglycan seems to become stronger during incubation of proteolytically active granule remnants with LDL. In the early stage of incubation, all the granule remnant–bound LDL can be released by 0.3 *M* NaCl (the "salt-sensitive" fraction of LDL). With time, an increasing proportion of the granule remnant-bound LDL requires 0.5 *M* NaCl for release (the "salt-resistant" fraction of LDL) (46). Thus on the granule surface there appears to be a flux of LDL particles from the salt-sensitive to the salt-resistant fraction. Chemical analyses show that about 20% of the apoB of LDL is lost from the salt-sensitive fraction and 60% from the salt-resistant fraction, without any change in the composition of the lipid portion. As a result, there is a decrease in density as compared with native LDL particles. The mean densities of the native, salt-sensitive, and salt-resistant LDL fractions are 1.040, 1.026, and < 1.015 g/mL, respectively (46). Electron microscopic analysis reveals that large fused particles of LDL reside in the highly proteolyzed salt-resistant fraction, but no fused particles can be found in the less proteolyzed salt-sensitive fraction (46).

The exact mechanism of the fusion process is unknown. It seems evident that both binding and extensive degradation of LDL by mast cell granule remnants is required for fusion of LDL particles to take place on the granule surface. The data presented here suggest that the degree of proteolysis per LDL particle predisposing to fusion must be over 20%. Breakdown in the apoB moiety of LDL appears to result in loss of integrity of the polar shell, and the consequent instability of the lipoprotein particles leads to fusion of their apolar cores.

The fused LDL particles in the salt-resistant LDL fraction represent a hitherto uncharacterized form of lipoprotein modification, the mast cell–modified LDL. In addition to increased particle size, the properties of this modified LDL include selective loss of apoB, with a consequent decrease in density and a stron-

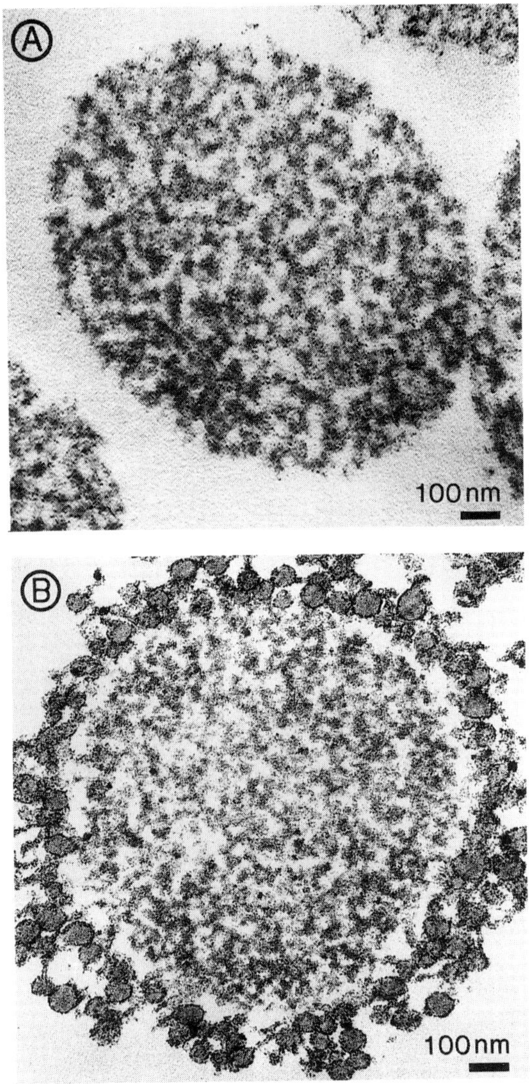

Figure 6 Electron micrographs of granule remnants incubated without LDL (A) and with LDL (B). (A) Control granule remnants incubated at 37°C for 18 h without LDL. (B) 40 μg of granule remnants was incubated at 37°C with 200 μg of LDL for 18 h. After incubation the granules were sedimented by centrifugation and prepared for electron microscopy as described in more detail in Ref. 46. Black rims encircle the fused particles of LDL on the granule surface. Bar. 100 nm.

ger ionic interaction between the apoB component and the heparin proteoglycan of the granule remnant. Such mast cell–modified LDL particles, unlike native LDL, are taken up by macrophages as granule remnant–LDL complexes at a greatly enhanced rate (42).

III. EFFECT OF MAST CELLS ON LIPOPROTEIN METABOLISM

A. Granule Remnant–Mediated Uptake of LDL by Macrophages In Vitro

As described in Section I.A, after mast cell stimulation, a fraction of the exocytosed granule remnants are phagocytosed by scavenging cells such as macrophages. If proteolytically inactive granule remnants are co-cultured with mouse macrophages in vitro in the presence of LDL, the uptake of LDL by cultured mouse macrophages increases markedly (approximately 10-fold) (42). Calculations show that an LDL uptake of 40 μg of cholesteryl esters per milligram of cell protein requires phagocytosis of approximately 110 granule remnants per macrophage (39).

If proteolytically active granule remnants are used instead of proteolytically inactive granule remnants, the rate of LDL uptake increases 50-fold, whereas with proteolytically inactive granule remnants the rise is approximately 10-fold (44). The fivefold greater capacity of the proteolytically active granule remnants to enhance the uptake of LDL by macrophages results from their greater capacity to bind LDL due to fusion of LDL particles on the granule surface (see Section II.C). Thus the two major components of mast cell granule remnants, the proteases and heparin proteoglycans, act in concert to promote the uptake of LDL by macrophages in vitro.

The most critical step in the mast cell-mediated uptake of LDL by macrophages is the binding of LDL to the granule remnants. Thus treatment of the granule remnants with avidin or protamine or treatment with LDL with 1,2-cyclohexanedione, all of which inhibit the binding of LDL to granule remnants, totally prevents granule remnant–mediated uptake of LDL (42).

The uptake process itself appears to be phagocytosis of LDL-bearing granule remnants (42,44). The trigger for the uptake of LDL-bearing granule remnants is not known. However, the presence of native LDL or modified LDL on the granule surfaces neither triggers nor enhances the phagocytosis. Thus it seems evident that the recognition site is not an LDL receptor but that the granule remnants themselves trigger phagocytosis.

When engulfed by macrophages, the LDL particles are hydrolyzed within the lysosomes, as demonstrated by the inhibition of granule remnant–mediated stimulation of cholesteryl ester synthesis by the lysosomotropic agents chloroquine and ammonium chloride (42). Lysosomal hydrolysis of the cholesteryl esters of LDL and subsequent reesterification of the cholesterol lead to formation of the cholesteryl ester droplets characteristic of foam cells. Thus the process parallels the pathway described in macrophages incubated with acetylated LDL (21).

1. *Uptake and Degradation of Insoluble Complexes of Mast Cell Soluble Heparin Proteoglycans and LDL by Macrophages*

As described in Section II.A.1, the soluble heparin proteoglycans released on stimulation of rat serosal mast cells are capable of forming insoluble complexes with LDL particles. These complexes are avidly taken up by cultured mouse macrophages, uptake being 45-fold higher than that of native LDL (7). As with granule remnants, the mode of uptake is phagocytosis. But in contrast to granule remnant–mediated uptake of LDL, uptake of these complexes is inhibited by acetylated LDL and polyinosinic acid in excess, indicating that uptake is mediated by scavenger receptors (7). Indeed, cultured mouse macrophages avidly take up insoluble heparin proteoglycan–LDL complexes by scavenger receptor-–mediated phagocytosis. The soluble heparin proteoglycan-mediated LDL uptake represents about 50% of the total (granule remnants + soluble heparin proteoglycans) mast cell–mediated LDL uptake by macrophages.

B. Granule Remnant–Mediated Uptake of LDL by Macrophages In Vivo

Since binding of LDL to mast cell granule remnants is critical for mast cell-–mediated uptake of LDL to occur, it was of special importance to test whether mast cell stimulation in vivo, in the presence of natural binding inhibitors (which include fibronectin and fibrinogen), would augment LDL uptake by macrophages (53,54). Such experiments were conducted in the peritoneal cavity of the rat, a site known to contain both macrophages and mast cells (45). Figure 7 shows an experiment in which the role of exocytosed granule remnants as carriers of LDL to the macrophages was visualized with gold-labeled LDL. The results demonstrate that stimulated mast cells do promote uptake of LDL by macrophages in the peritoneal cavity of the rat, a site where natural binding inhibitors are present. The observations that interaction of LDL with mast cell granule remnants or with the soluble heparin proteoglycans markedly increases the uptake of LDL by macrophages and leads to foam cell formation means that mast cells are potentially atherogenic.

C. Granule Remnant–Mediated Inhibition of Reverse Cholesterol Transport

According to the cholesterol balance hypothesis of atherogenesis (27), accumulation of LDL-derived cholesterol in the intima is prevented by HDL particles

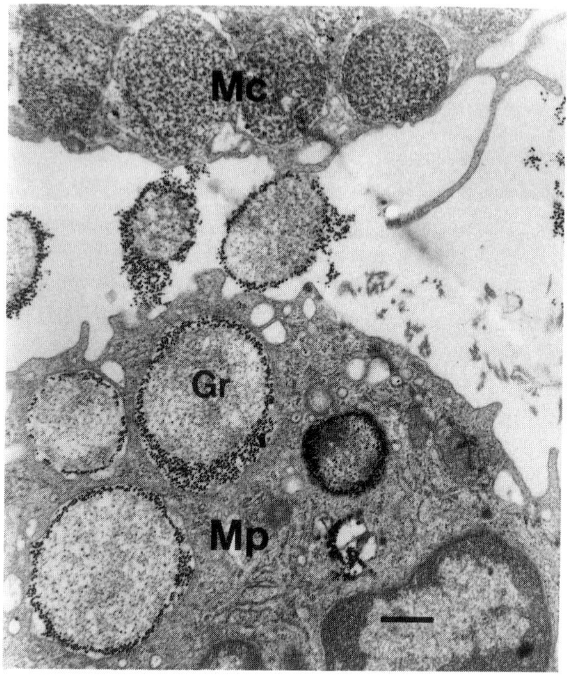

Figure 7 Electron micrograph of mast cell granule remnant–mediated uptake of LDL by macrophages in vivo. Each rat was given an intraperitoneal injection of 500 μL of phosphate-buffered saline (PBS) containing 200 μg of gold-labeled LDL and 100 μg of compound 48/80. After 1 h, the cells of the peritoneal cavities were collected and prepared for transmission electron microscopy as described in more detail in Ref. 45. The upper part of the figure shows a stimulated mast cell with three extracellularly located granule remnants coated with gold-LDL (small black dots). Inside the macrophage three phagosomes can be seen, each containing a LDL-bearing granule remnant. Bar, 0.5 μm. Mc, mast cell; Mp, macrophage; Gr, mast cell granule remnant. (From Ref. 50.)

in a reverse cholesterol transport process (28). In this process, HDL carries accumulated cholesterol from the macrophage foam cells back into the bloodstream.

In the presence of rat serosal mast cells stimulated to trigger exocytosis of their cytoplasmic secretory granules, the ability of HDL to induce cholesterol efflux from cholesterol-loaded macrophages is largely lost. This loss is due to the proteolytic action of chymase, which degrades the apolipoproteins of HDL, rendering them unable to mediate cholesterol efflux from the foam cells (51). Degradation of less than 4% of the apolipoproteins of HDL reduces cholesterol efflux to about half. Such minimal degradation of HDL has no effect on the mobility of HDL when analyzed by agarose gel electrophoresis (51). The observation that mast cell chymase prevents cholesterol efflux from the macrophages, thus reversing the cholesterol balance toward intracellular cholesterol accumulation, an early stage of atherogenesis, is a potentially atherogenic effect.

D. Internalization and Modification of LDL in Mast Cells Recovering from Degranulation

After stimulation of rat serosal mast cells, about 75% of the granule remnants remain in the degranulation channels and are ultimately internalized by the recovering mast cells (9). Like the granule remnants expelled into the "free" extracellular space, the granule remnants located in the degranulation channels of stimulated mast cells also bind LDL (9). Moreover, such remnant-bound LDL is rapidly (within 10 to 20 min) internalized by the mast cells along with the granule remnants (Fig. 8). As expected, the uptake of LDL particles by mast cells depends on their ability to bind to exocytosed granule remnants. Acetylated LDL and HDL_3, two classes of lipoprotein particles unable to bind to granule remnants (43), are not internalized by recovering mast cells (9). After a single stimulation in the presence of LDL, mast cells are filled with material staining positively with Oil Red O (i.e., with neutral lipid). This change is accompanied by a 30-fold increase in the cellular content of cholesteryl esters (9).

During subsequent recovery from degranulation, the apoB of the intracellularly located remnant-bound LDL becomes effectively (up to 70%) degraded by the proteolytic enzymes of the granules. But because the granules lack cholesteryl esterase activity, the LDL cholesterol is not made available for cellular nutrition. Instead, selective proteolytic degradation of the bound LDL leads to fusion and enlargement of the LDL particles on the granule surface similar to that observed on isolated granule remnants (48).

Rat serosal mast cells recover rapidly from stimulation. In vitro, the rat serosal mast cells were ready for a second stimulation after a 90-min recovery period (48). This restimulation affected about 50% of the intracellularly modified LDL particles. Of these, about one-third were expelled into the incubation medium with the granules. The other two-thirds of the affected granules, which remained in the degranulation channels, bound and internalized more LDL from the extracellular fluid. After recovery from the second stimulation the intracellular cholesterol content typically had increased about threefold (48). Thus during the second internalization more LDL was taken up than had been secreted, indicating a mechanism by which LDL could accumulate in mast cells. Moreover, the results demonstrated that mast cells, when stimulated, are able to take up, modify, and secrete LDL repeatedly.

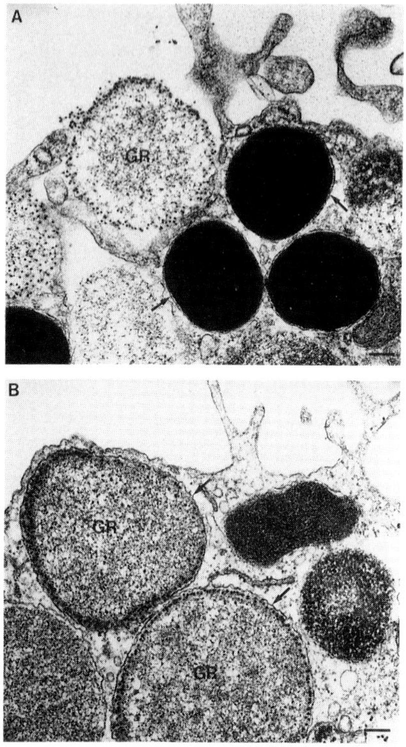

Figure 8 Electron micrographs of uptake of gold-labeled LDL by recovering mast cells. Rat serosal mast cells were stimulated with compound 48/80. After incubation for 2 min at 37°C, the reaction was stopped by adding ice-cold buffer, and the cells were sedimented and resuspended in warm (37°C) buffer containing gold-LDL. After incubation for 1 min (A) or for 30 min (B) at 37°C, the cells were prepared for electron microscopy as described in more detail in Ref. 48. (A) The electron-dense granules surrounded by granule membranes (arrows) represent intact secretory granules that were not involved in degranulation. An electron-transparent granule remnant (GR) that is covered with gold-LDL (small black dots) is seen within the degranulation channel opening. (B) After a 30-min recovery period channel openings are seen to be closed, and the LDL-containing granule remnants are again surrounded by membranes (arrows). Bar, 200 nm. GR, mast cell granule remnant. (From Ref. 48.)

Since the critical factor in the granule remnant–dependent modification of LDL is the proteolytic activity of chymase, inhibitors of proteolysis, if present in the extracellular fluid, will block LDL modification. High concentrations (1 mg/mL) of soybean trypsin inhibitor or high concentrations (up to 20 mg/mL of protein) of rat or human lipoprotein–deficient serum containing natural neutral protease inhibitors will totally inhibit the extracellular granule remnant-mediated degradation of LDL (J. O. Kokkonen and P. T. Kovanen, unpublished observations). In contrast, intracellular LDL degradation is not inhibited by inhibitors present in the extracellular fluid, suggesting that proteolytic inhibitors are excluded from the granule compartment (48). In terms of proteolysis, the granule compartment of stimulated mast cells appears to be a "protected" environment.

This round trip of LDL in mast cells exposed to repeated stimulation constitutes a hitherto unknown intracellular pathway of LDL (9,48). Rat serosal mast cells take up LDL by a highly specific process in which LDL is bound to the heparin proteoglycan component of the granule remnants. After internalization of the granule remnants, the LDL is selectively degraded. This contrasts with the fate of LDL particles in typical lysosomes, in which LDL disintegrates completely as a result of complete hydrolysis of all the components (i.e., not only apoB but also the various lipids). In terms of the metabolic fate of LDL, a critical difference between typical lysosomes and the secretory granules, which are considered to be modified lysosomes, is that the granules contain only one major pair of proteolytic enzymes, chymase and carboxypeptidase A, and are totally without lipolytic activity (8,41,46). Because cholesteryl ester hydrolase activity is absent from the secretory granules, no cholesterol can be released from the LDL particles. Accordingly, even massive uptake of LDL by stimulated mast cells will not result in transfer of cholesterol from the granule compartment to the cytosol. This, again, implies that all the LDL cholesterol present in the granule compartment of mast cell is metabolically inert. In this respect, the metabolic fate of LDL cholesterol in the mast cell granule pathway would contrast sharply with that in the LDL receptor pathway (17).

We have shown previously that rat serosal mast cells do not express LDL receptors even under in vivo conditions that greatly enhance LDL receptor activity in other rat cell types, such as those in the liver and adrenal gland (40). Accordingly, binding of LDL by the granule remnants in the degranulation channels appears to be the only specific and regulated pathway that mast cells possess for uptake of LDL.

E. Inhibition of Oxidation of LDL by Mast Cells

As described in Section I.B.1, LDL modification is generally held to be a prerequisite for cholesterol accumulation in atherosclerosis. Recent findings that oxidatively modified LDL particles are indeed present in atherosclerotic lesions of both human and animal arteries (23,24) has led to intensive research on this modification.

Transition metal ions (e.g., copper) play an important role in LDL oxidation. Indeed, LDL is avidly oxidized in the presence of micromolar concentrations of copper ions (55). Moreover, this oxidative modification of LDL is fully inhibited by chelators of transition metal ions [e.g., ethylenediaminetetra-acetic acid (EDTA)]. The LDL particles undergo several changes in structural and biological properties during the oxidative process and are recognized and avidly taken up by macrophages (56). Such copper ion–oxidized LDL particles have actually been used as the "prototype" in the investigations on oxidatively modified LDL particles.

In vitro, not only copper ions but also several cell types present in the arterial intima are able to modify LDL oxidatively into a form that is recognized and taken up avidly by macrophages. These cells include endothelial cells (57), smooth muscle cells (58), lymphocytes (59), and the macrophages themselves (60). The cell-mediated changes in the LDL particles also require the presence of transition metal ions (55,56,61). But in contrast to the micromolar concentration of copper ions needed for the copper ion–mediated oxidation of LDL, only nanomolar concentrations of copper ions are needed to propagate the lipid peroxidation process induced by the cells (56). Whatever the mechanism by which LDL becomes oxidatively modified in vivo, the oxidized LDL seems to have properties similar to those of LDL oxidized in vitro either by cells or with copper ions.

1. *Inhibition of Copper Ion–Mediated Oxidation of LDL by Mast Cells*

Stimulation of rat serosal mast cells with compound 48/80 in the presence of LDL and copper sulfate leads to total inhibition of the copper ion–mediated oxidation of LDL (47). The inhibitory effect depends on the ability of the neutral protease chymase of exocytosed mast cell granule remnants to degrade the apoB of LDL. This proteolysis releases degradation products consisting of peptides (M_r < 5000) and amino aids capable of binding any copper ions present in the aqueous phase, with formation of apparently redox-inactive copper ion complexes (47). Moreover, the apoB degradation products formed in the granule compartment of mast cells recovering from stimulation are released into the extracellular fluid on restimulation and effectively inhibit copper ion–mediated oxidation of LDL (62).

In addition, it was shown that copper ions bound to the apoB of LDL are sufficient to cause LDL oxidation and that removal of such copper ion–containing apoB peptides from LDL particles inhibits oxidation of LDL (47). Hence the protective effect of stimulated mast cells against copper ion–mediated oxidation of LDL depends on two mechanisms: (a) binding of free copper ions to the peptides and amino acids released from the apoB of LDL by the proteolytic action of the granule chymase, and (b) release of copper ion–containing apoB peptides from the LDL particles by the granule chymase.

The amino acids histidine, cysteine, and tryptophan have been shown to be capable of forming complexes with copper ions (63–65). Since these amino acids are present among the peptides (M_r < 5000) released from the apoB of LDL by the proteolytic action of mast cell granule remnants (41), they appear likely to be the breakdown products responsible for the observed binding of copper ions. The frequent occurrence of these amino acids in proteins suggests further that other proteolytic peptides besides those derived from apoB of LDL can act as copper ion chelators. As suspected, HDL degradation products have also been found to be capable of inhibiting the copper ion–mediated oxidation of LDL (K. A. Lindstedt and P. T. Kovanen, unpublished observations). The ability of the small apoB-derived degradation products to bind copper ions is the first biological function assigned to them.

2. *Inhibition of Macrophage-Mediated Oxidation of LDL by Mast Cells*

The effect of the apoB degradation products on macrophage-mediated LDL oxidation, which represents a more "biological" model of LDL oxidation, is rather weak, suggesting that in this assay the copper ions bound to them were still redox active. However, stimulation of the mast cells with compound 48/80 resulted in about 80% inhibition of LDL oxidation by co-cultured macrophages (49). The inhibitory effect depended on the degree of mast cell stimulation and resided in the histamine fraction. Histamine, being a histidine derivative, effectively binds copper ions and totally inhibits the macrophage-mediated oxidation of LDL (49). Thus mast cells, if sufficiently stimulated, can prevent macrophage-mediated oxidation of LDL and may play a preventive role in atherosclerosis.

IV. ROLE OF MAST CELLS IN ATHEROSCLEROSIS

A. Evidence from Experiments in Rats

The main observations relating to lipoprotein metabolism by rat serosal mast cells are summarized in Fig. 9. Panel A shows the potentially atherogenic effects and panel B shows the anti-atherogenic effects. Mast cell stimulation in the presence of LDL leads to binding of LDL to the exocytosed granule remnants, whether they are expelled into the "free" extracellular space or remain associated with the mast cells. The LDL molecules bound by the granule remnants are eventually either transported to macrophages or internalized by the recovering mast cells. The concomitant degradation of the apoB moiety of the bound LDL by the neutral proteases of the granule remnants located in the free extracellular space leads to modification of the LDL in such a way that large, fused LDL particles are formed on each granule remnant surface. The significance of this modification is that it increases the capacity of the granule remnants to bind LDL approximately fivefold, enabling them to carry five times as much LDL into macrophages as granule remnants in which proteolytic activity has been inhibited. Moreover, when stimulated, rat serosal mast cells secrete granules from

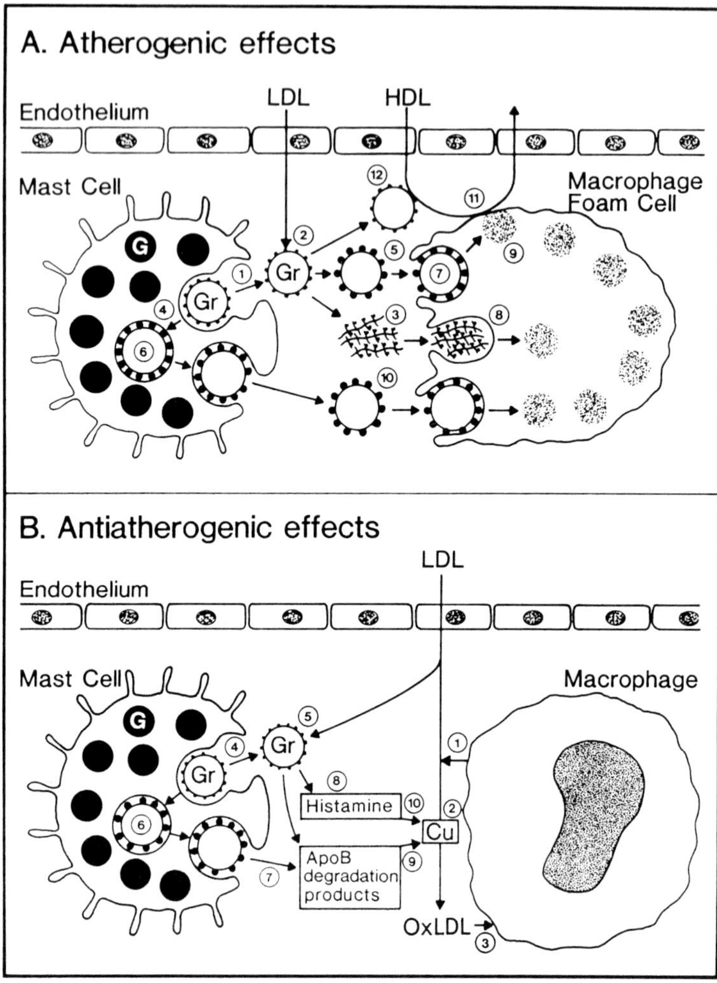

Figure 9 Graphical representation of atherogenic (A) and antiatherogenic (B) effects of stimulated mast cells on lipoprotein metabolism in the arterial intima. (A) Stimulation of mast cells leads to formation of two pools of exocytosed granule remnants (Gr). One pool remains in the degranulation channels of the parent mast cell. A second pool is secreted into the "free" extracellular space (1). Large black circles (G) represent secretory granules not involved in degranulation. Low-density lipoprotein (LDL) particles are transported into the arterial intima through the endothelium. Mast cell stimulation in the presence of LDL (small black dots) leads to binding of LDL to both pools of exocytosed granule remnants (2) and to the soluble heparin proteoglycan fraction released from the granules on stimulation (3). The mast cell–associated LDL-bearing granule remnants are rapidly internalized by the recovering mast cell (4). Concomitant proteolytic degradation of bound LDL by the neutral proteases of the granule remnants located in the free ex-

which soluble heparin proteoglycans dissolve in the surrounding fluid; these mediators are capable of forming insoluble complexes with LDL particles. These complexes are avidly taken up by cultured mouse macrophages. Phagocytosis by macrophages of LDL-coated granule remnants or insoluble complexes between LDL and the soluble heparin proteoglycan fraction leads to accumulation of non-membrane-bound cholesteryl ester droplets in the macrophages, which thus become typical foam cells (Fig. 9A).

The LDL molecules taken up by the recovering mast cells are degraded by the neutral proteases of the granule remnants within the granule compartment. This leads to formation of fused LDL particles on the surface of the internalized mast cell granule remnants similar to those described for the extracellular granule remnants. However, this mechanism does not produce a foam cell, for no non-membrane-bound cholesteryl ester droplets, such as those present in macrophage foam cells, are formed. On restimulation of the mast cell, some of the modified LDL particles are resecreted along with the granules, thereby creating an intracellular LDL loop. Whether or not LDL cholesterol accumu-

---

tracellular space (5) or inside the mast cell (6) leads to modification of LDL in such a way that large fused LDL particles (enlarged black dots) are formed on the granule surface. Phagocytosis by macrophages of the granule remnants containing fused LDL particles (7) or insoluble complexes between the soluble heparin proteoglycan fraction and LDL (8) leads to accumulation of nonmembrane-bound cholesteryl ester droplets (large gray dots) in macrophages, typical of foam cells (9). After restimulation of the mast cell, the granule remnants containing modified LDL particles are expelled into the free extracellular space and ultimately phagocytosed by macrophages (10). HDL particles transported through the endothelium into the intima remove cholesterol from macrophage foam cells and carry it back to the bloodstream (11). Proteolytic degradation of HDL particles by the granule remnants prevents HDL-mediated cholesterol efflux from foam cells (12). (B) Macrophages oxidatively modify LDL particles in the presence of copper ions (1). Active copper ions (Cu) are also able to modify LDL oxidatively in the absence of cells (2). Oxidized LDL (OxLDL) is taken up by macrophages leading to accumulation of cholesterol in macrophages, typical of foam cells (3). Stimulation of mast cells leads to formation of two pools of exocytosed granule remnants, which both bind extracellular LDL particles (4). Proteolytic degradation of bound LDL by the neutral proteases of the granule remnants located in the free extracellular space (5) or inside the mast cell (6) leads to formation of degradation products (small apoB-derived peptides and amino acids). After restimulation of the mast cell, the apoB degradation products contained in the granule compartment are released into the free extracellular space (7). Histamine is released from the granules on stimulation (8). The apoB peptides and amino acids (9) together with the histamine (10) chelate the copper ions and inhibit the macrophage- and copper ion–mediated LDL oxidation. Consequently, cholesterol accumulation in macrophages and foam cell formation are inhibited.

lates in mast cells depends on the balance between the amounts of LDL secreted and internalized by the mast cell on repeated stimulation. The LDL-bearing granules released into the free extracellular space are eventually phagocytosed by macrophages. The fate of the granule remnant–modified LDL particles seems ultimately to be phagocytosis by scavenger cells, regardless of whether they are modified extracellularly or inside the mast cell (Fig. 9A). Proteolytic degradation by the exocytosed granule remnants renders HDL, a cholesterol acceptor, unable to mediate cholesterol efflux from the macrophage, thus promoting cellular accumulation of cholesterol in these cells (Fig. 9A).

In contrast to actions that favor cholesterol accumulation, stimulation of rat serosal mast cells may lead to inhibition of the oxidation of LDL. The inhibitory effect depends on the extracellular and intracellular generation of copper-binding apoB peptides and amino acids from LDL by chymase. Extracellularly, the granule remnant–bound LDL fraction acts as the source of copper ion–binding peptides and amino acids. Intracellularly, copper ion–binding apoB degradation products are produced in the granule compartment containing internalized LDL. In addition, removal of copper ion–containing apoB peptides from LDL particles by granule chymase inhibits oxidation of LDL (not shown in panel B). Furthermore, histamine released from the mast cells on stimulation binds copper ions and totally inhibits macrophage-mediated oxidation of LDL (Fig. 9B).

B. Prerequisites for Human Mast Cells to Affect Lipoprotein Metabolism

In humans, the most critical site where mast cells, macrophages, and LDL coexist is the arterial intima. Under what conditions would mast cells act on lipoproteins in the arterial intima?

1. *Stimulation of Mast Cells*

For mast cells to affect the metabolism of lipoproteins, the cells must first be stimulated to degranulate. Since the process of degranulation is unaffected by the nature of the stimulus, it appears likely that any stimulus leading to mast cell degranulation will initiate the sequence of events that we have described. Recent studies suggest that mast cells are stimulated by activated cells located in their microenvironment. Such cells include macrophages and activated T lymphocytes (66), which are also to be found in the atherosclerotic human intima. In addition, activated complement, formed after local tissue damage and potentially capable of stimulating mast cells, is also present in the atherosclerotic human arterial intima (67). Thus the mast cells of atherosclerotic lesions reside in immunologically active areas containing many factors capable of triggering their degranulation.

2. *Binding of LDL to Granule Remnants*

The most critical prerequisite for mast cell granule remnant–induced uptake of LDL by macrophages and by recovering mast cells is the binding of LDL to the exocytosed granule remnants. This binding occurs after interaction between the apoB component of LDL and the heparin proteoglycan component of the granule remnants. The proteoglycan component of human mast cell granules is composed of highly sulfated molecules such as heparin and chondroitin sulfate E (68). Among glycosaminoglycans tested in vitro, the one with the highest affinity for LDL is heparin (69). Moreover, observations in vivo demonstrate that TC mast cells of the human skin (70) may extrude granule remnants similar in appearance to those of rat serosal mast cells. Thus binding of LDL to the extruded granule remnants of human TC mast cells can be expected to occur under conditions similar to those that permit binding to LDL to the granule remnants of rat serosal mast cells in vivo. Rat serosal mast cells also secrete soluble heparin proteoglycans, and these form insoluble complexes with LDL that are eventually phagocytosed by macrophages. Whether such soluble proteoglycans are liberated on stimulation of human TC mast cells is not known.

Two critical conditions determining the magnitude of granule binding (and hence the efficiency of the granule remnant–mediated LDL uptake by macrophages) are the concentration of LDL and the presence in the extracellular fluid of other molecules that have an affinity for heparin. In the presence of natural binding inhibitors, high concentrations of LDL favor LDL binding to rat mast cell granule remnants (45). The molecules with affinity for heparin that are present in the human arterial intima include fibronectin and fibronigen (53,54). In the extracellular fluid of most tissues, the LDL, because of its large size, is present at concentrations much lower than those of the low-molecular-weight molecules listed above. Indeed, the concentration of LDL in lymphatic fluid is only one-tenth of that in the circulating blood (71). In the human arterial intima, in contrast, the concentration of LDL is high, equaling or even exceeding that in the blood circulation (72). Hence, compared with mast cells in other tissue sites, the mast cells in the human arterial intima reside in an environment that favors binding of LDL to the extracellularly located mast cell granule remnants with ensuing uptake of LDL by macrophages and stimulated mast cells.

3. *Degradation of LDL and HDL by Granule Remnants*

Since the critical factor in granule remnant–dependent modification of LDL, generation of antioxidative peptides, and proteolytic degradation of cholesterol acceptor HDL is the proteolytic activity of chymase, the presence of inhibitors of proteolysis in the extracellular fluid may block these effects. However, stimulated mast cells take up and degrade LDL in the presence of high concentrations of protease inhibitors. Thus, in terms of the enzymatic activity of mast cell neutral proteases, the granule compartment of stimulated mast cells appears to be a "protected" environment, which suggests that these are the sites of production of peptides and amino acids.

V. CONCLUSIONS

Some of the multiple actions of mast cells on lipoprotein metabolism appear to favor the accumulation of cholesterol in macrophages, the stimulated mast cells being potentially powerful inductors of foam cell formation, which is the hallmark of early atherosclerosis. These actions include proteolytic modification of LDL both extracellularly and within the recovering mast cell, with ensuing uptake of LDL-containing granule remnants by macrophages. In addition, phagocytosis by macrophages of insoluble complexes between LDL and the soluble heparin proteoglycan fraction leads to foam cell formation. Furthermore, proteolytic degradation of HDL particles may disturb the cholesterol balance in macrophages by preventing efflux of cholesterol from them. On the other hand, there are other actions suggesting that mast cells may prevent oxidation of LDL, implying a potentially preventive role for the mast cell in atherosclerosis.

Inevitably, mast cell stimulation and release of mediators cause many effects on lipoproteins, both direct and indirect. Some of the effects may be atherogenic and others anti-atherogenic. Consequently, the net effect of mast cell stimulation on atherosclerosis will depend on a balance between these effects. The net effect of mast cells is not known and further studies are needed to elucidate the impact of these cells in lipoprotein metabolism and atherosclerosis.

REFERENCES

1. D. Lagunoff, T. W. Martin, and G. Read, Agents that release histamine from mast cells, *Annu. Rev. Pharmacol. Toxicol. 23*:331 (1983).
2. M. C. Liu, D. Proud, L. M. Lichtenstein, D. W. MacGlashan, R. P. Schleimer, N. F. Adkinson, A. Kagey-Sobotka, E. S. Schulman, and M. Plaut, Human lung macrophage-derived histamine-releasing activity is due to IgE-dependent factors, *J. Immunol. 136*:2588 (1986).
3. E. Brzezinska-Blaszczyk, M. Czuwaj, and P. Kuna, Histamine release from mast cells of various species induced by histamine releasing factor from human lymphocytes, *Agents Actions 21*:26 (1987).
4. M. V. White, A. P. Kaplan, M. Haak-Frendscho, and M. Kaliner, Neutrophils and mast cells: comparison of neutrophil-derived histamine-releasing activity with other histamine-releasing factors, *J. Immunol. 141*:3575 (1988).
5. L. M. Zheutlin, S. J. Ackerman, G. J. Gleich, and L. L. Thomas, Stimulation of basophil and rat mast cell histamine release by eosinophil granule–derived cationic proteins, *J. Immunol. 133*:2180 (1984).
6. S. J. Galli, A. M. Dvorak, and H. F. Dvorak, Basophils and mast cells:

morphologic insights into their biology, secretory patterns, and function, *Prog. Allergy 34*:1 (1984).
7. K. A. Lindstedt, J. O. Kokkonen, and P. T. Kovanen, Soluble heparin proteoglycans released from stimulated mast cells induce uptake of low density lipoproteins by macrophages via scavenger receptor–mediated phagocytosis, *J. Lipid Res. 33*:66 (1992).
8. L. B. Schwartz and K. F. Austen, Structure and function of the chemical mediators of mast cells, *Prog. Allergy 34*:271 (1984).
9. J. O. Kokkonen and P. T. Kovanen, Accumulation of low density lipoproteins in stimulated rat serosal mast cells during recovery from degranulation, *J. Lipid Res. 30*:1341 (1989).
10. U. Lindahl, H. Pertoft, and R. Seljelid, Uptake and degradation of mast-cell granules by mouse peritoneal macrophages, *Biochem. J. 182*:189 (1979).
11. M. Baggiolini, U. Horisberger, and U. Martin, Phagocytosis of mast cell granules by mononuclear phagocytes, neutrophils and eosinophils during anaphylaxis, *Int. Arch. Allergy Appl. Immunol. 67*:219 (1982).
12. B. Slutsky, D. Jarvis, P. Bibb, R. S. Feldgerg, and D. E. Cochrane, Viability and recovery from degranulation of isolated rat peritoneal mast cells, *Exp. Cell Res. 168*:63 (1987).
13. A. M. Dvorak, R. P. Schleimer, E. S. Schulman, and L. M. Lichtenstein, Human mast cells use conservation and condensation mechanisms during recovery from degranulation, *Lab. Invest. 54*:63 (1986).
14. S. Margolis, Structure of very low and low density lipoproteins, in *Structural and Functional Aspects of Lipoproteins in Living Systems* (E. Tria and A. M. Scanu, eds.), Academic Press, London, 1969, p. 369.
15. T. J. Knott, R. J. Pease, L. M. Powell, S. C. Wallis, S. C. Rall, T. L. Innerarity, B. Blackhart, W. H. Taylor, Y. Marcel, R. Milne, D. Johnson, M. Fuller, A. J. Lusis, B. J. McCarthy, R. W. Mahley, B. Levy-Wilson, and J. Scott, Complete protein sequence and identification of structural domains of human apolipoprotein B, *Nature 323*:734 (1986).
16. C.-Y. Yang, S.-H. Chen, S. H. Gianturco, W. A. Bradley, J. T. Sparrow, M. Tanimura, W.-H. Li, D. A. Sparrow, H. DeLoof, M. Rosseneu, F.-S. Lee, Z.-W. Gu, A. M. Gotto, and L. Chan, Sequence, structure, receptor-binding domains and internal repeats of human apolipoprotein B-100, *Nature 323*:738 (1986).
17. M. S. Brown, P. T. Kovanen, and J. L. Goldstein, Regulation of plasma cholesterol by lipoprotein receptors, *Science 212*:628 (1981).
18. E. B. Smith and R. S. Slater, Relationship between low density lipoprotein in aortic intima and serum lipid levels, *Lancet i*:463 (1972).
19. M. D. Haust, The natural history of human atherosclerotic lesions, in *Vascular Injury and Atherosclerosis* (S. Moore, ed.), Marcel Dekker, New York, 1981, p. 1.
20. M. J. Mitchinson and R. Y. Ball, Macrophages and atherosclerosis, *Lancet ii*:146 (1987).
21. J. L. Goldstein, Y. K. Ho, S. K. Basu, and M. S. Brown, Binding site on macrophages that mediates uptake and degradation of acetylated low density lipoprotein,

producing massive cholesterol deposition, *Proc. Natl. Acad. Sci. USA 76*:333 (1979).
22. D. Steinberg, S. Parthasarathy, T. E. Carew, J. C. Khoo, and J. L. Witztum, Beyond cholesterol: modifications of low density lipoprotein that increase its atherogenicity, *N. Engl. J. Med. 320*:915 (1989).
23. M. E. Haberland, D. Fong, and L. Cheng, Malondialdehyde-altered protein occurs in atheroma of Watanabe heritable hyperlipidemic rabbits, *Science 241*:215 (1988).
24. S. Ylä-Herttuala, W. Palinski, M. E. Rosenfeld, S. Parthasarathy, T. E. Carew, S. Butler, J. L. Witztum, and D. Steinberg, Evidence for the presence of oxidatively modified low density lipoprotein in atherosclerotic lesions of rabbit and man, *J. Clin. Invest. 84*:1086 (1989).
25. E. B. Smith, C. Ashall, and J. E. Walker, High density lipoprotein (HDL) subfractions in interstitial fluid from human aortic intima and atherosclerotic lesions, *Biochem. Soc. Trans. 12*:843 (1984).
26. S. Stender and H. Hjelms, In vivo transfer of cholesteryl ester from high and low density plasma lipoproteins into human aortic tissue, *Arteriosclerosis 8*:252 (1988).
27. P. T. Kovanen, Atheroma formation: defective control in the intimal roundtrip of cholesterol, *Eur. Heart J. 11 (Suppl. E)*:238 (1990).
28. J. A. Glomset, The plasma lecithin:cholesterol acyltransferase reaction, *J. Lipid Res. 9*:155 (1968).
29. Y. K. Ho, M. S. Brown, and J. L. Goldstein, Hydrolysis and excretion of cytoplasmic cholesteryl esters by macrophages: stimulation by high density lipoprotein and other agents, *J. Lipid Res. 2*:391 (1980).
30. M. Aviram, E. L. Bierman, and J. F. Oram, High density lipoprotein stimulates sterol translocation between intracellular and plasma membrane pools in human monocyte-derived macrophages, *J. Lipid Res. 30*:65 (1989).
31. P. K. Vadiveloo and N. H. Fidge, Studies on the interaction between apolipoprotein A-II–enriched HDL_3 and cultured bovine aortic endothelial (BAE) cells, *Biochim. Biophys. Acta 1045*:135 (1990).
32. Y. Nagano, H. Arai, and T. Kita, High density lipoprotein loses its effect to stimulate efflux of cholesterol from foam cells after oxidative modification, *Proc. Natl. Acad. Sci. USA 88*:6457 (1991).
33. J. F. Oram, A. J. Mendez, J. P. Slotte, and T. F. Johnson, High density lipoprotein apolipoproteins mediate removal of sterol from intracellular pools but not from plasma membranes of cholesterol-loaded fibroblasts, *Arterioscler. Thromb. 11*:403 (1991).
34. A. Barkia, P. Puchois, N. Ghalim, G. Torpier, R. Barbaras, G. Ailhaud, and J.-C. Fruchart, Differential role of apolipoprotein Al–containing particles in cholesterol efflux from adipose cells, *Atherosclerosis 87*:135 (1991).
35. D. L. Graham and J. F. Oram, Identification and characterization of a high density lipoprotein-binding protein in cell membranes by ligand blotting, *J. Biol. Chem. 262*:7439 (1987).
36. R. W. Robinson, I. N. Likar, and L. J. Likar, Arterial mast cells, *Monogr. Atheroscler. 8*:73 (1978).
37. A. A. Trillo, The cell population of aortic fatty streaks in African green monkeys

with special reference to granulocytic cells: an ultrastructural study, *Atherosclerosis 43*:259 (1982).
38. H. C. Stary, The sequence of cell and matrix changes in atherosclerotic lesions of coronary arteries in the first forty years of life, *Eur. Heart J. 11 (Suppl. E)*:3 (1990).
39. J. O. Kokkonen, The effect of mast cells on the metabolism of low density lipoproteins: an experimental study with rat serosal mast cells, Academic dissertation, University of Helsinki, 1990.
40. J. O. Kokkonen, and P. T. Kovanen, Low density lipoprotein degradation by rat mast cells: demonstration of extracellular proteolysis caused by mast cell granules, *J. Biol. Chem. 260*:14756 (1985).
41. J. O. Kokkonen, M. Vartiainen, and P. T. Kovanen, Low density lipoprotein degradation by secretory granules of rat mast cells: sequential degradation of apolipoprotein B by granule chymase and carboxypeptidase A, *J. Biol. Chem. 261*:16067 (1986).
42. J. O. Kokkonen and P. T. Kovanen, Stimulation of mast cells leads to cholesterol accumulation in macrophages in vitro by a mast cell granule–mediated uptake of low density lipoprotein, *Proc. Natl. Acad. Sci. USA 84*:2287 (1987).
43. J. O. Kokkonen and P. T. Kovanen, Low density lipoprotein binding by mast-cell granules: demonstration of binding of apolipoprotein B to heparin proteoglycan of exocytosed granules, *Biochem. J. 241*:583 (1987).
44. J. O. Kokkonen and P. T. Kovanen, Proteolytic enzymes of mast cell granules degrade low density lipoproteins and promote their granule-mediated uptake by macrophages in vitro, *J. Biol. Chem. 264*:10749 (1989).
45. J. O. Kokkonen, Stimulation of rat peritoneal mast cells enhances uptake of low density lipoproteins by rat peritoneal macrophages in vivo, *Atherosclerosis 79*:213 (1989).
46. P. T. Kovanen and J. O. Kokkonen, Modification of low density lipoproteins by secretory granules of rat serosal mast cells, *J. Biol. Chem. 266*:4430 (1991).
47. K. A. Lindstedt J. O. Kokkonen, and P. T. Kovanen, Inhibition of copper-mediated oxidation of low density lipoproteins by rat serosal mast cells: a novel cellular antioxidative mechanism involving proteolysis of the substrate under oxidative stress, *Arterioscler. Thromb. 13*:23 (1993).
48. J. O. Kokkonen, K. A. Lindstedt, and P. T. Kovanen, Metabolism of low density lipoprotein in mast cells recovering from degranulation: description of a novel intracellular pathway leading to proteolytic modification of the lipoprotein, *Arterioscler. Thromb. 13*:276 (1993).
49. K. A. Lindstedt, Inhibition of macrophage-mediated LDL oxidation by stimulated rat serosal mast cells, *J. Biol. Chem. 268*:7741 (1993).
50. J. O. Kokkonen and P. T. Kovanen, The metabolism of low density lipoproteins by rat serosal mast cells, *Eur. Heart J. 11 (Suppl. E)*:134 (1990).
51. M. Lee, L. K. Lindstedt, and P. T. Kovanen, Mast-cell-mediated inhibition of reverse cholesterol transport, *Arterioscler. Thromb. 12*:1329 (1992).
52. P. T. Kovanen, Mast cell granule–mediated uptake of low density lipoproteins by macrophages: a novel carrier mechanism leading to the formation of foam cells, *Ann. Med. 23*:551 (1991).

53. E. B. Smith and C. Ashall, Fibronectin distribution in human aortic intima and atherosclerotic lesions: concentration of soluble and collagenase-releasable fractions, *Biochim. Biophys. Acta 880*:10 (1986).
54. E. B. Smith, K. M. Alexander, and I. B. Massie, Insoluble "fibrin" in human aortic intima: quantitative studies on the relationship between insoluble "fibrin," soluble fibrinogen and low density lipoprotein, *Atherosclerosis 23*:19 (1976).
55. U. P. Steinbrecher, S. Parthasarathy, D. S. Leake, J. L. Witztum, and D. Steinberg, Modification of low density lipoprotein by endothelial cells involves lipid peroxidation and degradation of low density lipoprotein phospholipids, *Proc. Natl. Acad. Sci. USA 81*:3883 (1984).
56. S. Parthasarathy, L. G. Fong, M. T. Quinn, and D. Steinberg, Oxidative modification of LDL: comparison between cell-mediated and copper-mediated modification, *Eur. Heart J. 11 (Suppl. E)*:83 (1990).
57. T. Henriksen, E. M. Mahoney, and D. Steinberg, Enhanced macrophage degradation of low density lipoprotein previously incubated with cultured endothelial cells: recognition by receptors for acetylated low density lipoproteins, *Proc. Natl. Acad. Sci. USA 78*:6499 (1981).
58. T. Henriksen, E. M. Mahoney, and D. Steinberg, Enhanced marophage degradation of biologically modified low density lipoprotein, *Arteriosclerosis 3*:149 (1983).
59. D. J. Lamb, G. M. Wilkins, and D. S. Leake, The oxidative modification of low density lipoprotein by human lymphocytes, *Atherosclerosis 92*:187 (1992).
60. S. Parthasarathy, D. J. Printz, D. Boyd, L. Joy, and D. Steinberg, Macrophage oxidation of low density lipoprotein generates a modified form recognized by the scavenger receptor, *Arteriosclerosis 6*:505 (1986).
61. J. W. Heinecke, H. Rosen, and A. Chait, Iron and copper promote modification of low density lipoprotein by human arterial smooth muscle cells in culture, *J. Clin. Invest. 74*:1890 (1984).
62. K. A. Lindstedt, The effect of mast cells on oxidation of low density lipoproteins: an experimental study with rat serosal mast cells, Academic dissertation, Åbo Akademi University, Turku, 1993.
63. J. Porath, J. Carlsson, I. Olsson, and G. Belfrage, Metal chelate affinity chromatography, a new approach to protein fractionation, *Nature 258*:598 (1975).
64. N. Kalant, S. McCormick, and M. A. Parniak, Effects of copper and histidine on oxidative modification of low density lipoprotein and its subsequent binding to collagen, *Arterioscler. Thromb. 11*:1322 (1991).
65. H. Masuda, T. Sugimori, A. Odani, and O. Yamauchi, Structural evidence for the intramolecular charge–transfer interaction involving an indole ring in ternary copper(II) complexes with L-tryptophan and aromatic diamines, *Inorg. Chim. Acta 180*:73 (1991).
66. L. Jonasson, J. Holm, O. Skalli, G. Bondjers, and G. K. Hansson, Regional accumulations of T cells, macrophages, and smooth muscle cells in the human atherosclerotic plaque, *Arteriosclerosis 6*:131 (1986).
67. H. G. Rus, F. Niculescu, E. Constantinescu, A. Cristea, and R. Vlaicu, Immunoelectron-microscopic localization of the terminal C5b-9 complement complex in human atherosclerotic fibrous plaque, *Atherosclerosis 61*:35 (1986).

68. R. L. Stevens, C. C. Fox, L. M. Lichtenstein, and K. F. Austen, Identification of chondroitin sulfate E proteoglycans and heparin proteoglycans in the secretory granules of human lung mast cells, *Proc. Natl. Acad. Sci. USA 85*:2284 (1988).
69. P.-H. Iverius, The interaction between human plasma lipoproteins and connective tissue glycosaminoglycans, *J. Biol. Chem. 247*:2607 (1972).
70. K. B. Yancey, C. H. Hammer, L. Harvath, L. Renfer, M. M. Frank, and T. J. Lawley, Studies of human C5a as a mediator of inflammation in normal human skin, *J. Clin. Invest. 75*:486 (1985).
71. D. Reichl, N. B. Myant, and J. J. Pflug, Concentration of lipoproteins containing apolipoprotein B in human peripheral lymph, *Biochim. Biophys. Acta 489*:98 (1977).
72. E. B. Smith and C. Ashall, Low density lipoprotein concentration in interstitial fluid from human atherosclerotic lesions: relation to theories of endothelial damage and lipoprotein binding, *Biochim. Biophys. Acta 754*:249 (1983).

13

Physiological Regulation of C3a Anaphylatoxin Activity by Mast Cell Chymase

Tony E. Hugli
The Scripps Research Institute, La Jolla, California

I. INTRODUCTION

Anaphylatoxins are a group of biologically active protein fragments generated from blood components C3, C4, and C5 during complement activation. The fragments are denoted as C3a, C4a, and C5a, respectively, and several reviews provide detailed descriptions of the biological and biochemical properties of these molecules (1–4). A property that is shared by each of these complement fragments is their ability to induce smooth muscle contraction (i.e., spasmogenic function). Another property of the anaphylatoxins is that they enhance vascular permeability. These two functions have been associated with anaphylatoxin-induced histamine release, and hence activation of tissue mast cells has been implicated. Early observations that dramatic cutaneous responses were elicited by injection of minute quantities of anaphylatoxins, in both animals (5) and humans (6,7), led to speculation that tissue mast cells were largely responsible for this skin reaction. The skin wheal and flare reaction to anaphylatoxins occurs without pain and has a short duration of 10 to 15 min. The hypothesis that mast cells were directly involved in this skin reaction was further supported by evidence that degranulated mast cells were observed in biopsies taken immediately after subcutaneous injection of the anaphylatoxins.

More recent investigations have attempted to identify receptors for the anaphylatoxins on mast cells (8–10). It was this line of investigation that led not only to studies of anaphylatoxin degradation by mast cell proteases but also to our realization that rat peritoneal mast cells (RMCs) were apparently activated by C3a/C4a primarily via a non-receptor-mediated mechanism (i.e., involving the physical properties of the C3a/C4a molecule). Therefore, we revisited observations made almost 20 years ago by Morrison et al. (11), who described "two distinct mechanisms of activation of mast cells by cationic molecules." There are the type I mediators, such as compound 48/80 and polymyxin B, representing fast releasers (i.e., 1 min), and the type II mediators, such as C3a and a cationic cobra venom factor (CVA), which release vasoamines from these cells more slowly, over 20 to 30 min. In both cases the release was noncytotoxic, but type I release was inhibited (i.e., mast cells were desensitized) by 20 μM ATP, whereas activation by the type II mediator CVA (and presumably C3a) was enhanced by ATP. The distinction between various cationic molecules and their differential ability to activate mast cells remains an enigma that may be related to different types or classes of surface anionic molecules with which the mediators undergo charge–charge interaction. In this chapter we focus on the mechanism of the non-receptor-mediated activation of mast cells by C3a or C4a.

C3a is a highly cationic molecule ($pK > 9.0$) with a charge density similar to that of compound 48/80, protamine, polymyxin B, or other positively charged molecules capable of polycationic activation of mast cells (12). Although the physical characteristics of C3a may explain why the mast cell is activated by C3a, it fails to identify how activation is accomplished. I will discuss in greater detail the current view of C3a interactions with mast cells, including how the activation process may occur and what role the proteases chymase and tryptase may have in regulating C3a levels in situ. More important, the processes of mast cell activation by C3a, and subsequent degradation of C3a by protease(s) released from these activated cells, promise to be important physiological and pathophysiological mechanisms. The emphasis in this chapter is on the C3a anaphylatoxin molecule, and to some extent C4a, which is only slightly less cationic than C3a (13). We do not consider mast cell interactions with factor C5a because this factor appears to be neither sufficiently abundant nor cationic to induce this type of mast cell activation. Unlike C3a, C5a does not attain plasma or tissue levels adequate for activating mast cells via the polycationic pathway. Entirely different mechanisms appear to exist in vivo for the regulation of C3a and C5a. C5a (and $C5a_{des\ Arg}$) have high-affinity receptors on neutrophilic, eosinophilic, and basophilic leukocytes as well as on monocytes and tissue macrophages. These high-affinity cellular receptors essentially extract the C5a from circulation. This does not mean that C5a interactions with mast cells are physiologically less important than those of C3a, but it does suggest that mast cells may be much more important in regulating extravascular C3a levels than C5a levels. The impact of C3a under pathophysiologic conditions is yet to be defined. However, there are clinical situations where recruitment of mast cell products may be particularly harmful, such as in acute trauma, sepsis, multiorgan failure, and adult respiratory distress syndrome. The added contribution of C3a-induced mediator release from mast cells may be sufficient to promote life-threatening consequences in these patients.

II. FACTOR C3a

The C3a molecule is derived from component C3, the most abundant blood protein of the complement cascade. Blood contains between 1 and 1.6 g C3/L and as much as 60 mg/L of C3a can be generated in plasma. Since the molecular weight of C3a is approximately 9000 Da, as much as 6×10^{-6} *M* C3a can be released by total in vivo activation of either the classical or alternative complement pathways. Factor C4a is generated only during activation of the classical pathway, but quantities of C4a as high as 1 to 2×10^{-6} *M* C4a can be generated in plasma, and this factor could augment the total polycation level in circulation. Consequently, large quantities of these two factors can be generated in plasma, which supports our hypothesis that C3a and/or C4a may play a significant role in cationic activation of mast cells under certain pathophysiological conditions. Therefore, proteolytic regulation of these factors could prevent involvement of anaphylatoxins in mediating undesirable cellular or tissue responses.

Human C3a and C4a contain 77 and 76 amino acid residues, respectively, and are cleaved from their parent molecules C3 and C4 by highly specialized plasma proteases called convertases. The convertases are serine proteases having trypsin-like specificity, which cause bond scission to occur in the respective α chains of the parent proteins, as illustrated in Fig. 1. Both intact C3a and C4a contain a C-terminal arginyl residue when generated in circulation by complement activation, and this intact form exhibits the characteristic spasmogenic and vascular permeability functions that define these biological factors. A serum carboxypeptidase (carboxypeptidase N) removes the C-terminal arginine from these factors rapidly and efficiently, forming a des Arg derivative that continues to circulate (14). Enzymatic conversion of C3a/C4a to the des Arg form occurs rapidly in blood, with a half-life of several seconds, and this modification renders these factors virtually inactive for receptor-mediated cell activation and tissue responses. The $C3a_{des\ Arg}$ molecule fails to compete with intact C3a for either activation of guinea pig platelets or contraction of guinea pig ileum. Studies using synthetic analog peptides modeled after the C-terminal amino acid sequence of C3a demonstrated that the C-terminal region is the essential effector site of this molecule (15,16). These synthetic peptide studies helped to prove that C3a has receptors on a variety of cell types, including macrophages and eosinophils. Platelets from certain animal species such as

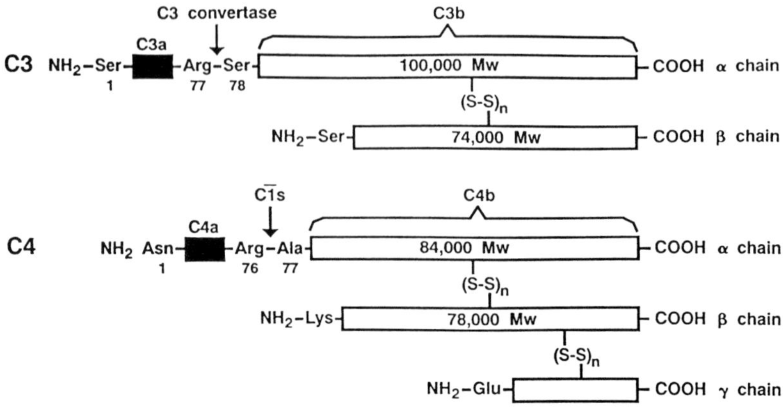

Figure 1 The human complement component C3 is a two-chain molecule. Convertases of the classical or alternative pathways of complement activation selectively cleave the C3 α chain between arginine 77 and serine 78 to generate the C3a anaphylatoxin. The human C4 molecule circulates as a three-chain protein. The protease C1s of the classical pathway of complement activation cleaves the C4 α chain between arginine 76 and alanine 77 to generate C4a.

guinea pigs have C3a receptors, but human platelets may not bear C3a receptors. Observation that removal of a C-terminal aginyl residue from C3a or C4a inactivates these factors for receptor-mediated functions elucidated the major control mechanism for regulating these anaphylatoxins in circulation. Therefore, it was assumed that carboxypeptidase N represents the primary method for regulating these factors in vivo. As we will see later, a non-receptor-mediated mechanism exists for mast cell activation by the des Arg form of C3a, indicating that another form of regulation is required to control or destroy this "inactive" $C3a_{des\ Arg}$ molecule.

The C3a and C4a molecules exist in compact and highly stable conformations composed of four helical regions interconnected by three disulfide bonds. The typical globular shape of C3a/C4a exposes a polar hydrophilic surface (17). The highly cationic human C3a molecule ($pK > 9.0$), with a net positive charge of 13 (12 for $C3a_{des\ Arg}$), is nearly as basic as such classical mast cell secretagogues as protamine, substance P, mastroparan, and the synthetic polycation 48/80 (Fig. 2). Human C4a is less cationic than C3a, having a net positive charge of 9 (8 for $C4a_{des\ Arg}$) and a pK value above 8.5. It is the unusual cationic nature of these factors and their charge density that is remarkable and which apparently relates to the ability of these molecules to activate mast cells via a nonreceptor polycation–mediated mechanism.

1 10
NH_2-Ser-Val-Gln-Leu-Thr-Glu-**Lys-Arg**-Met-Asn-**Lys**-Val-Gly-**Lys**-Tyr-Pro-**Lys**-Glu-

20 30
-Leu-**Arg-Lys**-Cys-Cys-Glu-Asp-Gly-Met-**Arg**-Gln-Asn-Pro-Met-**Arg**-Phe-Ser-Cys-

40 50
-Gln-**Arg-Arg**-Thr-**Arg**-Phe-Ile-Ser-Leu-Gly-Glu-Ala-Cys-**Lys-Lys**-Val-Phe-Leu-Asp-

60 70
-Cys-Cys-Asn-Tyr-Ile-Thr-Glu-Leu-**Arg-Arg**-Gln-**His**-Ala-**Arg**-Ala-Ser-**His**-Leu-Gly-

Leu-Ala-(**ARG**)-COOH

Figure 2 Sequence of the highly cationic human C3a molecule. The numerous cationic residues, indicated in bold, are well distributed throughout the entire sequence of this molecule. The rather uncharacteristic frequency of cationic residue repeats in the sequence, such as Lys–Arg, Arg–Lys, Arg–Arg, and Lys–Lys, may be related to the ability of this molecule to activate mast cells. The C-terminal arginyl residue (ARG) is rapidly removed in circulation to form the $C3a_{des\ Arg}$ derivative.

III. C3a INTERACTIONS WITH MAST CELLS

Many investigators have studied the effects of anaphylatoxins on mast cell histamine release. It is extremely important to identify both the species and tissue source of mast cells being used because this cell is heterogeneous depending on tissue location, and wide variations in their responses have been reported (18). Schulman et al. (19) reported that up to a level of 55 μg/mL, C3a was unable to activate isolated human lung mast cells, while human skin was reactive to sub-μg/mL levels of C3a. Although the human skin mast cell was degranulated after exposure to C3a, there is no evidence that C3a directly stimulated the human skin mast cells. The skin reaction in humans was dependent on intact C3a, because the response was greatly depressed for the des Arg form of C3a. However, evidence that C3a activates human mast cells directly is sparse to nonexistent in the literature.

Physical evidence for C3a binding to the rat peritoneal mast cell was provided by ter Laan et al. (20), who showed by an indirect immunofluorescence technique that C3a was bound to the surface of RMCs. The pattern of binding indicated an uneven distribution of the C3a as demonstrated by intense spots of fluorescence scattered over the membrane surface. A key observation by these investigators was that degranulated RMCs had little fluorescence (i.e., bound C3a) on the cell surface, while intact [i.e., ethylenediaminetetraacetic acid (EDTA) inhibited] RMCs were highly fluorescent. These data, coupled with evidence that C3a induced degranulation and activation of the isolated RMC (8,9,21), suggested that interactions between this factor and RMCs were both specific and receptor mediated. Ter Laan et al. (20) interpreted their data as indicating that the "receptors" for C3a were "sensitive to proteases released during mast cell degranulation." Although incorrect in assuming that C3a was bound to receptors on the RMCs, they did accurately suggest that degranulation resulted in a degradative event. They incorrectly interpreted the decrease in C3a binding to the activated RMCs as an attack on the "receptor" and not on the ligand C3a. Therefore, without knowing it, these investigators had produced the first experimental data indicating that C3a was degraded on the surface of activated RMCs, thus explaining the disappearance of fluorescent C3a on degranulated RMCs versus uptake by EDTA-stabilized RMCs.

IV. C3a DEGRADATION BY CHYMASE

In-depth studies of C3a–mast cell interactions (22) provided surprising results that challenged concepts and conclusions regarding the nature of this interaction as based on earlier observations. Uptake of [^{125}I]C3a (human) by RMCs was compared with uptake by a variety of other types of cells, and it was immediately apparent that much more C3a was taken up by mast cells than by any other cell type examined, including human endothelial cells, eosinophils, neutrophils, or the rat basophilic leukemia cell line (RBL) (Fig. 3). This uptake of C3a was nonsaturable at 0°C with $> 10^{-7}$ molecule of C3a bound per cell. The uptake behavior and extraordinarily large number of sites suggested that atypical binding sites were involved. Another curious result was that greater binding occurred at 0°C than at 37°C, a clear indication that degradation may influence the results. Furthermore, the number of C3a sites increased dramatically when a serine protease inhibitor such as phenylmethylsulfonyl fluoride (PMSF) was present. When radiolabeled C3a was recovered from the RMC supernatant and was analyzed by sodium dodecyl sulfate (SDS) gel electrophoresis, the mystery was partially solved (22). The C3a was extensively degraded in the presence of RMCs, indicating that a protease was present on the cell surface (Fig. 4).

C3a was covalently linked to predominantly one component of the RMC surface using a conventional homobifunctional cross-linking agent [bis(sulfosuccinimidyl) suberate]. The complex of C3a linked to a component on the RMCs was estimated by SDS-PAGE analysis to be 36,000 Da. Chymase is

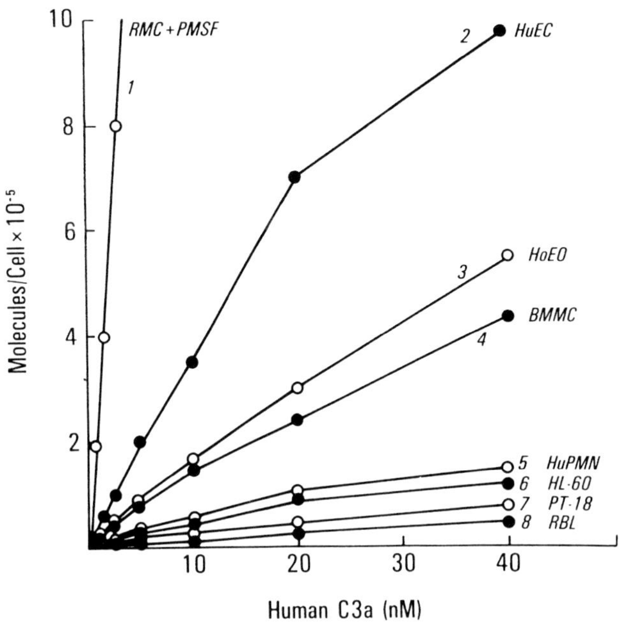

Figure 3 Uptake of [^{125}I]Hu C3a by a variety of isolated and cultured cells. Each cell type at 5 × 10^6 cells/mL was incubated with various quantities of [^{125}I]Hu C3a in 250 μL of RPMI-BSA buffer for 1 h at 0°C. The cell types include: 1, rat mast cell (RMC); 2, human endothelial cell (HuEC); 3, horse eosinophil (HoEO); 4, human eosinophil (HuEO); 5, bone marrow mast cell (BMMC); 6, human polymorphonuclear leukocyte (Hu PMN); 7, human promyelocytic leukemia cell line (HL-60); 8, mouse bone marrow cell line (PT-18); and 9, rat basophilic leukemia cell line (RBL). Note that PMSF (5 m*M*) was added to the RMC to prevent degradation. Addition of PMSF to the other cell types had little effect on total uptake. (Adapted from Ref 22.)

approximately a 29,000-Da protein and C3a is 9000 Da; therefore, the combined weights agree well with the estimated size of a covalently cross-linked complex of C3a and chymase (i.e., 36,000 Da). When the RMCs were washed in 1 *M* NaCl to extract surface chymase, before exposure to [^{125}I]C3a in the presence of the cross-linking agent, the 36,000-Da band was eliminated. Specific inhibitors of chymase such as chymostatin and tosyl-L-phenylalanine chloromethyl ketone (TPCK) were effective in preventing C3a degradation, while nonserine protease inhibitors such as azide, EDTA, and elastinal were not. At this point the evidence suggested that chymase on the RMC surface was respon-

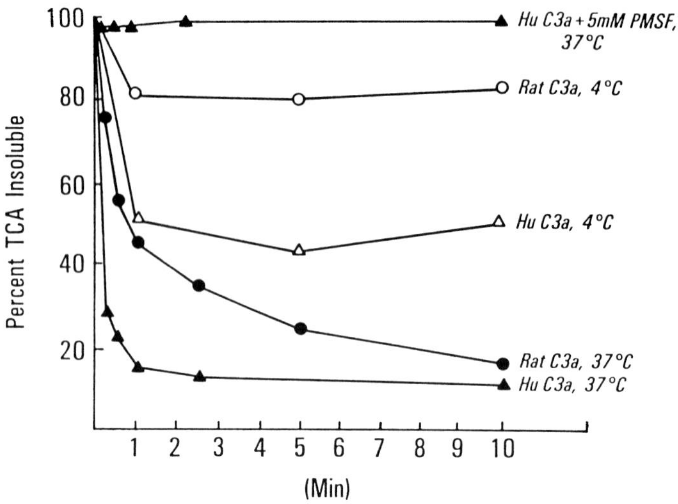

(a)

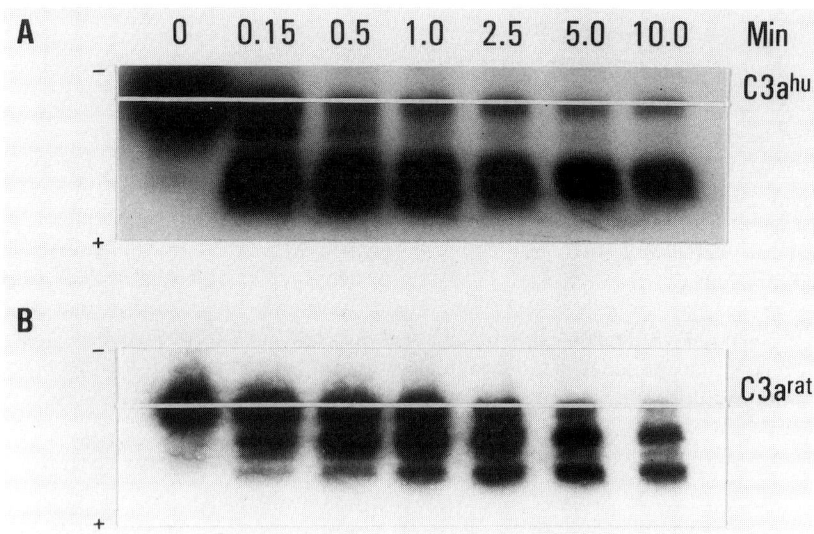

(b)

Figure 4 (a) Time course for the breakdown of [^{125}I]Hu C3a and [^{125}I]rat C3a by RMC. [^{125}I]Hu Ca3 or [^{125}I]rat C3a were incubated with 10^6 RMCs in 200 μL of buffer for various lengths of time at 4° or 37°C. At each time point aliquots were taken and 5 m*M* PMSF was added before centrifugation to remove the cells. The quantity of undigested ligand was estimated by counting the TCA precipitate. Rates of degradation in incubates containing [^{125}I]Hu C3a at 4°C (Δ) and at 37°C (▲) were compared with incubates of [^{125}I]rat C3a at 4°C (○) and 37°C (●). Note that degradation of [^{125}I]Hu C3a was prevented even at 37°C (▲) by adding 5 m*M* PMSF. (b) I. SDS-PAGE gel analysis

sible for the extensive C3a degradation. This observation also suggested that tissue C3a ($C3a_{des\ Arg}$) levels could be regulated by the degradative action of surface proteases on the RMCs.

We do not propose that cationic C3a binds to the cationic chymase but rather to the sulfated proteoglycans. Polyanionic proteoglycans such as heparin compose the insoluble complexes of the RMC secretory granules to which the chymase is adsorbed (24). When C3a binds to the polyanionic complexes on the surface of exposed granules, the chymase that is embedded in the matrix can attack the C3a. The degradative effect of the chymase proceeds as it would with any polypeptide substrate. Cross-linking of the C3a to the chymase is believed possible only because of the relative abundance and proximity of these two proteins. The larger question that remains is whether binding of C3a to the proteoglycan is related to the cationic activation mechanism. What needs to be examined is the nature of the interaction between C3a and cellular components at the cell surface that result in stimulation of the cell. Cellular activation appears to be independent of the C3a–granular interaction or the degradation mechanism. This conclusion is based on the fact that the proteoglycan–chymase complexes are not exposed to the cell surface until after the activation event occurs (25). Recent evidence shows that C3a can directly stimulate the mast cell G protein (G_i subtype) GTPase, which introduces a novel concept of how these cells may be stimulated (26). The argument against direct C3a stimulation of G protein is the fact that C3a appears not to gain access to the inside of the cell. Alternatively, it is conceivable that C3a binding to surface sialic acid groups results in activation of G protein via an indirect signal mechanism. This aspect of C3a action requires further investigation to resolve how this polycation stimulates mast cell activation.

Polycationic activation by C3a does not eliminate the possibility that specific C3a receptors also exist on RMCs. Functional evidence against C3a receptor–mediated stimulation of RMCs is discussed in the following section. However, there is physical evidence that RMCs are devoid of C3a receptors based on the same cross-linking experiments that demonstrated the protease. The cross-linkage pattern obtained using $[^{125}I]C3a_{des\ Arg}$ was identical to the pattern obtained with intact C3a. Also, unlabeled C3a failed to compete with the labeled material, indicating that this interaction was nonspecific. The interpretations based on binding and the SDS-PAGE gel patterns remain inconclusive evidence

---

of supernatants recovered from mixtures of 10^6 RMCs and $[^{125}I]$Hu C3a (A) or $[^{125}I]$rat C3a(B) in 200 μL of Tyrode buffer containing 1% (w/v) gelatin incubated at 37°C vs. time. The autoradiographic results indicate extensive degradation of both rat and human C3a. Conversion of intact C3a to smaller fragments is complete in 5 to 10 min. The smallest visualized fragments of C3a migrate with the dye front and are estimated to be less than 2000 to 3000 Da. (Adapted from Ref. 22.)

for claiming that RMCs are devoid of C3a receptors because (a) the receptor numbers may be too small to observe by these techniques, (b) binding may be too weak to obtain efficient cross-linkage between C3a and the receptor, or (c) the cross-linking reagent is not suitable for connecting the C3a-receptor molecules. On the other hand, evidence that C3a receptors exist on other cells, such as guinea pig platelets, has been obtained successfully using these techniques (23).

V. C3a ACTIVATION OF MAST CELLS

Data collected from isolated RMCs demonstrated that these cells could be stimulated to release both histamine and β-glucuronidase by exposure to either C3a and $C3a_{des\ Arg}$ (22). Intact C3a was actually less potent than was the des Arg derivative in these experiments. Release occurred only at levels above 10^{-6} *M*. We interpreted our data as indicating that the RMC release response to C3a/$C3a_{des\ Arg}$ was non-receptor mediated and that activation was consistent with a polycationic mechanism. An indirect measure of skin mast cell activation was obtained by eliciting a cutaneous response to C3a in rats. Both the diameter of the wheal and the volume of edema fluid at the site of C3a injection indicated that the in vivo response to C3a and $C3a_{des\ Arg}$ in rat was nearly identical. As with the isolated mast cells, the quantities of material required to elicit a response were in the range 1 to 10 μ*M*. This result is in direct opposition to the skin response in humans, where cutaneous reactions to C3a are markedly greater (10 to 100 times) than to the des Arg form and C3a elicits a response in human skin at concentrations 10 to 100 lower than in rat skin (27). We interpret these results as further evidence that RMCs may be devoid of C3a receptors.

Since C3a was degraded in the presence of isolated and "unstimulated" RMCs, we concluded that some quantity of chymase was either normally exposed on the surface of RMCs or a portion of these isolated cells were activated. The cells were examined in vivo by introducing labeled C3a into the rat peritoneum, where direct contact could be made between the undisturbed RMCs and C3a (25). If unactivated mast cells in vivo have chymase on their surface, C3a would be degraded; otherwise, the factor would remain intact. Results of these experiments clearly demonstrated that (a) unstimulated RMCs express no chymase on their surface in vivo, (b) at elevated levels C3a or $C3a_{des\ Arg}$ can activate RMCs in situ, and (c) after RMC activation C3a was degraded extensively by the chymase.

VI. IN VIVO DEGRADATION OF C3a

Ten milliliters of Tyrode buffer containing 10^{-8} *M* $[^{125}I]C3a_{des\ Arg}$ was injected into the peritoneal cavity of adult rats and aliquots were recovered over a 1-h period (25). No degradation of the C3a occurred under these conditions, indicating that in vivo unstimulated RMCs have no chymase on their surface. When compound 48/80 was introduced at 0.1 and 1.0 μg/mL along with the [^{125}I]C3a$_{des\ Arg}$, the degradation versus time was extensive. These results indicated that compound 48/80 activated the RMCs as expected, and chymase (i.e., secretory granules) was transported to the cell surface. Injection of anti-IgE into the peritoneum of BN rats or to passively sensitized SD rats followed by introduction of nonstimulating levels (i.e., of 2×10^{-8} *M*) of [^{125}I]C3a$_{des\ Arg}$ also led to extensive degradation of the labeled material. When only the factor C3a or C3a$_{des\ Arg}$ (data not shown) was injected into the peritoneum, degradation of the labeled material also occurred, but only when levels greater than 10^{-6} *M* were introduced (Fig. 5). This result indicated that both C3a and C3a$_{des\ Arg}$ were capable of activating the RMCs in vivo, and that it required micromolar levels of cationic C3a to elicit chymase release. In all cases the degradation slowed versus time, indicating that the enzyme either became inactivated rapidly on the cell surface or was taken up again by the cells (Fig. 6). Similar experiments were

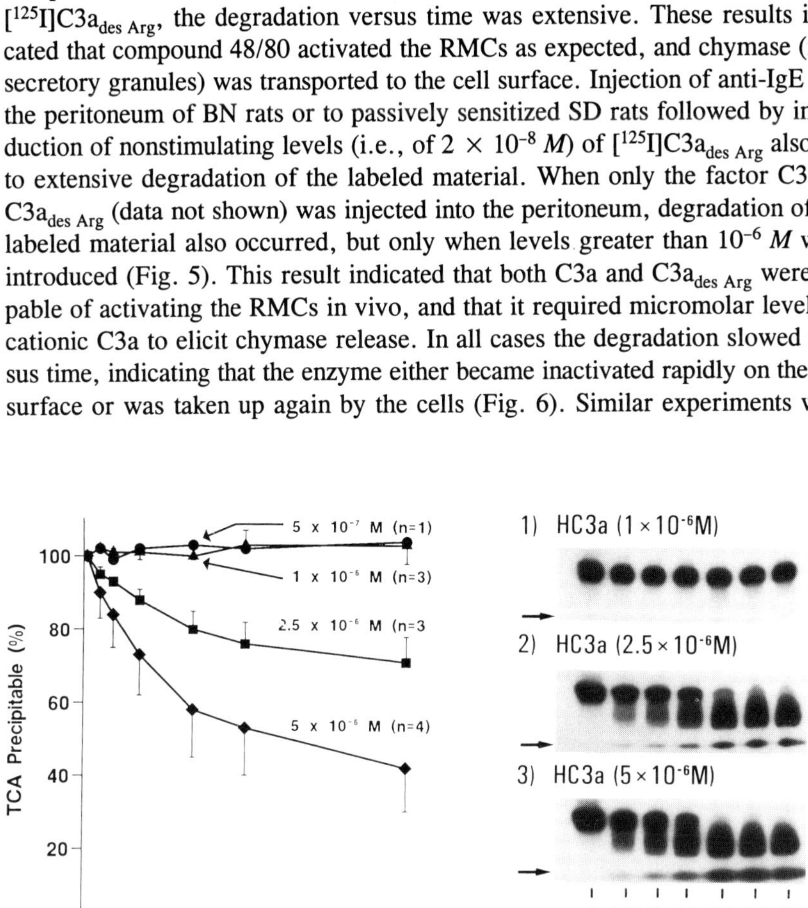

Figure 5 Intraperitoneal injection of various concentrations of Hu C3a leads to degradation of [^{125}I]Hu C3a$_{des\ Arg}$ in vivo (A). Ten milliliters of buffer containing 1×10^{-8} mol/L [^{125}I]Hu C3a$_{des\ Arg}$ and 0.5, 1, 2.5, or 5×10^{-6} mol/L Hu C3a was injected in the peritoneal cavity of Sprague-Dawley rats. After gentle massage for 1 min, samples were withdrawn, centrifuged, and the supernatants were analyzed for degradation. Significant degradation of [^{125}I]Hu C3a$_{des\ Arg}$ was observed when greater than 1.0×10^{-6} mol/L Hu C3a was introduced to the peritoneum (B2 and B3). We concluded that Hu C3a at elevated levels is an activator of peritoneal mast cells. (Adapted from Ref. 25.)

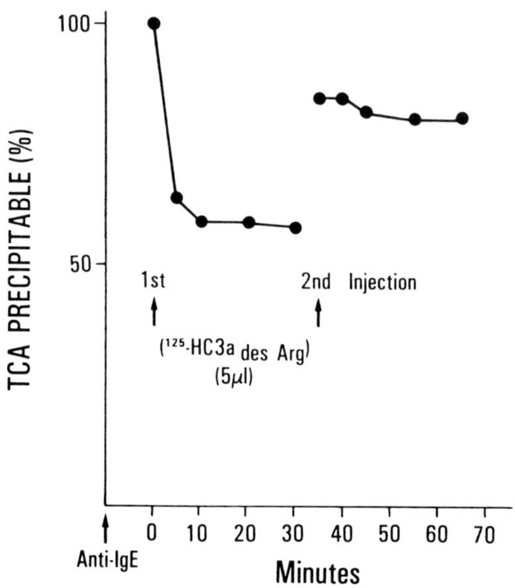

Figure 6 When $C3a_{des\ Arg}$ is injected into the peritoneum of a Brown-Norway rat challenged with anti-IgE to activate the mast cells, the anaphylatoxin is rapidly degraded. Degradation falls off after 10 min and a second injection of the $C3a_{des\ Arg}$ fails to elicit further degradation. This result indicates that the secreted chymase is either rapidly inactivated or removed from the cell surface so that in the absence of further stimulation the mast cell is no longer able to regulate C3a levels.

performed in both normal and mast cell–deficient [WBB6F1/J-WW^V (WW)] mice to prove that C3a degradation was mast cell (i.e., chymase) dependent. Intraperitoneal injection of compound 48/80 or ionophore A23187 with [^{125}I]$C3a_{des\ Arg}$ resulted in degradation of the C3a in normal mice, whereas no degradation was observed for C3a recovered from the peritoneum of mast cell–deficient animals. Intraperitoneal injection of 5×10^{-5} *M* $C3a_{des\ Arg}$ into normal mice resulted in only partial degradation versus time, suggesting that C3a is less effective in activating mast cells in mice than in rats (25). This observation may point to the existence (between species) of major variations in mast cell responsiveness to cationic activation.

VII. CONCEPTS/HYPOTHESIS/SPECULATIONS

It is tempting to speculate that tissue mast cells act as sentinals capable of regulating local concentrations of polycationic factors such as the anaphylatoxins by a degradative mechanism. It was earlier shown that the C3a molecule was rapidly degraded by tryptase in the presence of heparin (28). Therefore, both mucosal (MMC) and connective tissue (CTMC) mast cells may participate as regulators of bioactive peptides through surface-disposed proteases that are expressed as a result of cell activation. Anaphylatoxins generated either systemically, followed by extravasation of the factors into the tissue, or produced locally at a tissue site may have pathophysiological consequences. Should levels of these and other cationic materials [i.e., histones, protamine, and cell-derived basic proteins (29)] attain local concentrations capable of activating mast cells, proteases such as chymase and tryptase will appear on the cell surface. These enzymes are capable of degrading the activating factors, which would further limit activation of mast cells and thereby regulate overstimulation of these cells. Mast cells activated by immune mechanisms such as IgE-induced stimulation would be forearmed with surface proteases that could prevent the buildup of anaphylatoxin levels capable of secondary activation of these mast cells. Under pathologic conditions or extreme trauma, complement activation may proceed rapidly and exceed the capacity of mast cell proteases to regulate tissue concentrations of the anaphylatoxins C3a and C4a. In such cases these polycationic molecules could have harmful effects by overstimulating mast cells and inducing high levels of mediator release. It is mainly these extreme cases that suggest possible pathophysiologic involvement of C3a/C4a in human disease. Clinical treatments such as protamine titration of heparin following complicated cardiovascular procedures and conditions such as sepsis, acute organ rejection, multiorgan failure, and respiratory distress syndrome, where complement activation may be unusually high, exemplify situations that may involve activation of mast cells via the polycationic pathway.

VIII. CONCLUSIONS

Evidence indicates that the polycationic anaphylatoxins C3a and C4a are capable of activating mast cells via a nonreceptor mechanism. The physical charge--charge interactions between the polycationic C3a/C4a molecules and putative polyanionic molecules on the mast cell surface results in cellular activation with release of vasoamines and proteases. The expression of proteases on the mast cell surface after stimulation represents a potential regulatory mechanism for limiting the levels of C3a/C4a in the surrounding environment of the mast cell. These proteases rapidly and effectively degrade the anaphylatoxins, regulating their effectiveness as activators of the mast cells (see Fig. 7). Polycationic activation of mast cells by C3a/C4a may have a clinical relevance under conditions whereby rapid complement activation occurs in patients already exposed to high levels of circulating mediators. Addition of C3a-induced mast cell mediators in compromised patients could have severe consequences. The pos-

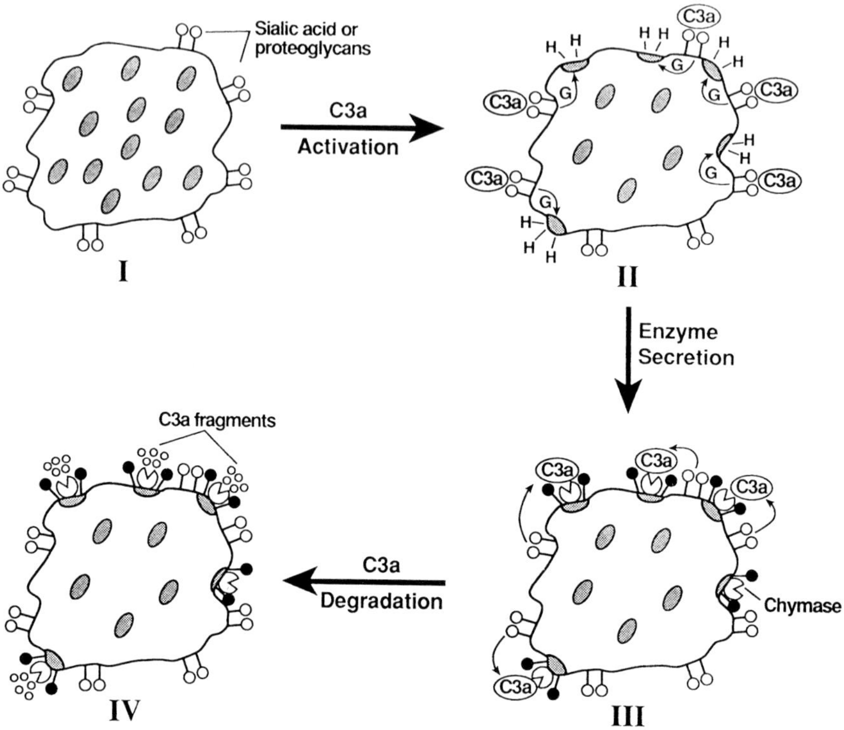

Figure 7 Schematic view of rat mast cell activation by C3a (C4a) and eventual degradation or regulation of these cationic factors. I. The unstimulated mast cell has anionic groups (○) distributed on the cell surface, including sialic acid and sulfated proteoglycans to which the C3a can bind. It is proposed that this binding somehow leads to activation of the cell. II. The activated mast cells release vasoamines (histamine) and various enzymes when the secretory granules fuse with the outer membrane. This process is believed to be mediated via the GTpase (G proteins) activation pathway. III. The granules also expose chymase in complex with heparin (●) on the cell surface. C3a/C4a binds to the anionic heparin in the complex. IV. The C3a is degraded by chymase in these complexes, thereby reducing levels of this polycation in the vicinity of the cell.

sibility that non-receptor-mediated actions of the anaphylatoxins play a significant clinical role has yet to be established; however, evidence that mast cells respond to polycationic C3a in vitro and in vivo is now well documented. The implications of these results indicate that further investigation is needed to explore the role of nonreceptor activation of mast cells by C3a (and C4a) in humans.

ACKNOWLEDGMENTS

The author wishes to thank Mary Glass for her assistance in preparing the manuscript. This work was supported in part by grants from the National Institutes of Health: AI32834, DK42717, AI17354, and HL25658.

REFERENCES

1. T. E. Hugli and H. J. Muller-Eberhard, Anaphylatoxins: C3a and C5a, *Adv. Immunol. 26*:1 (1978).
2. T. E. Hugli, Structure and function of the anaphylatoxins, *Springer Semin. Immunopathol. 7*:193 (1984).
3. T. E. Hugli and E. L. Morgan, Mechanisms of leukocyte regulation by complement-derived factors, *Contemp. Top. Immunobiol. 14*:109 (1984).
4. T. E. Hugli, Biochemistry and biology of anaphylatoxins, *Complement 3*:111 (1986).
5. T. E. Hugli, E. H. Vallota, and H. J. Muller-Eberhard, Purification and partial characterization of human and porcine C3a anaphylatoxin, *J. Biol. Chem. 250*:1472 (1975).
6. K. D. Wuepper, V. A. Bokisch, H. J. Muller-Eberhard, and R. B. Stoughton, Cutaneous responses to human C3 anaphylatoxin in man, *Clin. Exp. Immunol. 11*:13 (1972).
7. I. H. Lepow, R. A. Willms-Kretschmer, R. A. Patrick, and F. S. Rosen, Gross and ultrastructural observations of lesions produced by intradermal inject of human C3a in man, *Am. J. Pathol. 61*:13 (1970).
8. A. R. Johnson, T. E. Hugli, and H. J. Muller-Eberhard, Release of histamine from rat mast cells by the complement peptides C3a and C5a, *Immunology 28*:1067 (1975).
9. M. M. Glovsky, T. E. Hugli, T. Ishizaka, L. M. Lichtenstein, and B. W. Erickson, Anaphylatoxin-induced histamine release with human leukocytes: studies of C3a leukocyte binding and histamine release, *J. Clin. Invest. 64*:804 (1979).
10. J. E. Gervasoni, Jr., D. H. Conrad, T. E. Hugli, L. B. Schwartz, and S. Ruddy, Degradation of human anaphylatoxin C3a by rat peritoneal mast cells: a role for the secretory granule enzyme chymase and heparin proteoglycan, *J. Immunol. 136*:285 (1986).
11. D. C. Morrison, J. F. Roser, C. G. Cochrane, and P. M. Henson, Two distinct mechanisms for the initiation of mast cell degranulation, *Int. Arch. Allergy Appl. Immunol. 49*:172 (1975).
12. J. C. Foreman and L. M. Lichtenstein, Induction of histamine secretion by polycations, *Biochim. Biophys. Acta 629*:587 (1980).
13. J. P. Gorski, T. E. Hugli, and H. J. Muller Eberhard, Characterization of human C4a anaphylatoxin, *J. Biol. Chem. 256*:2707 (1981).
14. V. A. Bokisch and H. J. Muller-Eberhard, Anaphylatoxin inactivator of human plasma: its isolation and characterization as a carboxypeptidase, *J. Clin. Invest. 49*:2427 (1970).

15. L. H. Caporale, P. S. Tippett, B. W. Erickson, and T. E. Hugli, The active site of C3a anaphylatoxin, *J. Biol. Chem. 255*:10758 (1980).
16. C. G. Unson, B. W. Erickson, and T. E. Hugli, Active site of C3a anaphylatoxin: contributions of the lipophilic and orienting residues, *Biochemistry 23*:585 (1984).
17. R. Huber, H. Scholze, E. P. Paques, and J. Deisenhofer, Crystal structure analysis and molecular model of human C3a anaphylatoxin, *Hoppe-Seyler's Z. Physiol. Chem. 361*:1389 (1980).
18. K. E. Barrett and F. L. Pearce, Mast cell heterogeneity, in *The Handbook of Immunopharmacology* (C. Page, series ed.), Vol. *Immunopharmacology of Mast Cells and Basophils* (J. C. Foreman, ed.), Academic Press, London, 1993, p. 29.
19. E. S. Schulman, J. J. Post, P. M. Henson, and P. C. Giclas, Differential effects of the complement peptides, C5a and $C5a_{desArg}$ on human basophils and lung mast cell histamine release, *J. Clin. Invest. 81*:918 (1988).
20. B. ter Laan, J. L. Molenaar, T. M. Feltkamp-Vroom, and K. W. Pondman, Interaction of human anaphylatoxin C3a with rat mast cells demonstrated by immunofluorescence, *Eur. J. Immunol. 4*:393 (1974).
21. C. T. Harman, Jr., and M. M. Glovsky, Complement activation requirements for histamine release from human leukocytes: influence of purified $C3a_{hu}$ and $C5a_{hu}$ on histamine release, *Int. Arch. Allergy Appl. Immunol. 66*:274 (1981).
22. Y. Fukuoka and T. E. Hugli, Anaphylatoxin binding and degradation by rat peritoneal mast cells: mechanisms of degranulation and control, *J. Immunol. 145*:1851 (1990).
23. Y. Fukuoka and T. E. Hugli, Demonstration of a specific C3a receptor on guinea pig platelets, *J. Immunol. 140*:3496 (1988).
24. R. J. Harvima and L. B. Schwartz, Mast cell-derived mediators, in *The Handbook of Immunopharmacology* (C. Page, series ed.), Vol. *Immunopharmacology of Mast Cells and Basophils* (J. C. Foreman, ed.), Academic Press, London, 1993, p. 115.
25. T. Kajita and T. E. Hugli, Evidence for in vivo degradation of C3a anaphylatoxin by mast cell chymase. I. Nonspecific activation of rat peritoneal mast cells by $C3a_{des\ Arg}$, *Am. J. Pathol. 138*:1359 (1991).
26. M. Mousli and T. E. Hugli, A mechanism of action for anaphylatoxin C3a stimulation of mast cells, *J. Immunol. 148*:2456 (1992).
27. T. E. Hugli and B. W. Erickson, Synthetic peptides with the biological activities and specificity of human C3a anaphylatoxin, *Proc. Natl. Acad. Sci. USA 74*:1826 (1977).
28. J. E. Gervasoni, Jr., D. H. Conrad, T. E. Hugli, L. B. Schwartz, and S. Ruddy, Degradation of human anaphylatoxin C3a by rat peritoneal mast cells: a role for the secretory granule enzyme and heparin proteoglycan, *J. Immunol. 136*:285 (1986).
29. J. C. Foreman, Non-immunological stimuli of mast cells and basophil leucocytes, in *The Handbook of Immunopharmacology* (C. Page, series ed.), Vol. *Immunopharmacology of Mast Cells and Basophils* (J. C. Foreman, ed.), Academic Press, London, 1993, p. 57.

14

Mast Cell Chymases and Tryptases: Phylogeny, Family Relations, and Biogenesis

George H. Caughey
Cardiovascular Research Institute, University of California, San Francisco, School of Medicine, San Francisco, California

I. INTRODUCTION

The mast cell proteases have rewarded the attentions of recent investigators in the field by revealing levels of genetic and catalytic complexity not imagined 10 or even 5 years ago. Some of these proteases are genetically, structurally, and functionally closely akin to enzymes made and released by other cells of bone marrow origin. In this chapter we place the mast cell serine proteases in the broader context of the granule-associated serine proteases of leukocytes as a whole and provide glimpses of the forest beginning to emerge from the trees of individual proteases. We also review recent advances in the understanding of mast cell protease biogenesis, which is shared in many of its particulars by other leukocyte serine proteases.

II. CHYMOTRYPSIN-LIKE MAST CELL PROTEASE FAMILY

A. Diversity of Mammalian Mast Cell Chymases

The chymase extended family is large and still growing, as new isoforms, species variants, and relatives expressed in hematopoietic cells are discovered and characterized. As in many areas of biology, the discovery of new proteins has accelerated with the application of DNA cloning and sequencing technology, resulting in a temporary surfeit of cDNA-derived protease sequences relative to functional data revealing the biological importance of the observed structural diversity.

An alignment of amino acid sequences of mast cell chymases and cathepsin G's and a phylogenetic tree generated from the aligned sequences are shown in Figs. 1 and 2, respectively. These figures illustrate several points. First, they reveal the considerable sequence variation among individual members of the chymase family, with many pairs of chymases exhibiting $<60\%$ amino acid identity in the multiple sequence alignment. Although not all of these chymases have been characterized chemically and catalytically they demonstrate wide variation in such properties as charge, solubility, proteoglycan binding, glycosylation, catalytic efficiency, and regulated expression in mast cell subsets. These differences predict that different chymases play different roles, a speculation at this point supported almost exclusively by data gathered in vitro. Presently, there is neither an adequate functional explanation of the sequence dissimilarity among chymases nor an explanation of the multiplicity of chymases in some individual species: five or more chymases in the mouse alone and at least two in the rat. Recently reported strain differences in mouse chymase expression add an additional layer of complexity (1). In humans, however, the situation may be simpler. Current genetic and enzymological data are most consistent with the existence of just one major chymase (2–5), the expression of which is confined to a subset of mast cells (6–8) and, perhaps, to certain vascular endothelial cells (9,10). It is possible that additional human mast cell chymases await discovery.

B. Human Chymase, Dog Chymase, and Mouse Chymase 5 (MMCP-5): A Separate Branch of the Chymase Tree

The closest known relatives of human chymase are dog chymase and mouse chymase 5 (3,18,20). As shown in Fig. 2, they appear to form a distinct branch of chymases. Surprisingly, neither of the two rat chymases, although they are the major granule-associated proteases of rat mast cells and have been characterized extensively, lies on the branch to which the dog and human chymases

---

Figure 1 Alignment of chymotrypsin-like mast cell proteinases. Complete published amino acid sequences of mast cell chymases and cathepsin G's were subjected to the multiple sequence alignment and scoring algorithm implemented in GeneWorks 2.0 (Intelligenetics). Amino acid rather than nucleic acid sequences were aligned to allow inclusion of rat chymase 1 (rat mast cell protease I), for which only the amino acid sequence is available (11). Residues that are shared by all of the aligned proteases are boxed. The sequence was obtained from the following sources: mouse cathepsin G (12), human cathepsin G (13), mouse chymases 1 (14,15), 2 (16), 4 (17), 5 (18), and L (17), rat chymase 2 (19), dog chymase (20), and human chymase (3).

```
Mouse cathepsin G   MQ-----PLL LLLTFILLQG DEAGKIIGGR EARPHSYPYM AFLLIQSPEG   45
Human cathepsin G   MQ-----PLL LLLAFLLPTG AEAGEIIGGR ESRPHSRPYM AYLQIQSPAG   45
Mouse chymase 1     MQ-----ALL FLMALLLPSG AGAEEIIGGV EARPHSRPYM AHLKIITDRG   45
Rat chymase 2       MQ-----ALL FLMALLLPSG AGAEEIIGGV ESIPHSRPYM AHLDIVTEKG   45
Mouse chymase 4     MQ-----ALL FLMALLLPSG AGAEEIIGGV ESRPHSRPYM AHLEITTERG   45
Rat chymase 1       ---------- ---------- -----IIGGV ESRPHSRPYM AHLEITTERG   25
Mouse chymase L     MQ-----ALL FLMALLLPSE AGAEEIIGGV ESEPHSRPYM AYVNTFRRKG   45
Dog chymase         MHCLPLTLLL LLLC----SR AEAEEIIGGT ESKPHSRPYM AHLEILTLRN   46
Human chymase       MLLLPLPLLL FLLC----SR AEAGEIIGGT ECKPHSRPYM AYLEIVTSNG   46
Mouse chymase 5     MHLLALHLLL LLLG----SS TKAGEIIGGT ECIPHSRPYM AYLEIVTSEN   46
Mouse chymase 2     MQ-----ALL FLMALLLPSG AGAEEIIGGV EAKPHSRPYM AYLKFTTKNG   45

Mouse cathepsin G   -LSACGGFLV REDFVLTAAH CLGSSINVIL GAHNIQMRER TQQLITVLRA   94
Human cathepsin G   -QSRCGGFLV REDFVLTAAH CWGSNINVIL GAHNIQRREN TQQHITARRA   94
Mouse chymase 1     SEDRCGGFLI APQFVLTAAH CKGREITVIL GAHDVSKSES TQQRIKVEKQ   95
Rat chymase 2       LRVICGGFLI SRQFVLTAAH CKGREITVIL GAHDVRKRES TQQKIKVEKQ   95
Mouse chymase 4     FTATCGGFLI TRQFVMTAAH CSGREITVIL GAHDVSKTES TQQKIKVEKQ   95
Rat chymase 1       YKATCGGFLV TRQFVMTAAH CKGRETTVIL GVHDVSKTES TQQKIKVEKQ   75
Mouse chymase L     YVAICGGFLI TPQFVMTAAH CRGRRMTVIL GAHNVRKREC TQQKIKVEKY   95
Dog chymase         HLASCGGFLI RRNFVLTAAH CAGRFIMVIL GAHNIQKKED TWQKLEVIKQ   96
Human chymase       PSKFCGGFLI RRNFVLTAAH CAGRSITVIL GAHNITEEED TWQKLEVIKQ   96
Mouse chymase 5     YLSACSGFLI RRNFVLTAAH CAGRSITVLL GAHNKTSKED TWQKLEVEKQ   96
Mouse chymase 2     SKERCGGFLI APQFLMTAAH CRGSEISVIL GAHNINKNEP TQQIIKTEKT   95

Mouse cathepsin G   IRHPDYNPQN IRNDIMLLQL RRRARRSGSV KPVALPQASK KLQPGDLCTV  144
Human cathepsin G   IRHPQYNQRT IQNDIMLLQL SRRVRRNRNV NPVALPRAQE GLRPGTLCTV  144
Mouse chymase 1     IIHKNYNVSF NLYDIMLLKL EEKAELTPTV DVIPLPGPSD FIDPGKMCWT  145
Rat chymase 2       IIHESYNSVP NLHDIMLLKL EKKVELTPAV NVVPLPSPSD FIHPGAMCWA  145
Mouse chymase 4     IVHPKYNFYS NLHDIMLLKL QKKAKETPSV NVIPLPRPSD FIKPGKMCRA  145
Rat chymase 1       IVHPNYNFYS NLHDIMLLKL QKKAKVTPAV DVIPLPQPSD FLKPGKMCRA  125
Mouse chymase L     ILPPNYNVSS KFNDIMLLKL KKQANLTSAV DVVPLPGPSD FAKPGTMCWA  145
Dog chymase         FPHPKYDDLT LRHDIMLLKL KEKANLTLAV GTLPLSPQFN FVPPGRMCRV  146
Human chymase       FRHPKYNTST LHHDIMLLKL KEKASLTLAV GTLPFPSQFN FVPPGRMCRV  146
Mouse chymase 5     FLHPKYDENL VVHDIMLLKL KEKAKLTLGV GTLPLSANFN FIPPGRMCRA  146
Mouse chymase 2     FVHPKFQYLS GFNDIMLLKL QKKAELNSDV DVISLPSSSD FIKPGKMCWT  145

Mouse cathepsin G   AGWGRVSQSR -GTNVLQEVQ LRVQMDQMCA NRFQFYNSQT QICVGNPRER  193
Human cathepsin G   AGWGRVSMRR -GTDTLREVQ LRVQRDRQCL RIFGSYDPRR QICVGDRRER  193
Mouse chymase 1     AGWGKTGEKE PTSETLREVE LRIMDKEACK MYKHYDY-NF QMCVGSSTKL  194
Rat chymase 2       AGWGKTGVRD PTSYTLREVE LRIMDEKACV DYRYYEY-KF QMCVGSPTTL  194
Mouse chymase 4     AGWGRTGVTE PTSDTLREVK LRIMDKEACK NYWHYDY-NL QMCVGSPRKK  194
Rat chymase 1       AGWGQTGVTK PTSNTLREVK QRIMDKEACK NYFHYNY-NF QMCVGSPRKI  174
Mouse chymase L     AGWGRTGVKK IISHTLREVE LKIVGEKACK IFRHYKD-SL QICVGSSTKV  194
Dog chymase         AGWGKRQVNG SGSDTLQEVK LRLMDPQACR HYMAFDH-NL QLCVGNPRKT  195
Human chymase       AGWGRTGVLK PGSDTLQEVK LRLMDPQACS HFRDFDH-NL QLCVGNPRKT  195
Mouse chymase 5     VGWGRTNVNE PASDTLQEVK MRLQEPQACK HFTSFRH-NS QLCVGNPKKM  195
Mouse chymase 2     AGWGKTGKNN PLSVTLREVE LRIMDQEACK DHNDYDY-QL QMCAGSPTTL  194

Mouse cathepsin G   KSAFRGDSGG PLVCSNVAQG IVSYGSNNGN PPAVFTKIQS FMPWIKRTMR  243
Human cathepsin G   KAAFKGDSGG PLLCNNVAHG IVSYGKSSGV PPEVFTRVSS FLPWIRTTMR  243
Mouse chymase 1     KTAYMGDSGG PLLCAGVAHG IVSYGDSHGK PPAVFTRISA YVPWIKTVIN  244
Rat chymase 2       RAAFMGDSGG PLLCAGVAHG IVSYGHPDAK PPAIFTRVST YVPWINAVIN  244
Mouse chymase 4     RSAYKGDSGG PLLCAGVAHG IVSYGRGDAK PPAVFTRISS YVPWINRVIK  244
Rat chymase 1       RSAYKGDSGG PLVCAGVAHG IVSYGRGDAK PPAVFTRISP YVPWINKVIK  224
Mouse chymase L     ASVYMGDSGG PLLCAGVAHG IVSSGRGNAK PPAIFTRISP HVPWINRVIK  244
Dog chymase         KSAFKGDSGG PLLCAGVAQG IVSYGQNDAK PPAVFTRISH YRPWINKVLK  245
Human chymase       KSAFKGDSGG PLLCAGVAQG IVSYGRSDAK PPAVFTRISH YRPWINQILQ  245
Mouse chymase 5     QNVYKGDSGG PLLCAGIAQG IASYVHRNAK PPAVFTRISH YRPWINKILR  245
Mouse chymase 2     KSIGQGDSGG PLVCDGVAHG IASSYEAKA- -PAVFTRISY YLPWIYKVLK  242

Mouse cathepsin G   RFAPRYQRPA NSLSQAQTX                                   262
Human cathepsin G   SF-----KLL DQMETPL--                                   255
Mouse chymase 1     ---------- ------GK-                                   246
Rat chymase 2       ---------- -----TSS-                                   247
Mouse chymase 4     ---------- -----GE--                                   246
Rat chymase 1       ---------- -----GKD-                                   227
Mouse chymase L     ---------- -----GK--                                   246
Dog chymase         ---------- -----QNKA                                   249
Human chymase       ---------- -----AN--                                   247
Mouse chymase 5     ---------- -----EN--                                   247
Mouse chymase 2     ---------- -----SK--                                   244
```

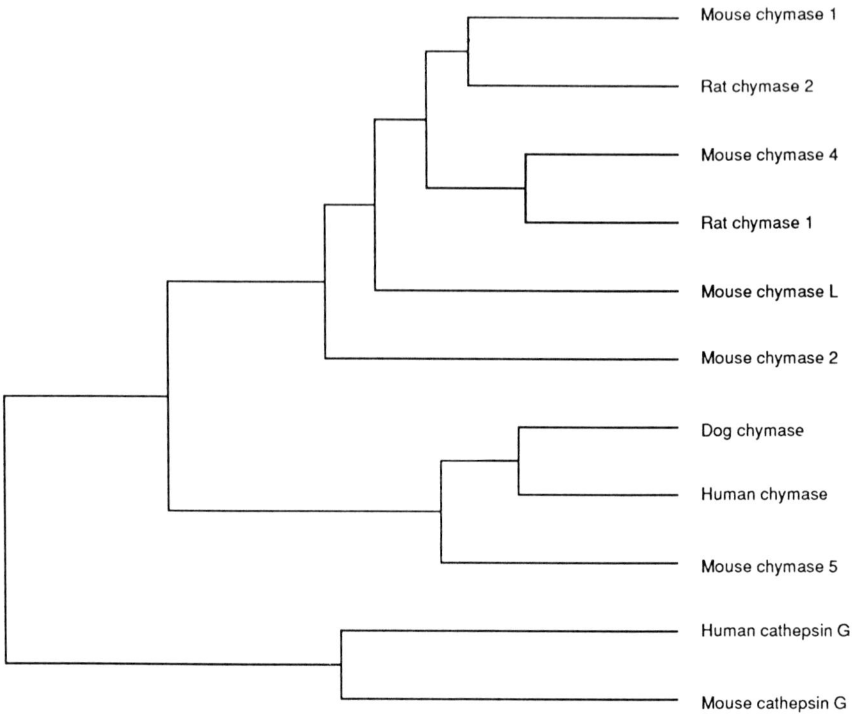

Figure 2 Tree of chymotrypsin-like mast cell proteinases. This phylogenetic tree was constructed from aligned amino acid sequences (see Fig, 1) using the Unweighted Pair Group with Arithmetic Mean Method, as implemented in GeneWorks. Distance from the right margin of the horizonatal lines to a branch point reflects directly the level of sequence variation between pairs or groups of proteases assessed by the multiple sequence alignment and reflects indirectly the evolutionary distance between the proteases and their putative most recent shared ancestor.

and mouse chymase 5 belong. This suggests that caution should be applied in using the results of experiments involving rat chymases to predict the role of chymase in human biology. Human chymase and its close relatives in the dog and the mouse share more than a high level of amino acid sequence identity. Dog and human chymases, for example, are very similar in the kinetics of hydrolysis of peptide-based substrates (21,22). Both are highly cationic [with predicted net charge of 16 and 13 for the dog (20) and human (3) enzymes, respectively] and both bind to heparin proteoglycan and are released from mast cells in the form of a soluble high-molecular-weight complex (23,24). Also, both are modified posttranslationally by N-linked glycosylation (8,20), although many chymases, including rat chymases 1 and 2, are not altered in this manner.

The enzymology of mouse chymase 5 is unknown; thus it is not known whether its substrate preferences, inhibitor susceptibilities, and other key attributes of protease function are closely related to those of dog and human chymase. Nonetheless, its deduced amino acid sequence predicts a net charge (+12) similar to that of its dog and human counterparts and also predicts that it is glycosylated at a site shared with human chymase (18). Additionally, the pattern of mast cell expression of human chymase and mouse chymase 5 is similar, both being stored in serosal and perivascular "connective tissue" mast cells (e.g., of the dermis) but not in "mucosal" mast cells (e.g., of the gut). On these grounds, it is highly likely (though not proven) that dog and human chymases and mouse chymase 5 are functionally as well as structurally orthologous.

C. Chymase and Cathepsin G: Siblings or Distant Cousins?

The same MC_{TC} subset of human mast cells that expresses chymase also contains cathepsin G (or an enzyme enzymatically and antigenically very similar to cathepsin G) (5,8,25), which has chymase-like activity and is similar to chymase in predicted activation sequence, N-glycosylation, charge, and intramolecular disulfide bonds (13), as well as in the intron–exon organization and chromosomal location of its gene (3,4,26) (Fig. 3). Thus chymase and cathepsin G are related, and both could be considered "chymases." However, as shown in Fig. 1, human and mouse cathepsin G lie on a branch separate from that of any of the classically defined mast cell chymases. Furthermore, there is no

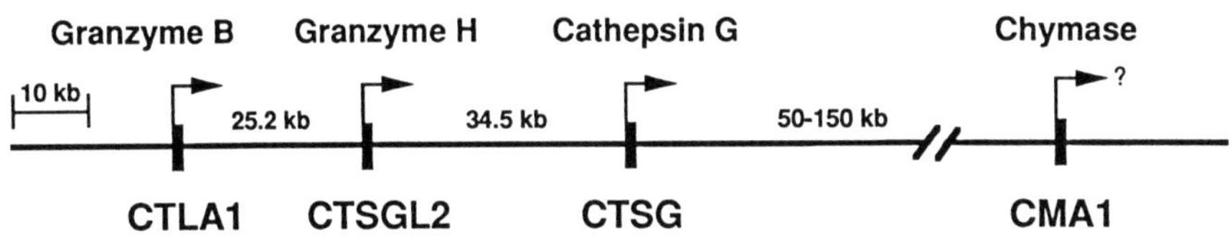

Figure 3 Serine protease gene cluster on human chromosome 14. This cluster, which localizes to band 11.2 of chromosome 14's long arm, contains genes encoding two proteases (i.e., cathepsin G and chymase) expressed in MC_{TC} mast cells (4). Cathepsin G is also expressed in myelomonocytic cells. Granzyme B is an "asp-ase" expressed in cytolytic T lymphocytes. It appears to be an important effector of apoptosis and DNA fragmentation in the targets of cytolytic T lymphocytes (27). Granzyme H mRNA is expressed in natural killer (NK) cells and cytolytic T lymphocytes. It is not yet clear whether granzyme H mRNA is translated into an active protease. In the figure the arrows indicate the direction of transcription, which is uncertain in the case of chymase. The black rectangles indicate the approximate size and position of each gene. The appropriate genome data base symbol (CTLA1, etc.) is identified below the position of each gene.

present evidence of cathepsin G expression in mast cells of mice. Therefore, human mast cell expression of cathepsin G may well be idiosyncratic. Also, as shown in Fig. 4, cathepsin G is no more closely related to chymase than it is to granzyme B. Thus the tradition of a separate nomenclature for chymases and cathepsin G's remains appropriate. Nonetheless, the nomenclature remains somewhat misleading. The "cathepsin" label traditionally is affixed to lysosomal proteases, regardless of mechanistic class (aspartyl, thiol, metallo, serine) or pH optimum, although most cathepsins are maximally active at the acidic pH of the lysosome. The cathepsin label is probably appropriate for cathepsin G in the neutrophil, where it resides in the azurophil granule, whose function is considered to be primarily lysosomal. However, like neutrophil elastase (which resides in the same neutrophil granules and which, by the same logic, could be labeled as a cathepsin), cathepsin G is released from neutrophils under circumstances such as "frustrated phagocytosis" or cell death. However, the "cathepsin" label is probably inappropriate for cathepsin G in mast cell granules, which are thought to function primarily as secretory granules rather than as lysosomes.

D. Relation of Chymase and Cathepsin G to Other Granule-Associated Serine Proteases

Chymase and cathepsin G belong to a broader group of granule-associated serine proteases made by various cells whose precursors originate in the bone marrow. The predicted phylogenetic relationship between various human versions of these proteases, which include enzymes made by myelomonocytic cells (cathepsin G, elastase, proteinase-3, and azurocidin), cytolytic $CD4^+$ T lymphocytes and natural killer cells (granzymes A, B, and H and metase), and mast cells (chymase, cathepsin G, and tryptases), is shown in Fig. 4. With the exception of tryptases, which are the most distant relatives of chymase and cathepsin G in this group, all share certain features which declare their shared evolutionary origins. One of these features is gene organization (i.e., pattern of intron phase and placement in the genes corresponding to the individual proteases). A gene encoding a given mammalian serine protease falls easily into one of six or seven basic organizational patterns. The four-intron phase-I,II,0,0 pattern of the granule-associated serine protease genes, exclusive of tryptase genes (28,29), is one such pattern. As shown in Fig. 3, the genes corresponding to four members of the group (chymase, cathepsin G, and granzymes B and H) are clustered in humans on a fragment of the genome mapping to band 11.2 of the long arm of chromosome 14. The reason for the proximity of these genes, each of which has a different pattern of cellular expression, is not known. Possibly it reflects gene duplication events too recent for scattering of the genes to different chromosomal sites to have taken place. More intriguing is the possibility that the genes

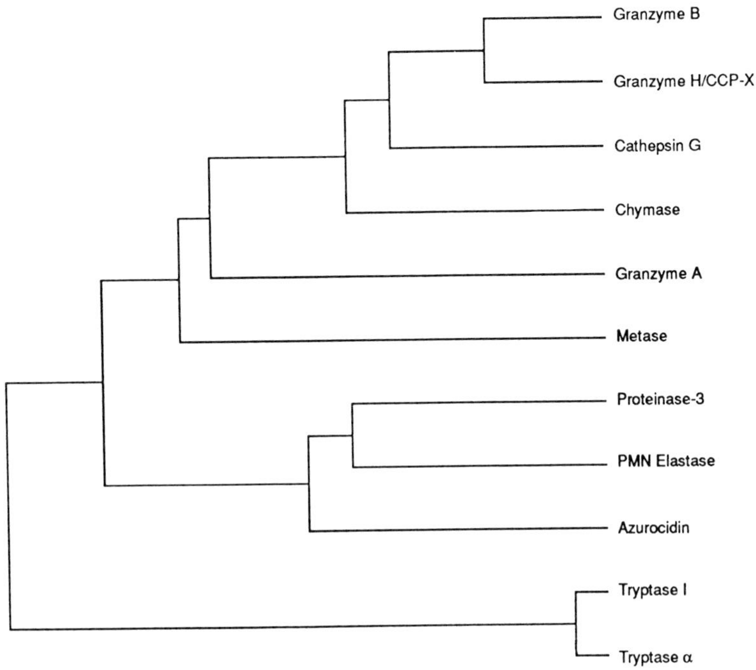

Figure 4 Tree of human granule-associated serine proteinases. As in Fig. 2, this phylogenetic tree was constructed from published, computer-aligned amino acid sequences of proteases, in this case from cytolytic T lymphocytes, natural killer cells, myelomonocytic cells, and mast cells. Each of the proteins listed has been purified and characterized, with the exception of granzyme H/CCP-X, whose amino acid sequence is deduced from a cloned gene clustered with the genes of granzyme B, cathepsin G, and chymase (4) (see Fig. 3). Each of the proteins characterized is catalytically active, with the exception of azurocidin. More detailed tree analysis of granzyme H/CCP-X and its closest relatives among granzymes, chymases, and cathepsin G (31) suggests that the granzyme H/CCP-X gene was generated recently in mammalian evolution by interlocus recombination between the ancestor genes of granzymes B and H/CCP-X, with the result that a portion of granzyme H/CCP-X gene is very closely related to the granzyme B gene, whereas the remainder is related more closely to the cathepsin G gene than to the granzyme B gene.

retain their proximity because of shared dependence on regulatory elements required by the group for orchestrated expression. Recent evidence suggests that the mouse genome contains a similar chymase-granzyme locus (30), which may be larger and more crowded, given the large number of closely related chymase and group B granzyme genes in the mouse relative to those in humans.

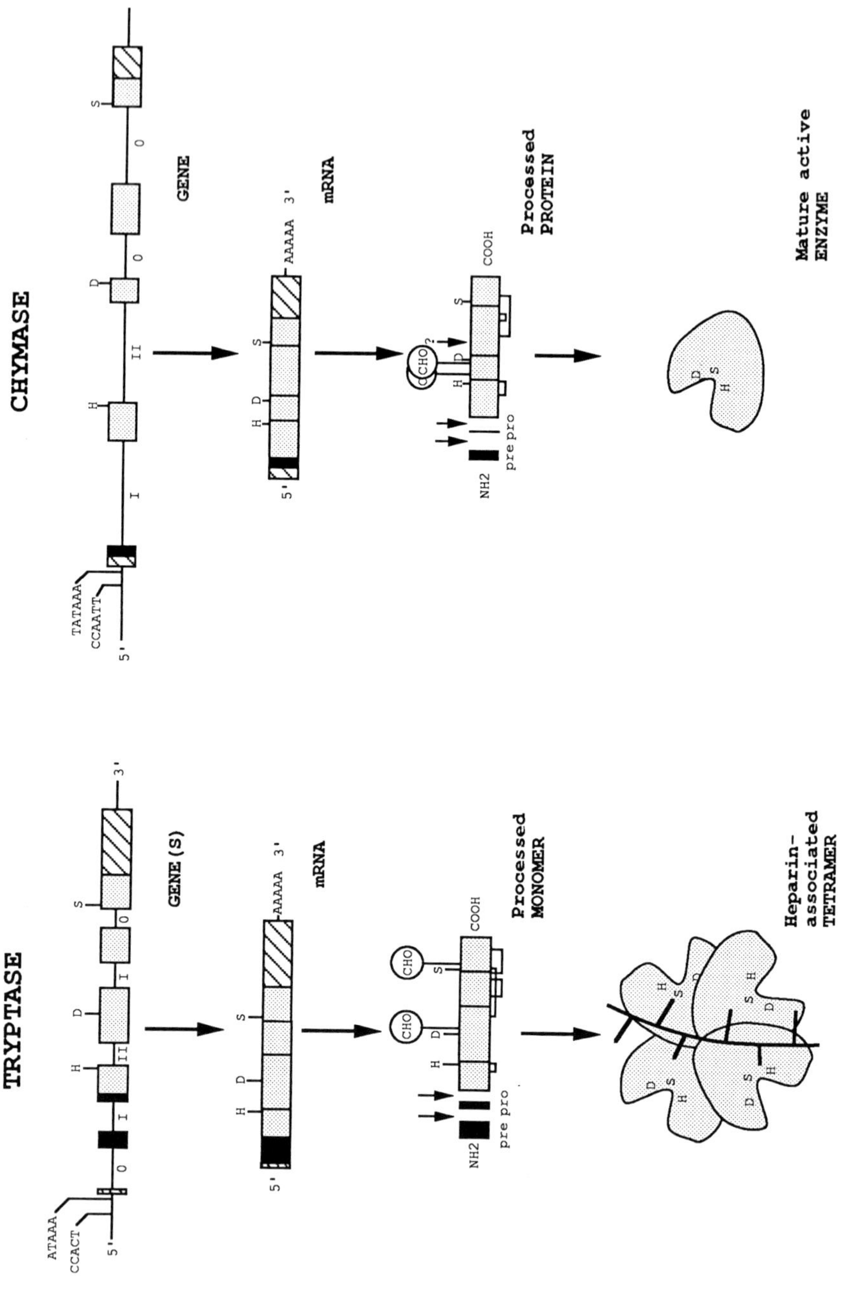
CHYMASE
TATAAA
CCAATT
5'
I
H
II
D
0
S
GENE
5'
H D
S
AAAAA 3'
mRNA
NH2
pre pro
CHO
?
COOH
Processed
PROTEIN
D
S
H
Mature active
ENZYME
TRYPTASE
ATAAA
CCACT
5'
0
I
H
II
D
S
3'
GENE(S)
5'
H
D
S
AAAAA 3'
mRNA
NH2
pre pro
CHO
COOH
Processed
MONOMER
Heparin-
associated
TETRAMER

The granule-associated serine proteases can also be grouped according to patterns of disulfide bond formation, which serve as crude indicators of structural relatedness. The mast cell and leukocyte serine proteases under discussion fall into two broad groups. One group, with just three predicted disulfide bonds (the fewest among known serine proteases of the trypsin family), includes group B granzymes, human granzyme H, neutrophil elastases, cathepsins G, and all known chymases. Crystal-based structures have been solved for two members of this group, chymase (32) and neutrophil elastase (33). The second group, which four predicted disulfide bonds [in positions similar to those in trypsins, with contain four to six disulfide pairs (34)], includes tryptases, group A granzymes, metases, and the recently described rat NK cell tryptic serine protease, RNK-Tryp-2 (35). A surprisingly large number of the proteases are predicted to contain unpaired cysteines in their activated catalytic domain. This includes dog tryptase, human chymase, mouse granzymes B and F, and mouse and human granzyme A. Among these enzymes, only granzyme A, which forms a homodimer, appears to use the "unpaired" cysteine to form intermolecular disulfide bonds (36). Dog mast cell protease-3 (see below), whose cDNA-derived amino acid sequence predicts the presence of 12 cysteines (only 8 of which correspond to those found in trypsins and tryptases) (37), may use one or more of its cysteines to stabilize the active, oligomeric form of the enzyme (38, and G. H. Caughey, unpublished results).

The biogenesis of mast cell chymase is depicted schematically in Fig. 5. The pathway of maturation of chymases from preproenzymes appears to be similar to that of the other, structurally related leukocyte serine proteases, such as cathepsin G, elastase, and granzymes. Like most secreted or granule-associated proteins, these proteases contain classic signal peptides (as predicted from

---

Figure 5 Biogenesis of human tryptase and chymase. Exons in the gene are represented by boxes. Nonprotein coding exons are marked with diagonal lines. Prepropeptides and catalytic domains are shaded black and gray, respectively. Intron size is shown to scale and intron phase (0, I, or II) is indicated below the strand. Codons or amino acids representing essential catalytic triad amino acids (His = H, Asp = D, Ser = S) are positioned in DNA, mRNA, and protein as indicated. Sites of N-linked carbohydrate attachment are designated "CHO." Positions of disulfide linkages are shown below the processed protein strand. Small arrows show presumed sites of proteolytic processing of preproenzymes. The tryptase gene and protein shown is human tryptase I (28). The chymase gene and protein shown is the human enzyme (2,3). The gene organization, processing, and disulfide linkages of cathepsin G are similar to those of chymase (13,26,39). The branchlike structure overlying the tryptase tetramer represents heparin proteoglycan. Human chymase, although it binds to heparin, does not seem to require it for catalytic activity or stability. (Modified from Ref. 41.)

sequenced cDNAs), which direct the nascent translation product into the endoplasmic reticulum and are removed during or soon after translation by components the signal peptidase complex. Removal of signal peptides from members of this group of enzymes are predicted to generate inactive proenzymes containing an N-terminal acidic dipeptide. Further removal of this putative "activation dipeptide," which, based on the results of N-terminal sequence analysis, is absent from the mature enzymes, is thought to generate the catalytically active form of chymase and related proteases (39). The probable intracellular processing enzyme responsible for removing the activation sequence is dipeptidyl peptidase I, the inhibition of which diminishes the generation of cathepsin G- and elastase-like catalytic activity in extracts of U937 cells and granzyme A–like activity in extracts of NK cells and stimulated $CD8^+$ T lymphocytes (40). Although the activation of prochymases by dipeptidyl peptidase I has not been addressed directly in the laboratory, the similarity of the activation peptides of chymase to those of cathepsin G and related enzymes predicts that similar machinery is used for their removal.

As direct evidence that the Gly–Glu prosequence of human chymase serves as an activation peptide, removal of the dipeptide by incubation of COS-1 cell–expressed, inactive, recombinant human prochymase with extracts of cultured lymphoma cells generates active chymase (42). COS-1 cells themselves are unable to activate prochymase, suggesting that not all cells possess the enzymatic wherewithal to activate chymase and related proteases from their proforms. The chymase activation sequence also may play other roles, such as organelle targeting, protein folding, or protection from intracellular degradation. This hypothesis is partially supported by the finding of a range of sizes of inactive immunoreactive proteins in COS-1 cells transfected with recombinant chymase constructs lacking the two codons encoding the Gly–Glu prosequence (42). In contrast, COS cell expression of active recombinant mouse granzyme B (which has an identical Gly–Glu activation dipeptide) has been reported using transfections with a construct in which the activation peptide codons are deleted (43). Therefore, a role for the Gly–Glu sequence other than as an activation peptide remains speculative. As discussed below, the activation pathways followed by tryptases may differ from those of chymases.

Within the catalytic domain itself, the most prominent conserved aspect of primary structure in the greater group of chymase–granzyme–elastase-related leukocyte granule proteases is an octapeptide, PHSRPYMA (44), which corresponds to residues of 9 to 16 of mature chymases. It is not a feature of tryptases or of other known groups of serine proteases. Its function is unknown, although one can speculate that it provides a targeting signal for these granule-associated enzymes. Antibodies raised against this peptide recognize members of this protease family and may be useful in identifying new members (43).

Posttranslational modification by glycosylation is an inconstant feature of mammalian mast cell chymases. Dog and human chymase, for example, are N-glycosylated at two sites, as suggested by the identification of consensus glycosylation sites in sequences deduced from cloned DNA, by reduction in molecular weight and collapse of heterogeneity on SDS-PAGE after exposure to glycosidases, and by the results of direct protein sequencing (3,8,20). Human and dog chymases, despite a high level of amino acid sequence conservation, do not share consensus glycosylation sites. Other chymases, including rat chymases 1 and 2, are not glycosylated. This suggests that the extent and exact location of this type of posttranslational modification is not critical for the most basic elements of chymase function. Nonetheless, variation in glycosylation may be responsible for some of the differences among chymases observed in properties such as stability, solubility, and binding to proteoglycans. In the particular case of human chymase, glycosylation at Asn^{82} (3) may prevent the expression of the intrinsic activity of a portion of the protein whose homologous region in cathepsin G and granzyme B has antibacterial properties (45). Neither of the latter two enzymes is glycosylated at this site.

Studies of granzyme A and B secretion from lymphocytes obtained from patients with I-cell disease (46) suggest that some glycosylated granule proteases are targeted to granules via interactions with mannose-6-phosphate receptor. Lymphocytes from such patients (who are unable to phosphorylate mannose) divert a large fraction of their granzymes to a constitutive secretory pathway, although they also retain the ability to direct a portion of the enzymes to a regulated, granule compartment by mannose-6-phosphate receptor–independent mechanisms. In cytolytic T cells, as in mast cells, there appears to be no separation between lysosomal and secretory granule compartments (41,46). With respect to mast cell chymase granule localization and secretion, it seems unlikely that a mannose-6-phosphate receptor–dependent pathway is critical, given the very high levels of some granule chymases (e.g., rat chymases 1 and 2) that cannot use the mannose-6-phosphate pathway because of the complete absence of N-linked sugars.

In addition to removal of signal and activation dipeptides and attachment of sugars, some (or perhaps most) chymases are subject to further posttranslational modifications. For example, rat chymases 1 and 2, in their mature forms, both appear to have been shortened at their C termini based on comparisons of the primary structures of the purified, mature proteins with those predicted from cloned cDNAs (11,19,47,48). Cathepsin G appears to be trimmed in a similar fashion, at least in the neutrophil (13). The significance of C-terminal processing of these enzymes is unknown. In addition, at high concentrations in vitro, human chymase is susceptible to autolytic cleavage at a single internal site, with preservation of catalytic activity (9). It remains to be determined whether this cleavage occurs in vivo.

The importance of granule proteoglycans in the biogenesis of chymases is not clear. Mast cell chymases vary greatly in surface charge and in their binding to negatively charged heparin- and chondroitin sulfate–containing proteoglycans. Several of these enzymes, especially dog and human chymases and rat chymase 1, bind strongly to heparin proteoglycan (49–51). However, the ability to express active chymase in COS cells, which probably contain little heparin, suggests that proteoglycan may not be essential for the generation of active enzyme. For human tryptase, the presence of heparin during granule biogenesis may be more critical (see below). By neutralizing positive charge, granule proteoglycans probably influence the amount and concentration of a highly cationic protease such as human chymase that can be stored in a mast cell secretory granule. The influence of granule proteoglycans is most clear following exocytosis. The size of proteoglycan, the solubility of the chymase–proteoglycan complex, and the strength of ionic interaction between chymase and proteoglycan are crucial determinants of the extent and rates of diffusion of enzyme from the site of mast cell degranulation, the size and types of targets with which chymases interact, and the susceptibility to inactivation by circulating antiproteases.

III. FAMILY OF TRYPSIN-LIKE MAST CELL SERINE PROTEASES

A. Diversity of Mammalian Mast Cell Tryptases

The major mast cell enzymes with the trypsin-like ability to hydrolyze target peptides and proteins after internal basic amino acids are tryptases. The sequence alignment and tree shown in Figs. 6 and 7 reveal substantial structural diversity among tryptase-related serine proteases characterized to date. As a group, tryptases lack a close phylogenetic relationship to granule-associated proteases such as elastase, granzymes, and mast cell chymases. The separation between branches in the tree in Fig. 4 reveals the structural and evolutionary dissimilarity between tryptases and other leukocyte and mast cell serine proteases. Tryptases also are not closely related to any other groups of mammalian serine proteases, including other enzymes of trypsin-like specificity, such as trypsin

---

Figure 6 Alignment of tryptase-like mast cell proteinases. As in Fig. 1, complete published amino acid sequences were aligned by computer. A consensus sequence is shown below each section. Regions of amino acid identity are boxed. Sequence was obtained from the following sources: dog mast cell protease-3 (37), human tryptases I (28), II/β (28,52), III (28), and α (53) (as modified in Ref. 54), dog tryptase (37), and mouse tryptases 1 (MMCP-6) (29) and 2 (MMCP-7) (55).

```
Dog mast cell protease-3   MAP---L-IR MLDL---LGT LSP-----KV GIVGGCKVPA RRYPWQVSLR   38
Human tryptase I           MLNLLLLALP VLASRA-YAA PAPGQALQRV GIVGGQEAPR SKWPWQVSLR   49
Human tryptase II/β        -LNLLLLALP VLASRA-YAA PAPGQALQRV GIVGGQEAPR SKWPWQVSLR   48
Human tryptase III         --------LP VLASRA-YAA PAPGQALQRV GIVGGQEAPR SKWPWQVSLR   41
Human tryptase α           MLSLLLLALP VLASRA-YAA PAPVQALQQA GIVGGQEAPR SKWPWQVSLR   49
Dog tryptase               MPSPLVLALA LLGSLV-PVS PAPGQALQRV GIVGGREAPG SKWPWQVSLR   49
Mouse tryptase 1           MLKRRLLLLW ALSLLASLVY SAPRPANQRV GIVGGHEASE SKWPWQVSLR   50
Mouse tryptase 2           ------LTLP LLSS---LVH AAPGPAMTRE GIVGGQEAHG NKWPWQVSLR   41

Consensus                  M.....L.LP .L.S.A-... PAPGQALQRV GIVGGQEAP. SKWPWQVSLR   50

Dog mast cell protease-3   FHGMGSGQWQ HICGGSLIHP QWVLTAAHCV ELEGLEAATL RVQVGQLRLY   88
Human tryptase I           VHG---PYWM HFCGGSLIHP QWVLTAAHCV GPDVKDLAAL RVQLREQHLY   96
Human tryptase II/β        VHG---PYWM HFCGGSLIHP QWVLTAAHCV GPDVKDLAAL RVQLREQHLY   95
Human tryptase III         VRD---RYWM HFCGGSLIHP QWVLTAAHCV GPDVKDLAAL RVQLREQHLY   88
Human tryptase α           VRD---RYWM HFCGGSLIHP QWVLTAAHCL GPDVKDLATL RVQLREQHLY   96
Dog tryptase               LKG---QYWR HICGGSLIHP QWVLTAAHCV GPNVVCPEEI RVQLREQHLY   96
Mouse tryptase 1           FKL---NYWI HFCGGSLIHP QWVLTAAHCV GPHIKSPQLF RVQLREQYLY   97
Mouse tryptase 2           AND---TYWM HFCGGSLIHP QWVLTAAHCV GPDVADPNKV RVQLRKQYLY   88

Consensus                  ...---.YWM HFCGGSLIHP QWVLTAAHCV GPDVKD.A.L RVQLREQHLY  100

Dog mast cell protease-3   DHDCLCNVTE IIRHPNFNMS WYGWDTADIA LLKLEAPLTL SEDVNLVSLP  138
Human tryptase I           YQDQLLPVSR IIVHPQFYTA QIG---ADIA LLELEEPVNV SSHVHTVTLP  143
Human tryptase II/β        YQDQLLPVSR IIVHPQFYTA QIG---ADIA LLELEEPVKV SSHVHTVTLP  142
Human tryptase III         YQDQLLPVSR IIVHPQFYTA QIG---ADIA LLELEEPVKV SSHVHTVTLP  135
Human tryptase α           YQDQLLPVSR IMVHPQFYII QTG---ADIA LLELEEPVNI SSRVHTVMLP  143
Dog tryptase               YQDHLLPVNR IVMHPNYYTP ENG---ADIA LLELEDPVNV SAHVQPVTLP  143
Mouse tryptase 1           YGDQLLSLNR IVVHPHYYTA EGG---ADVA LLELEVPVNV STHIHPISLP  144
Mouse tryptase 2           YHDHLMTVSQ IITHPDFYIV QDG---ADIA LLKLTNPVNI SDYVHPVPLP  135

Consensus                  YQDQLLPVSR IIVHP.FYT. Q.G---ADIA LLELE.PVNV S.HVH.V.LP  150

Dog mast cell protease-3   SPSLIVPPGM LCWVTGWGDI ADHTPLPPPY HLQEVEVPIV GNREQNCHYQ  188
Human tryptase I           PASETFPPGM PCWVTGWGDV DNDERLPPPF PLKQVKVPIM ENHICDAKYH  193
Human tryptase II/β        PASETFPPGM PCWVTGWGDV DNDERLPPPF PLKQVKVPIM ENHICDAKYH  192
Human tryptase III         PASETFPPGM PCWVTGWGDV DNDERLPPPF PLKQVKVPIM ENHICDAKYH  185
Human tryptase α           PASETFPPGM PCWVTGWGDV DNDEPLPPPF PLKQVKVPIM ENHICDAKYH  193
Dog tryptase               PALQTFPIGT PCWVTGWGDV HSGTPLPPPF PLKQVKVPIV ENSMCDVQYH  193
Mouse tryptase 1           PASETFPPGT SCWVTGWGDI DNDEPLPPPY PLKQVKVPIV ENSLCDRKYH  194
Mouse tryptase 2           PASETFPSGT LCWVTGWGNI DNGVNLPPPF PLKEVQVPII ENHICDLKYH  185

Consensus                  PASETFPPGM PCWVTGWGDV DNDE.LPPPF PLKQVKVPI. ENH.CD.KYH  200

Dog mast cell protease-3   TIL--EQDDE VIKQDMLCAG SEGHDSCQMD SGGPLVCRWK CTWIQVGVVS  236
Human tryptase I           LGAYTGDDVR IVRDDMLCAG NTRRDSCQGD SGGPLVCKVN GTWLQAGVVS  243
Human tryptase II/β        LGAYTGDDVR IVRDDMLCAG NTRRDSCQGD SGGPLVCKVN GTWLQAGVVS  242
Human tryptase III         LGAYTGDDVR IVRDDMLCAG NTRRDSCQGD SGGPLVCKVN GTWLQAGVVS  235
Human tryptase α           LGAYTGDDVR IIRDDMLCAG NSQRDSCKGD SGGPLVCKVN GTWLQAGVVS  243
Dog tryptase               LGLSTGDGVR IVREDMLCAG NSKSDSCQGD SGGPLVCRVR GVWLQAGVVS  243
Mouse tryptase 1           TGLYTGDDFP IVHDGMLCAG NTRRDSCQGD SGGPLVCKVK GTWLQAGVVS  244
Mouse tryptase 2           KGLITGDNVH IVRDDMLCAG NEGHDSCQGD SGGPLVCKVE DTWLQAGVVS  235

Consensus                  LG.YTGDDVR IVRDDMLCAG N..RDSCQGD SGGPLVCKV. GTWLQAGVVS  250

Dog mast cell protease-3   WGYGCGY-NL PGMYARVTSY VSWIHQHIPL SPGP                   269
Human tryptase I           WGEGCAQPNR PGIYTRVTYY LDWIHHYVPK KP--                   275
Human tryptase II/β        WGEGCAQPNR PGIYTRVTYY LDWIHHYVPK KP--                   274
Human tryptase III         WGEGCAQPNR PGIYTRVTYY LDWIHHYVPK KP--                   267
Human tryptase α           WDEGCAQPNR PGIYTRVTYY LDWIHHYVPK KP--                   275
Dog tryptase               WGEGCAQPNR PGIYTRVAYY LDWIHQYVPK EP--                   275
Mouse tryptase 1           WGEGCAQPNK PGIYTRVTYY LDWIHRYVPE HSX-                   277
Mouse tryptase 2           WGEGCAQPNR PGIYTRVTYY LDWIHHYVPK DF--                   267

Consensus                  WGEGCAQPNR PGIYTRVTYY LDWIHHYVPK .P--                   284
```

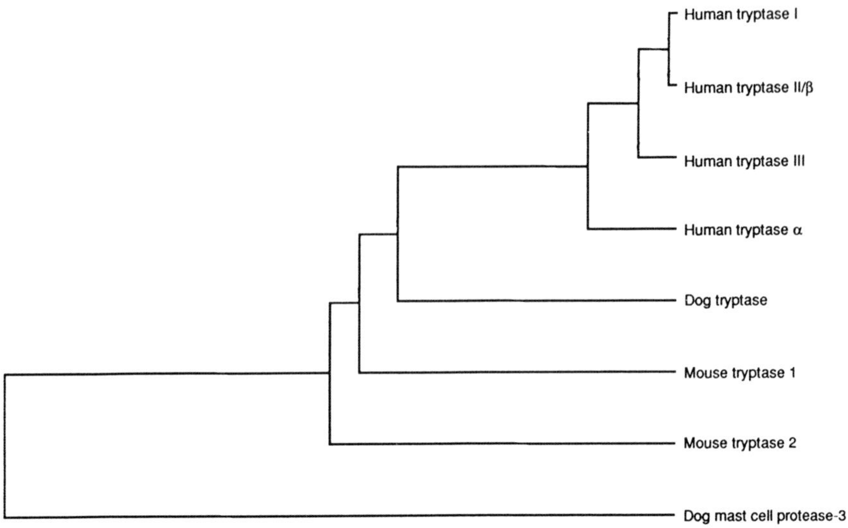

Figure 7 Tree of tryptase-like mast cell proteinases. As in Fig. 2, this phylogenetic tree was constructed from computer-aligned amino acid sequences (see Fig. 4). Dog mast cell protease-3 [originally termed *dog mastocytoma protease* (37)], although only 48 to 49% identical in amino acid sequence to human, dog, and mouse tryptases, is more closely related to tryptases than to other known proteases, and so is included in this grouping.

and the many tryptic serine proteases of the hemostatic, fibrinolytic, and complement cascades. Indeed, tryptases are functionally as well as structurally idiosyncratic. A number of possible biological functions for tryptases are discussed by other authors elsewhere in this volume. Like chymases, tryptases exhibit several isoforms, the variety and number of which vary in different mammalian species. Tryptases also manifest rather striking species differences in the amounts of enzyme expressed in specific mast cell subsets.

Two tryptase cDNAs and their corresponding genes have been identified in the mouse (see Chapter 9 in this volume). The enzymatic properties of mouse tryptases 1 and 2 [also known as mouse mast cell proteases (MMCP) 6 and 7] remain to be explored. However, their deduced amino acid sequences predict that their biophysical properties will be similar to those of more extensively characterized tryptases from dogs, rats, and humans. Protease-specific RNA analysis and immunohistochemical studies suggest that mouse tryptase expression is limited to a subpopulation of safranin-staining, connective tissue mast cells (29,55). Mouse tryptase 1 and 2 expression is absent in the safranin-negative, mucosal mast cells that proliferate in response to intestinal helminthosis. More is known of the enzymatic properties of the rat tryptase from skin (56) and peritoneal mast cells (57–59) than is known of mouse tryptases 1 and 2. Less is known, however, of their genes, transcription, and cell-specific expression. At the time this chapter was being written, rat tryptase cDNA had not been cloned, and it was not known whether more than one tryptase gene is expressed in this species. However, studies using antibodies raised against rat skin tryptase suggest that the pattern of tryptase expression in the rat is similar to that in the mouse, with immunoreactive tryptase being restricted to a fraction of safranin-staining mast cells. The percentage of safranin-staining mast cells that are tryptase positive varies dramatically with tissue location, being highest in airway tissues (60). Although there are differences in the reported behavior of tryptase isolated from skin and peritoneal cells, these difference are explained at least in part by variable association with an inhibitor, trypstatin, which appears to be a fragment of inter-α-trypsin inhibitor light chain that is taken up from serum by some preparations of peritoneal mast cells (61). A major difference between rat tryptase and its dog and human counterparts is its lack of binding to or stabilization by heparin (56,59). Although levels of tryptase in mast cell subsets have not been quantitated precisely, Lagunoff and colleagues estimate that the level of tryptase is only about 5% of that of chymase in the population of peritoneal mast cells as a whole (59). Thus current evidence suggests that the expression of tryptases in mice and rats is quite limited.

Humans contrast rather strikingly with mice and rats in that tryptases are expressed at high levels in virtually all mast cells in all tissues. Mast cell subset-specific differences in levels of human tryptase, and the more dramatic variations in levels of chymase, cathepsin G, and carboxypeptidase, are discussed in detail in Chapter 7 and by other authors in this volume. Although the cloning of tryptase cDNAs and genes predict the existence of at least four different tryptase monomers (α, I, II/β, and III) (28,52,53), the differences in deduced amino acid sequence among most of these (especially I, II/β, and III) are small. One or more of the identified cDNAs may encode allelic variants. At present, there is no strong evidence that different human tryptase genes are expressed preferentially in one or another tissue or mast cell subset. Indeed, there are no convincing differences in the properties of human tryptases purified from different tissue sites. Although initial comparison of tryptases suggested differences in the kinetics of substrate hydrolysis by preparations of enzymes extracted from skin and lung by various approaches in various laboratories (62), subsequent comparison of tryptases purified, titrated, and assayed under closely matched conditions failed to confirm the differences suspected on the basis of the earlier data (63). Thus even if different tryptase genes are expressed in varying ratios in different tissue microenvironments, it remains unclear whether these variations will be functionally consequential.

Recent studies show that some cultured tumor cell lines preferentially transcribe specific tryptase genes. U937 cells and KU812 cells, for example, primarily synthesize mRNA encoding tryptases I and II, respectively (54,64), although neither of these cell lines is known to produce active tryptase. Mono Mac 6 cells, some of which produce material recognized by antibodies to tryptase, predominantly or exclusively produce mRNA that is similar, but not identical, to mRNA encoding tryptase α (54); and HMC-1 cells, which produce active tryptase, synthesize mRNA that is mainly or exclusively of the β (i.e., I, II/β, or III) type (65). Although these observations in tumor-derived cells are consistent with the possibility that various tryptase genes are differentially transcribed in human mast cell subsets and basophils, their relevance to normal human cells remains to be established.

B. Dog Mast Cell Protease-3: A New Relative of Tryptase

As shown in the multiple sequence alignment in Fig. 6 and in the tree in Fig. 7, dog tryptase shares many structural features with human tryptases, to which it is slightly more closely related than are either of the two mouse tryptases. As established in studies published since the last review of dog tryptase (49), the dog enzyme exhibits several previously unrecognized similarities as well as differences with respect to the human enzyme (66–70). Like its human counterpart(s), the dog enzyme is an inhibitor-resistant heparin-stabilized tetramer of catalytically active subunits. Physical and functional characterization of dog tryptase to date has been accomplished using enzyme and cDNAs purified and cloned from the BR line of dog mastocytoma cells (37,71). Unlike in mice, rats, and humans, the mast cell subset distribution of tryptase in the dog has not been studied, nor has the presence of closely related dog tryptases been excluded. However, the search for a clone corresponding to dog tryptase did lead to the serendipitous discovery of a mastocytoma cDNA encoding a relative of tryptase, originally termed *dog mastocytoma protease* (37), but now termed *dog mast cell protease-3* (dMCP-3).

As demonstrated by the branching of the tree in Fig. 7, dMCP-3 is substantially different from dog tryptase and other known tryptases. Its prepro sequence is shorter, its catalytic domain is larger, it has many more cysteines, and it is predicted to be more anionic than known tryptases. Although dMCP-3 is predicted to be tryptic in specificity and is more closely related to tryptases than to any other group of serine proteases, one can argue that it should not be classified as a tryptase based on structure alone. Fortunately, recent characterization of the product of dMCP-3 mRNA further supports its classification as a tryptase. Antibodies raised against a synthetic peptide predicted from dMCP-3 cDNA recognize a product that appears to be expressed in the cytoplasmic granules of cultured dog mastocytoma cells and in widely scattered mononuclear, metachromatic cells in a variety of normal dog tissues (38). These antibodies do not cross-react with purified dog tryptase. Thus dMCP-3 appears to be a product of mast cells. The results of immunohistochemical staining raise the possibility that dMCP-3 is also expressed in neutrophils, although this is not supported by Western blots prepared from proteins extracted from circulating dog neutrophils. DMCP-3 appears to be the major serine protease in the G line of dog mastocytoma cells (38), from which relatively little tryptic activity is released in response to ionophore (72) but from which very high levels of tryptase-like activity are extracted at low ionic strength (73). Preliminary characterization of dMCP-3 activity suggests that it is a highly stable, heparin-binding, oligomeric, tryptic serine protease distinct from dog tryptase and from other known tryptases in its activity profile (G. Caughey, unpublished results). It is likely that a protease as distinct from other known proteases as dMCP-3 will be found to have orthologs in other species, although this remains to be validated.

C. Biogenesis of Tryptase-Related Enzymes

Figure 5 illustrates known steps in the biogenesis of tryptase, from gene transcription to assembly of a biologically active oligomer. The example is that of human tryptase I, whose gene and corresponding cDNA have been sequenced and characterized (28). It should be emphasized that basic issues concerning the contribution of different tryptase genes to the generation of active tryptases remain to be resolved. It is not known, for example, whether the products of separate tryptase genes can (or must) combine to form the active tetrameric form of the enzyme. Nonetheless, an outline of the process has emerged through the efforts of investigators in several laboratories.

Tryptase gene structure has one unusual feature, which is the insertion of an intron in phase with, and immediately preceding, the codon corresponding to the first methionine of the protein coding sequence (28). This intron separates the promoter regions and transcription start site from protein coding sequence. It is not known whether this intron carries information that influences genetic regulation of tryptase expression. During mRNA processing, this first intron is removed along with the downstream introns from the primary transcripts corresponding to the human tryptase I (28) and mouse tryptase 1 (MMCP-6) genes (29). Due to a point mutation at the acceptor splice junction in the gene encoding mouse tryptase 2 (MMCP-7), the first intron of this tryptase gene fails to be excised (55). However, because of its location upstream of the protein coding sequence, this "flaw" apparently is not fatal. The mouse tryptase 2 gene is transcribed and is presumably translated.

Tryptase mRNA is translated into preproprotein that contains a signal peptide and activation peptide linked in tandem to the inactive catalytic domain.

ACKNOWLEDGMENTS

This work is supported by Program Project Grant HL-24136 from the National Institutes of Health. The author is the recipient of a Career Investigator Award from the American Lung Association and of an Investigator Award from the Asthma and Allergy Foundation of America.

REFERENCES

1. R. L. Stevens, D. S. Friend, H. P. McNeil, V. Schiller, N. Ghildyal, and K. F. Austen, Strain-specific and tissue-specific expression of mouse mast cell secretory granule proteases, *Proc. Natl. Acad. Sci. USA 91*:128 (1994).
2. H. Urata, A. Kinoshita, D. M. Perez, K. S. Misono, F. M. Bumpus, R. M. Graham, and A. Husain, Cloning of the gene and cDNA for human heart chymase, *J. Biol. Chem. 266*:17173 (1991).
3. G. H. Caughey, E. H. Zerweck, and P. Vanderslice, Structure, chromosomal assignment, and deduced amino acid sequence of a human gene for mast cell chymase, *J. Biol. Chem. 266*:12956 (1991).
4. G. H. Caughey, T. H. Schaumberg, E. H. Zerweck, J. H. Butterfield, R. D. Hanson, G. A. Silverman, and T. J. Ley, The human mast cell chymase gene (CMA1): mapping to the cathepsin G/granzyme gene cluster and lineage-restricted expression, *Genomics 15*:614 (1993).
5. N. M. Schechter, W. Z. Mei, R. W. Blacher, S. R. Lessin, G. S. Lazarus, and H. Rubin, Determination of the primary structures of human skin chymase and cathepsin G from cutaneous mast cells of urticaria pigmentosa lesions, *J. Immunol. 152*:4062 (1994).
6. N. M. Schechter, J. K. Choi, D. A. Slavin, D. T. Deresienski, S. Sayama, G. Dong, R. M. Lavker, D. Proud, and G. S. Lazarus, Identification of a chymotrypsin-like proteinase in human mast cells, *J. Immunol. 137*:962 (1986).
7. A. A. Irani, N. M. Schechter, S. S. Craig, G. DeBlois, and L. B. Schwartz, Two types of human mast cells that have distinct neutral protease compositions, *Proc. Natl. Acad. Sci. USA 83*:4464 (1986).
8. N. M. Schechter, A. M. Irani, J. L. Sprows, J. Abernethy, B. Wintroub, and L. B. Schwartz, Identification of a cathepsin G–like proteinase in the MCTC type of human mast cell, *J. Immunol. 145*:2652 (1990).
9. H. Urata, A. Kinoshita, K. S. Misono, F. M. Bumpus, and A. Husain, Identification of a highly specific chymase as the major angiotensin II–forming enzyme in the human heart, *J. Biol. Chem. 265*:22348 (1990).
10. H. Urata, K. D. Boehm, A. Philip, A. Kinoshita, J. Gabrovsek, F. M. Bumpus, and A. Husain, Cellular localization and regional distribution of an angiotensin II–forming chymase in the heart, *J. Clin. Invest. 91*:1269 (1993).
11. H. Le Trong, D. C. Parmelee, K. A. Walsh, H. Neurath, and R. G. Woodbury, Amino acid sequence of rat mast cell protease I (chymase), *Biochemistry 26*:6988 (1987).
12. J. W. Heusel, E. M. Scarpati, N. A. Jenkins, D. J. Gilbert, N. G. Copeland, S.

D. Shapiro, and T. J. Ley, Molecular cloning, chromosomal location, and tissue-specific expression of the murine cathepsin G gene, *Blood 81*:1614 (1993).
13. G. Salvesen, D. Farley, J. Shuman, A. Przybyla, C. Reilly, and J. Travis, Molecular cloning of human cathepsin G: structural similarity to mast cell and cytotoxic T lymphocyte proteinases, *Biochemistry 26*:2289 (1987).
14. H. Le Trong, G. F. J. Newlands, H. R. P. Miller, H. Charbonneau, H. Neurath, and R. G. Woodbury, Amino acid sequence of a mouse mucosal mast cell protease, *Biochemistry 28*:391 (1989).
15. R. Huang, T. Blom, and L. Hellman, Cloning and structural analysis of MMCP-1, MMCP-4 and MMCP-5, three mouse mast cell-specific serine proteases, *Eur. J. Immunol. 21*:1611 (1991).
16. W. E. Serafin, D. S. Reynolds, S. Rogelj, W. S. Lane, G. A. Condor, S. S. Johnson, K. F. Austen, and R. L. Stevens, Identification and cloning of a novel mouse mucosal mast cell serine protease, *J. Biol. Chem. 265*:423 (1990).
17. W. E. Serafin, T. P. Sullivan, G. A. Conder, A. Ebrahimi, P. Marcham, S. S. Johnson, K. F. Austen, and D. S. Reynolds, Cloning of the cDNA and gene for mouse mast cell protease 4: demonstration of its late transcription in mast cell subclasses and analysis of its homology to subclass-specific neutral proteases of the mouse and rat, *J. Biol. Chem. 266*:1934 (1991).
18. H. P. McNeil, K. F. Austen, L. L. Somerville, M. F. Gurish, and R. L. Stevens, Molecular cloning of the mouse mast cell protease-5 gene: a novel secretory granule protease expressed early in the differentiation of serosal mast cells, *J. Biol. Chem. 266*:20316 (1991).
19. P. N. Benfey, F. H. Yin, and P. Leder, Cloning of the mast cell protease, RMCP II: evidence for cell-specific expression and a multigene family, *J. Biol. Chem. 262*:5377 (1987).
20. G. H. Caughey, W. W. Raymond, and P. Vanderslice, Dog mast cell chymase: molecular cloning and characterization, *Biochemistry 29*:5166 (1990).
21. G. H. Caughey, N. F. Viro, S. C. Lazarus, and J. A. Nadel, Purification and characterization of dog mastocytoma chymase: identification of an octapeptide conserved in chymotryptic leukocyte proteases, *Biochim. Biophys. Acta 952*:142 (1988).
22. J. C. Powers, T. Tanaka, J. W. Harper, Y. Minematsu, L. Barker, D. Lincoln, K. V. Crumley, J. E. Fraki, N. M. Schechter, G. G. Lazarus, K. Nakajima, K. Nakashino, H. Neurath, and R. G. Woodbury, Mammalian chymotrypsin-like enzymes. Comparative reactivities of rat mast cell proteases, human and dog skin chymases, and human cathepsin G with peptide 4-nitroanilide substrates and with peptide chloromethyl ketone and sulfonyl fluoride inhibitors, *Biochemstry 24*:2048 (1985).
23. S. M. Goldstein, J. Leong, L. B. Schwartz, and D. Cooke, Protease composition of exocytosed human skin mast cell protease–proteoglycan complexes: tryptase resides in a complex distinct from chymase and carboxypeptidase, *J. Immunol. 148*:2475 (1992).
24. C. P. Sommerhoff, S. J. Ruoss, and G. H. Caughey, Mast cell proteoglycans modulate the secretagogue, proteoglycanase, and amidolytic activities of dog mast cell chymase, *J. Immunol. 148*:2859 (1992).

25. R. C. Benyon, J. A. Enciso, and A. D. Befus, Analysis of human skin mast cell proteins by two-dimensional gel electrophoresis: identification of tryptase as a sialylated glycoprotein, *J. Immunol. 151*:2699 (1993).
26. P. A. Hohn, N. C. Popescu, R. D. Hanson, G. Salvesen, and T. J. Ley, Genomic organization and chromosomal localization of the human cathepsin G gene, *J. Biol. Chem. 264*:13412 (1989).
27. J. W. Heusel, R. L. Wesselschmidt, S. Shresta, J. H. Russell, and T. J. Ley, Cytotoxic lymphocytes require granzyme B for the rapid induction of DNA fragmentation and apoptosis in allogeneic target cells, *Cell 76*:977 (1994).
28. P. Vanderslice, S. M. Ballinger, E. K. Tam, S. M. Goldstein, C. S. Craik, and G. H. Caughey, Human mast cell tryptase: multiple cDNAs and genes reveal a multigene serine protease family, *Proc. Natl. Acad. Sci. USA 87*:3811 (1990).
29. D. S. Reynolds, D. S. Gurley, K. F. Austen, and W. E. Serafin, Cloning of the cDNA and gene of mouse mast cell protease-6: transcription by progenitor mast cells and mast cells of the connective tissue subclass, *J. Biol. Chem. 266*:3847 (1991).
30. M. F. Gurish, J. H. Nadeau, K. R. Johnson, H. P. McNeil, K. M. Grattan, K. F. Austen, and R. L. Stevens, A closely linked complex of mouse mast cell–specific chymase genes on chromosome 14, *J. Biol. Chem. 268*:11372 (1993).
31. P. Haddad, D. Jenne, J. Tschopp, M. V. Clement, D. Mathieu-Mahul, and M. Sasportes, Structure and evolutionary origin of the human granzyme H gene, *Int. Immunol. 3*:57 (1991).
32. S. J. Remington, R. G. Woodbury, R. A. Reynolds, B. W. Matthews, and H. Neurath, The structure of rat mast cell protease II at 1.9-Å resolution, *Biochemistry 27*:8097 (1988).
33. W. Bode, J. E. Meyer, and J. C. Powers, Human leukocyte and porcine pancreatic elastase: x-ray crystal structures, mechanism, substrate specificity, and mechanism-based inhibitors, *Biochemistry 28*:1951 (1989).
34. E. Emi, Y. Nakamura, M. Ogawa, T. Yamamoto, T. Nishide, T. Mori, and K. Matsubara, Cloning, characterization and nucleotide sequences of two cDNAs endoding human pancreatic trypsinogens, *Gene 41*:305 (1986).
35. T. J. Sayers, T. A. Wiltrout, M. J. Smyth, K. S. Ottaway, A. M. Pilaro, R. Sowder, L. E. Henderson, H. Sprenger, and A. R. Lloyd, Purification and cloning of a novel serine protease, RNK-Tryp-2,, from the granules of a rat NK cell leukemia, *J. Immunol. 152*:2289 (1994).
36. O. Krahenbuhl, C. Rey, D. Jenne, A. Lanzavecchia, P. Groscurth, S. Carrel, and J. Tschopp, Characterization of granzymes A and B isolated from granules of cloned human cytolytic T lymphocytes, *J. Immunol. 141*:3471 (1988).
37. P. Vanderslice, C. S. Craik, J. A. Nadel, and G. H. Caughey, Molecular cloning of dog mast cell tryptase and a related protease: structural evidence of a unique mode of serine protease activation, *Biochemistry 28*:4148 (1989).
38. M. J. Yezzi, I. E. Hsieh, and G. H. Caughey, Mast cell and neutrophil expression of dog mast cell proteinase-3, a novel tryptase-related proteinase, *J. Immunol. 152*:3064 (1994).
39. G. Salvesen and J. J. Enghild, An unusual specificity in the activation of neutrophil serine proteinase zymogens, *Biochemistry 29*:5304 (1990).

40. M. J. McGuire, P. E. Lipsky, and D. L. Thiele, Generation of active myeloid and lymphoid granule serine proteases requires processing by the granule thiol protease dipeptidyl peptidase I, *J. Biol. Chem. 268*:2458 (1993).
41. G. H. Caughey, The structure and airway biology of mast cell proteinases, *Am. J. Respir. Cell Mol. Biol. 4*:387 (1991).
42. H. Urata, S. S. Karnik, R. M. Graham, and A. Husain, Dipeptide processing activates recombinant human prochymase, *J. Biol. Chem. 268*:24318 (1993).
43. A. Caputo, R. S. Garner, U. Winkler, D. Hudig, and R. C. Bleackley, Activation of recombinant murine cytotoxic cell proteinase-1 requires deletion of an amino-terminal dipeptide, *J. Biol. Chem. 268*:17672 (1993).
44. G. H. Caughey, N. F. Viro, S. C. Lazarus, and J. A. Nadel, Purification and characterization of dog mastocytoma chymase: identification of an octapeptide conserved in chymotryptic leukocyte proteinases, *Biochim. Biophys. Acta 952*:142 (1988).
45. W. M. Shafer, J. Pohl, V. C. Onunka, N. Bangalore, and J. Travis, Human lysosomal cathepsin G and granzyme B share a functionally conserved broad spectrum antibacterial peptide, *J. Biol. Chem. 266*:112 (1991).
46. G. M. Griffiths and S. Isaaz, Granzymes A and B are targeted to the lytic granules of lymphocytes by the mannose-6-phosphate receptor, *J. Cell. Biol. 120*:885 (1993).
47. R. G. Woodbury, N. Katunuma, K. Kobayashi, K. Titani, and H. Neurath, Covalent structure of a group-specific protease from rat small intestine, *Biochemistry 17*:811 (1978).
48. A. Rouleau, M. Garbarg, J. C. Schwartz, and M. Ruat, Molecular cloning of rat mast cell protease 1 and development of specific probes for its gene transcript, *Biochem. Biophys. Res. Commun. 199*:593 (1994).
49. G. H. Caughey, Tryptase and chymase in dog mast cells, in *Neutral Proteases of Mast Cells* (L. B. Schwartz, ed.), S. Karger, Basel, 1990, p. 67.
50. S. Sayama, R. V. Iozzo, G. S. Lazarus, and N. M. Schechter, Human skin chymotrypsin-like proteinase chymase: subcellular localization to mast cell granules and interaction with heparin and other glycosaminoglycans, *J. Biol. Chem. 263*:6808 (1987).
51. H. Le Trong, H. Neurath, and R. G. Woodbury, Substrate specificity of the chymotrypsin-like protease in secretory granules isolated from rat mast cells, *Proc. Natl. Acad. Sci. USA 84*:364 (1987).
52. J. S. Miller, G. Moxley, and L. B. Schwartz, Cloning and characterization of a second complementary cDNA for human tryptase, *J. Clin. Invest. 86*:864 (1990).
53. J. S. Miller, E. H. Westin, and L. B. Schwartz, Cloning and characterization of complementary DNA for human tryptase, *J. Clin. Invest. 84*:1188 (1989).
54. R. Huang, M. Åbrink, A. E. Gobl, G. Nilsson, M. Aveskogh, L. G. Larsson, K. Nilsson, and L. Hellman, Expression of a mast cell tryptase in the human monocytic cell lines U-937 and Mono Mac 6, *Scand. J. Immunol. 38*:359 (1993).
55. H. P. McNeil, D. S. Reynolds, V. Schiller, N. Ghildyal, D. S. Gurley, K. F. Austen, and R. L. Stevens, Isolation, characterization, and transcription of the gene encoding mouse mast cell protease 7, *Proc. Natl. Acad. Sci. USA 89*:11174 (1992).
56. V. J. Braganza and W. H. Simmons, Tryptase from rat skin: purification and properties, *Biochemistry 30*:4997 (1991).

57. H. Kido, N. Fukusen, and N. Katunuma, Chymotrypsin- and trypsin-type serine proteases in rat mast cells: properties and functions, *Arch. Biochem. Biophys. 239*:436 (1985).
58. M. Muramatu, T. Itoh, M. Takei, and K. Endo, Tryptase in rat mast cells: properties and inhibition by various inhibitors in comparison with chymase, *J. Biol. Chem. 264*:4298 (1989).
59. D. Lagunoff, A. Rickard, and C. Marquardt, Rat mast cell tryptase, *Arch. Biochem. Biophys. 291*:52 (1991).
60. Z. Chen, A. A. Irani, T. R. Bradford, S. S. Craig, G. Newlands, H. Miller, T. Huff, W. H. Simmons, and L. B. Schwartz, Localization of rat tryptase to a subset of the connective tissue type of mast cell, *J. Histochem. Cytochem. 41*:961 (1993).
61. H. Kido, Y. Yokogoshi, and N. Katunuma, Kunitz-type protease inhibitor found in rat mast cells: purification, properties, and amino acid sequence, *Biochim. Biophys. Acta 967*:331 (1988).
62. T. Tanaka, B. J. McRae, K. Cho, R. Cook, J. E. Fraki, D. A. Johnson, and J. C. Powers, Mammalian tissue trypsin-like enzymes: comparative reactivities of human skin tryptase, human lung tryptase, and bovine trypsin with peptide 4-nitroanilide and thioester substrates, *J. Biol. Chem. 258*:13552 (1983).
63. I. T. Harvima, N. M. Schechter, R. J. Havima, and J. E. Fräki, Human skin tryptase: purification, partial characterization and comparison with human lung tryptase, *Biochim. Biophys. Acta 957*:71 (1988).
64. T. Blom and L. Hellman, Characterization of a tryptase mRNA expressed in the human basophil cell line KU812, *Scand. J. Immunol. 37*:203 (1993).
65. G. Nilsson, T. Blom, G. M. Kusche, L. Kjellén, J. H. Butterfield, C. Sundström, K. Nilsson, and L. Hellman, Phenotypic characterization of the human mast-cell line HMC-1, *Scand. J. Immunol. 39*:489 (1994).
66. E. K. Tam and G. H. Caughey, Degradation of airway neuropeptides by human lung tryptase, *Am. J. Respir. Cell Mol. Biol. 3*:27 (1990).
67. G. H. Caughey, F. Leidig, N. F. Viro, and J. A. Nadel, Substance P and vasoactive intestinal peptide degradation by mast cell tryptase and chymase, *J. Pharmacol. Exp. Ther. 244*:133 (1988).
68. S. J. Ruoss, T. Hartmann, and G. H. Caughey, Mast cell tryptase is a mitogen for cultured fibroblasts, *J. Clin. Invest. 88*:493 (1991).
69. T. Hartmann, S. J. Ruoss, W. W. Raymond, K. Seuwen, and G. H. Caughey, Human tryptase as a potent, cell-specific mitogen: role of signalling pathways in synergistic responses, *Am. J. Physiol. 262*:L528 (1992).
70. G. H. Caughey, W. W. Raymond, E. Bacci, R. J. Lombardy, and R. R. Tidwell, Bis(5-amidino-2-benzimidazolyl)methane and related amidines are potent, reversible inhibitors of mast cell tryptases, *J. Pharmacol. Exp. Ther. 264*:676 (1993).
71. G. H. Caughey, N. F. Viro, J. Ramachandran, S. C. Lazarus, D. B. Borson, and J. A. Nadel, Dog mastocytoma tryptase: affinity purification, characterization and amino-terminal sequence, *Arch. Biochem. Biophys. 258*:555 (1987).
72. G. H. Caughey, N. F. Viro, L. D. Calonico, D. M. McDonald, S. C. Lazarus, and W. M. Gold, Chymase and tryptase in dog mastocytoma cells: asynchronous ex-

pression as revealed by enzyme cytochemical staining, *J. Histochem. Cytochem.* *36*:1053 (1988).
73. G. H. Caughey, S. C. Lazarus, N. F. Viro, W. M. Gold, and J. A. Nadel, Tryptase and chymase: comparison of extraction and release in two dog mastocytoma lines, *Immunology 63*:339 (1988).
74. P. Vanderslice, C. S. Craik, J. A. Nadel, and G. H. Caughey, Molecular cloning of dog mast cell tryptase and a related protease: structural evidence of a unique mode of serine protease activation, *Biochemistry 28*:4148 (1989).
75. J. A. Cromlish, N. G. Seidah, M. Marcinkiewicz, J. Hamelin, D. A. Johnson, and M. Chretien, Human pituitary tryptase: molecular forms, NH_2-terminal sequence, immunocytochemical localization, and specificity with prohormone and fluorogenic substrates, *J. Biol. Chem. 262*:1363 (1987).
76. S. C. Alter, D. D. Metcalfe, T. R. Bradford, and L. B. Schwartz, Regulation of human mast cell tryptase: effects of enzyme concentration, ionic strength and the structure and negative charge density of polysaccharides, *Biochem. J. 248*:821 (1987).
77. L. B. Schwartz and T. R. Bradford, Regulation of tryptase from human lung mast cells by heparin: stabilization of the active tetramer, *J. Biol. Chem. 261*:7372 (1986).
78. L. B. Schwartz, J. W. Yunginger, J. Miller, B. Bokhari, and D. Dull, Time course of appearance and disappearance of human mast cell tryptase in the circulation after anaphylaxis, *J. Clin. Invest. 83*:1551 (1988).
79. L. B. Schwartz, D. D. Metcalfe, J. S. Miller, H. Earl, and T. Sullivan, Tryptase levels as an indicator of mast-cell activation in systemic anaphylaxis and mastocytosis, *N. Engl. J. Med. 316*:1622 (1987).
80. P. Vanderslice, C. S. Craik, and G. H. Caughey, Expression of recombinant human mast cell tryptase as an active monomer in *E. coli, FASEB J. 4*:A1945 (1990).

Appendix: Mast Cell Protease Nomenclature and Abbreviations

Chymases

Human chymase
Dog chymase
Rat chymase I = rat chymase 1 = rat mast cell protease-I (RMCP-I)
Rat chymase II = rat chymase 2 = rat mast cell protease-II (RCMP-II)
Mouse chymase 1 = mouse mast cell protease-1 (MMCP-1)
Mouse chymase 2 = mouse mast cell protease-2 (MMCP-2)
Mouse chymase 3 = mouse mast cell protease-3 (MMCP-3)
Mouse chymase 4 = mouse mast cell protease-4 (MMCP-4)
Mouse chymase 5 = mouse mast cell protease-5 (MMCP-5)
Mouse chymase L = mouse mast cell protease-like (MMCP-L)
Sheep chymase = sheep mast cell protease (SMCP)

Cathepsins G

Human cathepsin G (Cat G)
Mouse cathepsin G (Cat G)

Tryptases

Human tryptase α = α-tryptase
Human tryptase I
Human tryptase II = human tryptase β = β-tryptase

Human tryptase III
Human tryptase I, II/β, + III = "β-tryptases"
Dog tryptase
Dog mast cell protease-3 (dMCP-3) = dog mastocytoma protease
Rat tryptase
Mouse tryptase 1 = mouse mast cell protease-6 (MMCP-6)
Mouse tryptase 2 = mouse mast cell protease-7 (MMCP-7)
Bovine tryptase

Carboxypeptidases

Human mast cell carboxypeptidase A (MC-CPA) = human mast cell carboxypeptidase (HMC-CP)
Mouse mast cell carboxypeptidase (MC-CPA)

Aminopeptidase

Mouse mast cell aminopeptidase (MC-AP)

Index

About the Editor

GEORGE H. CAUGHEY is Chief of Research, Pulmonary Division, Department of Medicine, Parnassus Campus, University of California, San Francisco, School of Medicine; and Assistant Director, Cystic Fibrosis Research Center; Professional Faculty, Molecular Medicine Program; Associate Staff, Cardiovascular Research Institute; and Associate Professor of Medicine at the University of California, San Francisco, School of Medicine. The author or coauthor of over 90 professional papers, book chapters, and abstracts, he is a member of the American Thoracic Society, the American Physiological Society, and the American Society for Clinical Investigation. Dr. Caughey, a diplomate of the American Board of Internal Medicine, received the B.S. degree (1975) in chemistry from Arizona State University, Tempe, and the M.D. degree (1979) from Stanford University School of Medicine, California.